MICROMANIPULATORS AND MICROMANIPULATION

BY

DR. HAMED M. EL-BADRY

PROFESSOR AND HEAD OF THE DEPARTMENT OF MINING
AND PETROLEUM ENGINEERING, FACULTY OF ENGINEERING,
CAIRO UNIVERSITY, EGYPT, U. A. R.

WITH 177 FIGURES

1963

NEW YORK
ACADEMIC PRESS INC.

VIENNA
SPRINGER-VERLAG

This book is Volume 3 of
"Monographien aus dem Gebiete der qualitativen Mikroanalyse".
Herausgegeben von A. A. BENEDETTI-PICHLER, New York

SPRINGER-VERLAG

VIENNA

Published in the U.S.A. and Canada by
ACADEMIC PRESS INC.
111 Fifth Avenue, New York 3, New York

Library of Congress Catalog Card Number 63-21411

ISBN-13: 978-3-7091-5553-0   e-ISBN-13: 978-3-7091-5551-6
DOI: 10.1007/978-3-7091-5551-6

© BY SPRINGER-VERLAG VIENNA
Softcover reprint of the hardcover 1st edition 1963

1963

Preface

In the course of the years since H. D. SCHMIDT, in 1895, described his "microscopic dissector," a mechanical device for dissecting and studying biological materials, a great wealth of information has been published in the scientific and technical literature on methods involving the use of exceedingly delicate microtools mechanically guided under microscopic control for the investigation of microscopic structures and very small amounts of material. The operative tools used can be moved with considerable precision under various microscopic magnifications to perform the required tasks.

With the continuous progress in these methods, hundreds of micromanipulators and other mechanical manipulative devices, auxiliary equipment, and a great diversity of microtools have been described for performing varied operations on practically any type of materials and test objects. Thus, micromanipulative and related techniques have become innumerable and often intricate, and the applications, formerly confined to certain fields of biology and medicine, have been extended to the most diverse fields of science and technology as can be seen from a rapid glance at the contents of the present volume.

It follows, therefore, that pertinent information on the subject is widely scattered and must be sought in rather specialized scientific and technical journals difficultly accessible to a scientist or an engineer working in a particular field. Consequently, manipulative methods are still not fully appreciated as invaluable research and technical tools and, in spite of the fact that they have been extensively applied in many fields to solve a diversity of problems, they are generally overlooked by many workers who could benefit by their manifold uses. This is to be expected from a consideration of the highly specialized nature of these methods and their deviation from the more familiar practices.

In this volume, it has been my endeavour to present, as far as possible, a wide survey of the various instruments, manipulative and related techniques, and examples of applications in a variety of fields of science and technology. Compiling the fundamental information on the subject from such scattered sources makes it readily available in a single book. It is the author's hope that such a systematic collection of the various types of procedures will be of benefit to specialists in other fields. The volume is also intended to serve as a guide for investigators who are trying to find a suitable method for attacking a problem by enabling them to make a wise selection from the various alternatives presented or to formulate a workable procedure. It also may serve to introduce these methods to many workers who are not familiar with them and thus to contribute to a better understanding of their great potentialities and manifold uses. It is felt that this volume would satisfy real needs and fill a gap in the scientific and technical literature.

A general bibliography has been added, giving a list of articles not cited in the subject matter of the volume. It may provide some readers with additional references to papers dealing with their own particular field.

The author wishes to thank Professor A. A. BENEDETTI-PICHLER, New York, and Professor CECIL L. WILSON, Belfast, at whose suggestion the author undertook the task of preparing this volume. The work has been greatly influenced by the wholehearted co-operation of Professor BENEDETTI-PICHLER whose unfailing interest and encouragement greatly stimulated its progress. Furthermore, the generous co-operation of the publishers during the production of the volume is most gratefully acknowledged.

The author also acknowledges the courtesies extended by the National Research Center, Dokky, Cairo, Egypt, U.A.R., in furnishing photographic and documentation aids and in making available its library facilities. He especially wishes to express his deep obligation to the many writers, organizations, instrument manufacturers and other industrial firms who supplied him with personal information, reprints of original papers, microfilms, illustrative and pictorial materials, catalogues, and other pertinent information with permission to use them. These sources are duly credited in the subject matter of the volume.

It is with pleasure that the author acknowledges the untiring efforts of Sayed HASSAN M. EL-SABBAN who typed the final manuscript and the earlier drafts and who, actuated by a keen interest in the project, has rendered extensive and invaluable aid in checking the proofs, in assembling bibliographies and the subject index, and in other ways to an extent which places the author deeply in his debt.

The author would welcome criticisms and suggestions for improvements. He would also be grateful for any reprints and other pertinent material which may help in the preparation of a future revision and extension of the volume.

Cairo, Egypt, U.A.R., July, 1963

HAMED M. EL-BADRY

Contents

Part II

General Techniques

Part III

Applications

Contents

Introduction.

The term micromanipulation is used to include all operations performed in the microscopic field of vision with the aid of mechanical devices which guide the operating tools proper. The use of a microscope implies an optical magnification of object and tool points of at least ten diameters. The tool that actually contacts the object of interest is held by some device which aids the hand of the experimenter in performing the required task. The device may merely steady or guide the motions produced directly by the hand which grasps the handle of the tool, a principle used in simple manipulating devices, partly utilizing the motions provided by the microscope itself, as well as in gliding manipulators for medium to high magnifications. More often, the tool is held by the arm of a complicated mechanism which translates the orders imparted by the hand to a milled head, a lever, or a joy stick.

A clear evaluation of the advantages gained by the use of micromanipulators should first take into consideration that very delicate operations may be performed with the unaided hand, provided that the eye is assisted by optical magnification giving a distinct image of tool point and object. Thus it is possible to sort the particles of a powder with a needle guided by hand and a microscope magnification of up to about 150 to 200 diameters. It is also possible to insert, with the unaided hand, a textile fiber into the interior of the stinging hair of a nettle, *Urtica*, which may have an opening of not more than 30 μm = 0.03 mm. Such tasks are not at all difficult, especially when a stereoscopic binocular microscope is being used which, aside from the three-dimensional view, gives an upright image and does not necessitate to perform motions in directions opposite to those perceived in the image.

The common monocular compound microscope makes it necessary to reverse the directions of the motions, a difficulty which is by far not as formidable as one may imagine and is quickly overcome regardless whether the tool is operated by hand or by a mechanical device. Whenever practical, the image is projected on a screen which may be as small as 10 cm in diameter, and arrangements may be made so that an *upright* image is obtained.

The pronounced lack of depth, even with rather moderate magnifications, may be a help rather than a hindrance in manipulation. If the object is in sharp focus, it is only necessary to raise and lower the tool until its image becomes sharp; this will assure that it is very accurately in the same image plane as the object. When proceeding to the use of medium magnifications (100 to 500×), the lack of space between the object and the front lens of the microscope objective becomes the principal hindrance to manipulation, and manipulation by hand becomes definitely awkward.

Obviously, mechanical manipulation is a necessity when high magnifications are used. The use of manipulators, however, is also a great help for relatively simple tasks requiring only low magnification. Even a rather primitive manipulating device steadies the motion of the tool, reduces the nervous effort, and increases the efficacy of the experimenter. A task which, performed with the unaided hand, is somewhat of an art becomes an assured routine action. Equally important is that every tool held by a manipulator becomes an added hand. The experimenter's hand is often merely needed to get the tool into the required position, whereafter it becomes free for the performance of other tasks while the tool retains its position as long as desired. If one manipulator fixes the position of the object, the experimenter has still both hands available for the simultaneous handling of a second manipulator and an injection device or rheostat; switches may be thrown by means of a foot pedal. If either the object or the tool may be moved by means of the mechanical stage of the microscope, the latter may take the rôle of a second manipulator.

The purpose and merit of the micromanipulator is as obvious as the basic principle of its action. The invention of the compound microscope made it possible to see fine structures, and this led to curiosity concerning the nature, interrelation, and function of the parts. There is the natural desire to touch and test what is seen.

The object corresponding to the microscopic image—in contrast to that of the telescopic image—is always within reach. Consequently, attempts at manipulation of minute bodies may have been made as soon as the compound microscope became available,—or before that time with the help of magnifying lenses, transparent beads or rods. Details of microscopic structure, however, require very delicate feelers for their direct investigation. Lacking these, it was still possible to submit the whole object to some gross treatment and to observe the effect upon the detail of interest.

The treatment may be a variation of temperature which may produce revealing transitions occurring in parts of the structure. Decomposition, and oxidation may be observed in addition to vaporization, melting, and crystallization. The object may be exposed to reagents such as iodine vapor, ammonia gas, acid vapors, or to reagent solutions of various description including solutions of dyes. The interpretation of the changes observed under the microscope is often very simple and conclusive, but localized tests may not appear in the proper places if substances migrate during the treatment. Obviously, the direct approach, application of the test only to the detail to be investigated or mechanical separation of the detail from the rest of the object previous to the test, is more decisive. Regardless of method, however, the direct approach requires manipulation guided by optical magnification.

The sensitiveness of micromanipulation must grow with the fineness of the detail to be investigated and with the magnification needed to reveal it. Concerning the limits of possible accomplishment, it appears that the mechanical devices may be able to cope with all tasks that can be supervised by means of optical magnification. The ultimate limits may be determined by the inability to observe rather than by lack of discernment in the manipulation. If this is true, the field of micromanipulation could be expanded by the use of the electron microscope. Of course, the conditions to be satisfied by a preparation for electron microscopy seem to preclude the idea of manipulation simultaneous with observation. In August 1961, however, ERNST WIESENBERGER outlined and demonstrated during the Fourth International Symposium of Microchemistry some chemical experiments under the electron microscope, which permit to visualize the possibility of routes that circumvent the obstacles.

It has been attempted to present as complete as possible a survey of the instruments, methods, and uses of micromanipulation. This has been done with the hope of conveying a more generally correct outline of the subject and a large variety of detail, which may be useful for the specialist in a particular field as well as for the prospector searching for a way to the solution of a given problem. It cannot be helped that the methods, most of which have been developed by biologists, have deen described as seen with the eyes of a chemist or a chemical mineralogist. One may hope that the change of view point and perspective may contribute to directing attention toward neglected features.

The preoccupation with small dimensions suggested a table of units, which may be found on the inside of the back cover. For convenience, some simple equations have been added, which are useful in practical microscopy. The area of the field of vision is of interest since best illumination requires that all available light is collected inside it. Light received outside this area may be broken into the objective and add to haze derived from scattering within the field of vision.

Part I.

Apparatus.

1. Microscopes for Use in Micromanipulation.

Most of the principal types of optical microscopes have been used in the varied fields of micromanipulative work. Several instruments of different types, designs, and makes, which have been actually employed are indicated in appropriate places in this volume in connection with the particular assemblies and fields in which they have been tried.

The following pages contain a general account of the types of optical microscopes most commonly used in micromanipulation. The information given considers mainly the basic requirements of micromanipulative work. This may help the prospective worker in choosing the type of microscope suited for his purpose. Consequently, mainly the basic features are considered. No attempt is made to discuss in any detail the principles and theory of the varied methods of microscopy, or the various types and the internal construction of the different lens components. Detailed information on these are easily accessible in many specialized books on optical microscopy and also in catalogues, manuals, and other publications issued by the various microscope manufacturers so as not to merit any lengthly consideration in the present volume. The most useful single reference which deals with many essential aspects of microscopy is probably that to CHAMOT and MASON (64). Among the other useful books dealing with varied phases of the subject are those of VICKERS (232), PAYNE (183), SCHAEFFER (200), GAGE (103), and GIBB (108).

Instructions for general use of the microscope as well as procedures commonly used in the broad field of microscopy such as illumination of different specimens, micrometry, photomicrography, and microprojection are too well known to trained users of the microscope to be considered here. In general, micromanipulation in itself does not impose special requirements on such practices and procedures. The beginner is advised, however, in addition of studying one or more books on microscopy, to follow the instructions specified in the manual accompanying the microscope and other equipment he will employ. This will enable him to operate these instruments properly and to take full advantage of their potentialities.

The microscope may be mounted in a variety of ways depending upon the type of work and the manipulating device employed. Several microscope-micromanipulator assemblies are described in the present volume.

A few of the various possible arrangements are illustrated in Figs. 26, 31, 50. In a number of other set-ups, suitable mechanical manipulating devices are mounted directly on the microscope as illustrated in Figs. 10, 102. Stereoscopic microscopes, p. 16, may be set in a diversity of ways, p. 18, to suit various purposes and test objects. Certain testing microscopes may even be placed directly and securely upon the surface of large test objects which may be plane or cylindrical such as sheet metal, pipes, large castings, and machine beds. An example of these instruments is a special model of the Miniload microhardness tester (152), p. 85. This instrument, Figs. 59, 161, can be arranged in a variety of ways to suit test objects of practically any conceivable size and shape, p. 287.

Though it is the most extensively employed, the microscope is not the only optical magnifying device used in micromanipulation. A simple magnifier mounted on a suitable adjustable stand may provide magnification sufficient for the purpose.

a) Ordinary Compound Microscopes.

As is well known, the lens system of a compound or light microscope, consists of two essential elements, the objective and the eyepiece or ocular. Each of these two elements consists of a lens or a combination of lenses, and both elements function in producing the image of the specimen which is placed on the stage of the microscope. The objective forms a real magnified image of the object at the upper end of the body tube in the focal plane of the eyepiece, while the eyepiece acts as a simple magnifier for this real image and projects it into the eye of the observer. The objective and the eyepiece are usually mounted in a telescoping body tube, so that their optic axes coincide. Focusing on the object is achieved by means of suitably located coarse and fine adjustments. A third element, the condenser, ordinarily located below the stage, may be used to illuminate the object by concentrating light, usually reflected by a mirror from a suitable light source, and bringing it to a focus within the object. Condensers may also be used to provide oblique or darkfield illumination. The condenser is so supported that its optic axis coincides with that common to objective and eyepiece.

The above mentioned elements are incorporated in the comparatively simple instrument shown in Fig. 1, supplied by BAUSCH & LOMB OPTICAL Co. (20). This basic model as well as several other essentially similar instruments of different styles are available from a number of well known manufacturers and may be used for working with transparent objects, e. g., for many biological micromanipulative procedures performed on cells and tissues, p. 239.

The instrument, Fig. 1, is provided with a vertical monocular tube which may be fitted either with a $5\times$ or a $10\times$ Huygenian eyepiece and

with an attached rotating and precentered triple nosepiece permitting interchange of low, medium, and high power parfocal achromatic objectives
$10\times$, $43\times$, and $97\times$. Thus total magnifications in the wide range of
about 50 to about 970 diameters are available. Microscopes may also be
fitted with a quadruple nosepiece which, in certain essentially similar instruments of the same make, accomodates a $3.5\times$ low-power objective

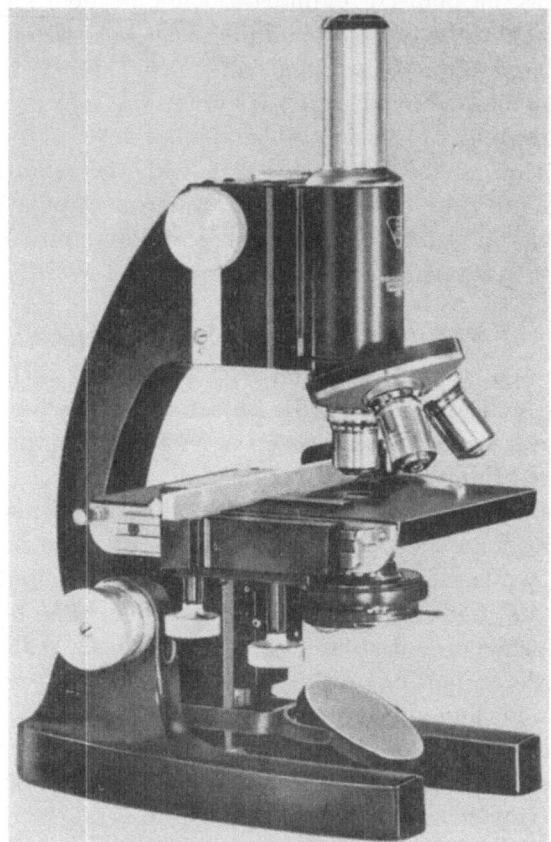

Fig. 1.   Ordinary Compound Microscope.   Courtesy of BAUSCH & LOMB (20).

in addition to the three objectives above indicated to supply also a total
magnification of less than 20 diameters. Nosepieces accomodating only
two objectives are also available. Microscopes may also be equipped with
a centering nosepiece for a single objective permitting precise interchange
of a number of individual standard objectives. The use of revolving nosepieces accomodating a number of objectives may be preferable, however,
when quick and frequent interchange of magnification is required.

Several types of eyepieces and objectives of varied magnifications,
numerical apertures, and degree of correction are also available for use
with standard instruments to suit various purposes. Objectives may be

classed according to the degree of optical correction into aplanatic, achromatic, apochromatic, and semiapochromatic or fluorite objectives. There are also the dry and the oil or water immersion objectives. Several types of oculars are available, among which the negative Huygenian and the positive Ramsden are in common use. The choice of suitable types of oculars and objectives (64, 108) is in most instances mainly dependent, among other factors, upon the material being investigated and the microscope magnifications employed rather than upon the type of micromanipulation, although low-power objectives with sufficiently long working distance have to be used in working with certain set-ups where the operating microtools are manipulated between the specimen and the objective as shown in Fig. 55. Certain ordinary compound microscopes may be fitted with an image-erecting eyepiece, p. 60, Fig. 31. This is desirable when the micromanipulator employed does not offer a reversal of movement which makes the operating microtool points appear to follow the natural movement of the hand when viewing the reversed image of an ordinary compound microscope, p. 26.

Ocular micrometers are used for accurate measurement of objects being investigated under the microscope. Usually, a micrometer disk with a transparent graduated scale is placed into the eyepiece, scale downwards, in such a position that the scale lies in the plane of image diaphragm. Scales having graduations of 100-μm and 50-μm divisions are in common use. Ocular micrometers should be calibrated against a standard scale or a stage micrometer. The stage micrometers are usually ruled in 10-μm and 100-μm divisions. A special type of eyepiece micrometers, the filar micrometer, offers improved accuracy of measurement. Other types of ocular micrometers are ruled for counting moulds, bacteria, dust particles, and similar material. For several purposes it may be necessary to use eyepieces which are equipped with cross hairs mounted in the plane of the diaphragm.

The vertical monocular tube, Fig. 1, has the conventional mechanical tube length of 160 mm and may be interchanged with an inclined or a vertical binocular body offering the user the comfort of binocular vision which is desirable especially when the work extends for long periods. Graduated adjustable monocular draw tubes are available and may be ordered, if required, in place of the fixed tube. The adjustable draw tube may be useful in the calibration of micrometer eyepieces to get simple numbers by making small changes in magnification. The graduated scale inscribed in the draw tube indicates the tube length. An inclined binocular body, similar to that shown in Fig. 3, allows the user to work in a natural, relaxed position, while the stage of the microscope remains horizontal. Such position of the stage is the most common in micromanipulative work and is often necessary as for instance when specimens are operated upon in oil immersion or in fluid cells. Consequently, the use of microscopes with a tilting mechanism is generally superfluous. The binocular body may be changed to a monocular one when micrometry, photomicrography, or micro-projection are required. In isolated instances, however, as for example in the manufacture of microtools by means of the de Fonbrune microforge, p. 152, micromanipulative operations are performed while the working stage, i. e., the circular disk carrying the vise which holds the material in preparation, is in an inclined position, Fig. 102.

The substage of the ordinary microscopes carries the illuminating system which usually consists of a condenser commonly fitted with a diaphragm, a mirror, and a light source. The condenser is the most important of these components and should be chosen to meet the aperture requirements of the microscope objective especially at higher magnifications in order to obtain maximum efficiency. For ordinary work in transmitted light with low magnifications, the use of condensers, however, is not necessary. With low-power objectives, a suitable light source integrated with the micro-scope or the microscope mirror reflecting the image of a suitable light source usually give satisfactory results without interposition of a con-denser. The room light or a bright sky may be adequate light sources. If a condenser is used with a low-power objective, the top component of the condenser is removed to illuminate the comparatively large field of vision. This also increases the working distance of the condenser.

A large variety of condensers and appropriate light sources are available to suit various purposes (64). This variety of condensers includes several types of bright-field, dark-field, and phase contrast condensers. The micro-scope shown in Fig. 1, is equipped with a substage Abbe condenser which is usually provided with an iris diaphragm for varying the illuminating aperture when working with medium and high power objectives. Abbe condensers are the simplest type of substage condensers and are suitable at various magnifications for most work where color correction is not neces-sary. Corrected condensers, especially the aplanatic condensers and the achromatic condersers are also in common use. Cardioid and paraboloid condensers for dark-field work, phase contrast condensers, and ultraviolet condensers are available to satisfy various requirements. The Panfocal Illuminator-Condenser (20) is a corrected achromatic condenser which has an integrated light source and fits into a ring mount in the substage of BAUSCH & LOMB Laboratory microscopes. This type of condenser is suitable for work in transmitted light under various magnifications, and permits quick interchange of bright-field, dark-field, and polarized light. A red, a green, a blue, or a daylight filter may also be brought into the condenser axis as required. For working with opaque objects using vertical illumi-nation, certain types of objectives, the vertical illuminating objectives, may be employed. In this type of objectives, which function as their own con-densers, the light from a near-by source enters the back of the objective which focuses it on the surface of the specimen. Several devices are available for illuminating opaque objects by reflected light. Of these may be mentioned the LEITZ Ultropak and the ZEISS Epi-condenser. All these and other devices together with a diversity of lamps and other light sources, which may be separate or integrated with the substage of the microscope and which may be used with suitable filters, are listed by microscope dealers and offer a wide choice of equipment for appropriate illumination of the desired type, intensity, and wave lengths, for different kinds of transparent and opaque objects under various magnifications.

Microscope stands may vary considerably in their design, but those having a sturdy construction are generally preferable since they offer greater stability and freedom from vibration. This is desirable for micromanipulation especially when the work is carried out under higher magnifications. The ordinary compound microscope shown in Fig. 1, similar to other ordinary microscopes, is equipped with a heavy stand of conventional design carrying the various components. The instrument has a plain spaceous rectangular object stage which is fitted with a built-on or integral mechanical stage permitting ample movements in two mutually perpendicular directions of the horizontal plane. Instead of the normal rectangular stage, some compound microscopes may be fitted with circular rotating and centering object stages similar to that shown in Figs. 4 and 5. These stages may have graduated peripheries and may be equipped with verniers. The revolving motion of the microscope stage, in addition to its usual applications in certain fields of microscopy such as in the examination of crystalline materials in polarized light, p. 13, is important for performing many operations in certain fields of micromanipulation. In carrying out chemical operations in capillary cones, p. 92, as well as in other fields of solution chemistry on the microgram and the nanogram scales, such rotating motion is always desirable for bringing capillary containers and micropipet in proper angular relation or into proper alignment, and it becomes indispensable when the micromanipulator itself does not offer such a rotating motion.

Suitable mechanical stages are essential in most fields of micromanipulative work for holding various specimens, glass slides, and operation chambers, which are retained securely in place by means of a spring finger, and also for getting the object to be operated upon into the field of the microscope. In addition, the mechanical stage often serves, together with the appropriate micromanipulator movements, to bring the specimen in the field against the operating microtool point. Attachable mechanical stages of various styles, both graduated and ungraduated, Fig. 2, are available for use with object stages of different designs. They may be clamped on the permanent stage of the microscope and enable the operator to guide the specimen in the two horizontal coordinates by means of two

rack-and-pinion systems. Mechanical stages provided with graduations and
verniers may be useful in certain cases in helping the operator to relocate
a given part of the preparation.

A compound microscope of a more elaborate design is shown in Fig. 3.
The instrument, supplied by E. LEITZ (152), is also shown in working
position in combination with the Leitz lever-activated high-power micromanipulator, p. 48, Figs. 25 and 26. The instrument is equipped with a
binocular body set on a Laborlux II stand.

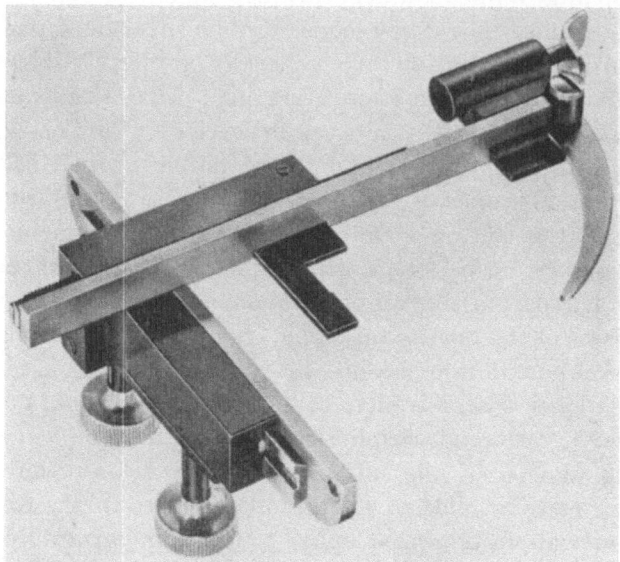

Fig. 2. Attachable Mechanical Stage, ungraduated, for holding and crossguiding of specimens.
Courtesy of BAUSCH & LOMB (20).

The Ortholux stand, of the same make, is essentially similar but is
equipped with low-position coarse and fine adjustments, both acting on
the stage of the microscope. Focusing is performed by raising or lowering
of the object stage while the microscope tube remains stationary. Low-
position focusing adjustments are probably more convenient to operate.
A focusing stage is typical of metallurgical microscopes and has the advantage that by lowering it an ample vertical space is provided for accomodating substantially thick objects while the ocular is maintained at a constant level. In micromanipulation, however, a stand with a vertically fixed
stage, where focusing is carried out with the tube, Fig. 3, is generally
preferable since this leaves the relative vertical positions of the object
and the operating microtool points unchanged during focusing. Another
advantage of stands with vertically fixed stages is that it prevents damage
to delicate microtools set just above the stage, which may occasionally
happen through oversight when the stage is being raised. If a microscope

stand with a lowering and raising stage is to be employed, an appropriate sequence of operations in the setting and use of the assembly should always be strictly followed to avoid such danger.

The Laborlux II stand, Fig. 3, with its widely curved tube carrier, different from that shown in Fig. 1, permits appropriate binocular or monocular inclined eyepieces to be so set that the stand is turned away from the observer while the object stage faces the operator, thus leaving an

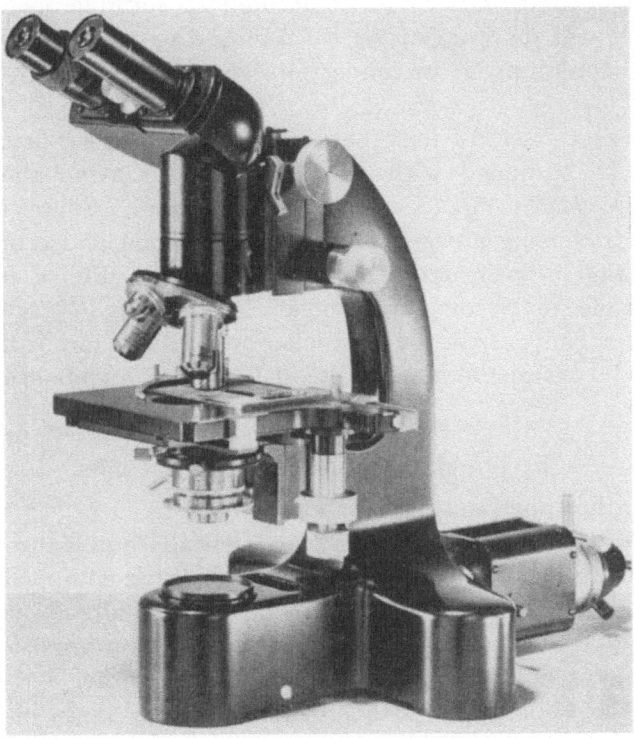

Fig. 3. Binocular Compound Microscope on a Laborlux II Stand with a Built-In Illuminating System.   Courtesy of E. Leitz (152).

unobstructed view of the stage as shown in Figs. 25 and 26.   The same feature, also illustrated in Fig. 29, which shows a Leitz stereoscopic binocular microscope of the Greenough type on a Laborlux II stand in working position in combination with a Leitz micromanipulator is rather preferable in micromanipulation since such arrangement permits an easy access of the object stage from all sides.

The stand, Fig. 3, is provided with a built-in illuminating system with a low-voltage filament lamp of adjustable intensity housed in the stand. The light source is adequate for various uses including dark-field, phase contrast, and photomicrography in black and white or color.   Change-over from visual observation to photomicrography is readily possible.   The high intensity source is also suitable for projecting bright microscopic images onto a screen. The microscope is normally equipped with a vertically adjustable two diaphragm bright-field condenser giving adequate and uniform illumination of the field of vision in all ranges of magnification. To illuminate the comparatively large field of low-power objectives, the top lens of the condenser is swung out of the system. The bright-field condenser can be interchanged with a suitable dark-field condenser or a phase contrast condenser. A brief account of suitable condensers which may be employed for working with hanging drop preparations in a moist chamber is given on p. 54, in connection with the LEITZ high-power micro-manipulator.

The Laborlux II stand may be fitted with a variety of object stages, which may be equipped with either built-on or attachable mechanical stages. This variety includes plain square stages, square sliding stages with ample traversing areas, and circular revolving and centering stages. For observation in polarized light the microscope, Fig. 3, may be fitted with a suitable revolving stage, polarizing filters, and other necessary attachments. The microscope may also be adapted for working with incident light by using the LEITZ Ultropak for the examination of opaque objects.

b) Special Compound Microscopes.

Several rather special types of microscopes have been used in micromanipulative work. Most of them are essentially similar to the ordinary compound microscopes and may differ only in being equipped with certain components, having various functions, which are usually absent in ordinary instruments. Some ordinary instruments, however, have provisions for change from one microscopic examination method to another by simply being fitted with the appropriate attachments. Certain types of instruments, however, are sufficiently different from the ordinary microscopes or are of particular importance in certain types of micromanipulative work to warrant special consideration. Of these the polarizing microscopes, the stereoscopic microscopes, and the inverted microscopes will be considered briefly.

Polarizing Microscopes. The simpler types of polarizing microscopes are often called chemical microscopes. The more elaborate types, petrological or petrographic microscopes, are extensively used in several fields of micromanipulation. All of them have nearly the same optical features.

A simple polarizing microscope consists essentially of an ordinary compound microscope which, in addition to the usual optical components of objective, eyepiece, and illuminating system, is fitted with two polarizing devices of similar construction. These two essential components, usually Nicol prisms made of long transparent cleavage rhombs of Iceland spar, cut and cemented so that only one beam of polarized light will pass through, are arranged in sequence and usually oriented so that when both are inserted in working position, their vibration directions are perpendicular to each other, one coinciding with the east-west and the other with the north-south directions. One of these, the polarizer, is mounted below the stage, whereas the other device, the analyzer, is located either in the microscope tube directly above the objective or above the eyepiece. In most modern instruments, Fig. 4, comparatively inexpensive and less bulky polarizing

Fig. 4.   Polarizing Microscope Standard GEL POL.   Courtesy of C. Zeiss, Oberkochen, Württemberg (244).

filters, manufactured of some oriented synthetic material, serve as polarizer and analyzer instead of the calcite prisms. The most commonly used of these recently developed filters are disks made of Polaroid. In addition to the polarizer and the analyzer which are the most important distinguishing components of the polarizing microscopes, the stands of these instruments, different from normal microscope stands, are invariably fitted with a rotating graduated stage, Fig. 4. The rotating motion of the stage, aside from offering additional flexibility of movement which may be necessary in certain fields of micromanipulative work, p. 92, permits efficient investigation of the interaction of the specimen with polarized light. The polarizing microscope is also equipped with, (a) cross-hair eyepieces fixed

so as to indicate accurately the vibration directions of polarizer and analyzer and (b) compensators, i. e., retardation plates and quartz wedge,
which can be inserted into a slot in the body tube. The instruments may
also be fitted with a Bertrand lens which can be inserted in the body tube
for conoscopic observations. Of course, the condenser should provide for
interchange of parallel and convergent light.

The incorporation of these additional features, however, while greatly
extending the capabilities of the compound microscope far beyond that
of a mere magnifier to be used only for the study of the outer form and for
revealing fine structures, usually does not decrease its potential value for
ordinary work. These accessories can be easily removed out of the system
as required. In addition to performing practically all the functions of an
ordinary compound microscope, the polarizing microscope yields much information on cyrstalline and pseudo-crystalline materials permitting the
determination of their optical properties (64, 74, 108, 119, 232, 233). Among
the useful optical properties of crystals most commonly determined with
the microscope are refractive indices, pleochroism, extinction angles, bire-
fringence, and several other characteristics. Between crossed polarizer and
analyzer, optically anisotropic substances can be observed clearly against
a dark field because of their polarization colors, whereas adjacent isotropic
substances are rendered invisible. The illumination of different types of
transparent and opaque objects and the use of appropriate methods for
differentiating them from the surrounding materials and for rendering
them clearly visible are discussed in detail by CHAMOT and MASON (64).

In addition to the extensive use of the polarizing microscope in the fields
of mineralogy, petrology, chemistry, and several other fields, the polarizing
microscope has become a valuable tool in biological investigations (13, 14,
64, 148, 166, 204, 232) since it often permits adjustment of contrast in the
specimen, which helps in studying the fine structures of biological materials.
Valuable knowledge can be obtained on the structure and orientation of
plant and animal cells, tissues, fibers, and membranes embedded in other
tissues as well as concerning the structure of long protein molecules and
the direction in which they orient themselves in a muscle cell when it is
subject to various reactions. Examination between polarizer and analyzer,
since it reveals fine structural features, is also very useful in the investigation of textile and paper fibers, plastics, colloidal aggregates, and other
materials.

Certain polarizing microscopes of more recent design may be fitted with
a binocular body. Fig. 5, shows a Standard GEL POL microscope, the
same as that shown in Fig. 4, but fitted with a binocular inclined tube.
The same type of instrument as well as certain other polarizing microscopes, in addition to being adaptable for bright-field, dark-field, and phase-
contrast microscopy, may also be fitted with an Epi-condenser permitting

observation in incident light to adapt the instrument to the examination of ores, coal, and other opaque objects. Finally, the same instrument may also be equipped for photomicrography and microprojection.

Polarizing microscopes of an even more versatile nature are available. The Ultraphot II POL (244), for instance, is equipped with an integral automatic camera and an adaptable illuminating apparatus in one instrument. The microscope can be adapted to practically all microscopic ex-

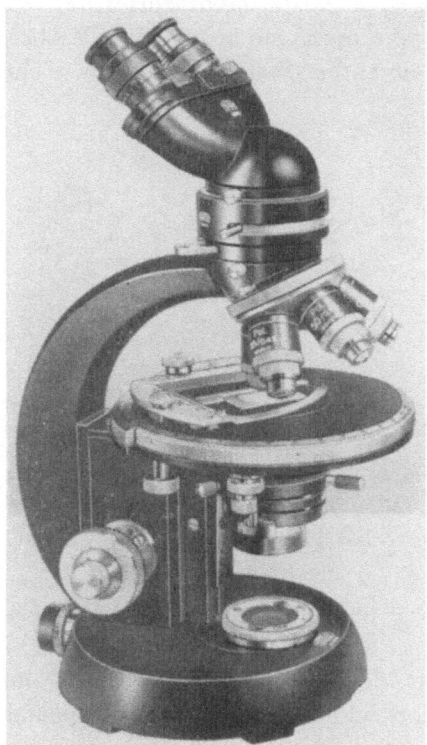

Fig. 5. Polarizing Microscope Standard GEL POL, with Inclined Binocular Tube. Courtesy of C. ZEISS, Oberkochen/Württemberg (244).

amination methods. With this instrument, objects of investigation can be photographed during examination and microscope images can be projected on a ground glass disk. Such universal instruments, however, are expensive. In most laboratories performing a diversity of microscopic work, it is generally more advantageous to procure a number of different types of microscopes of comparatively simple construction, each of which is suited and set up for a particular type of micromanipulative or other microscopic work. For instance, according to actual requirements in a given laboratory, it may be more beneficial to have a simple biological microscope, a chemical, petrological, or a metallographic microscope, and a stereoscopic microscope instead of one of these highly elaborate instruments. Several simple microscopes may be purchased at practically no greater total cost. Furthermore, it should be taken into consideration that ease of manipulation and simplicity of construction are features to be recommended particularly in micromanipulative work.

Stereoscopic Microscopes. The stereoscopic binocular microscopes, Figs. 6 to 8, have a number of remarkable features which render them particularly useful in performing a great diversity of low-power mechanical and other operations in scientific laboratories as well as in industry.

The Greenough-type stereomicroscopes enable the microscopic observation of objects in their three dimensional form. The stereoscopic vision,

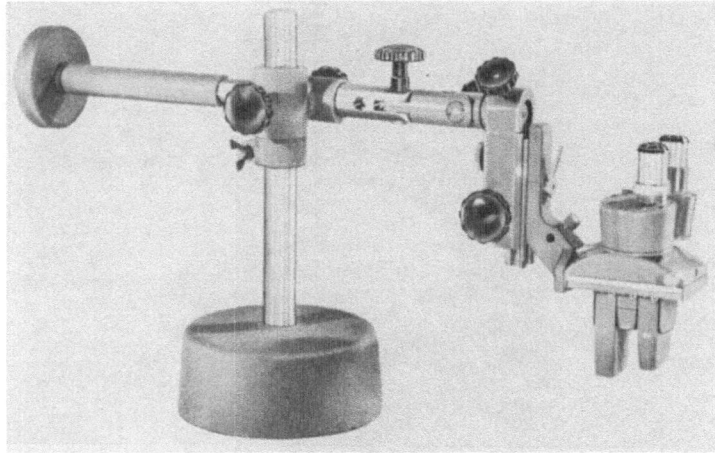

Fig. 6. Stereoscopic Microscope on a Heavy Pillar Stand for Examination of Large Objects. Courtesy of BAUSCH & LOMB (20).

coupled with a great depth of field, offers the depth of perception of ordinary human vision which gives a normal spatial image of the object and permits the observer to establish the true space relationship of different details in the object. The great penetrating power of these instruments permits the proper observation of relief and renders them especially suitable for the investigation of thick and irregular specimens. In the Greenough binocular microscope, such stereoscopic vision, which is always dependent on the production of a slightly different image in each eye, is achieved by utilizing two separate and complete compound microscopes, each having one objective and one eyepiece. The optic axes of the two microscopes are slightly inclined to one another. One microscope is used for each eye, and the two microscopes are directed at the same field from slightly different angles. In these instruments 18 mm objectives are the strongest objectives usually employed since it becomes increasingly more difficult to have the two objectives directed at the same field when focal length and working distance get very short.

In contrast to the Greenough-type microscopes, in monobjective binocular stereoscopic microscopes different images are formed in each eye by dividing the rays from a single objective between two oculars and modifying these rays in such a way so as to obtain a satisfactory stereoscopic image.

Another important asset of these instruments is that they offer an erect image. In other words, objects and movements are seen in their true, unreversed orientation. This feature is especially beneficial in micromanipulation since it enables the operative microtools to be guided easily and rapidly without confusion by avoiding the difficulty of interpreting reversed images as produced in conventional compound microscopes. When a micromanipulator is employed which does not offer a reversal of movement, p. 55, it may be used most conveniently in combination with a stereomicroscope. If a conventional microscope must be used, it should be equipped with an image-erecting eyepiece, p. 60.

Stereoscopic microscopes permit viewing large areas, which is another essential feature in many fields of application. A further asset is the exceptionally long free working distance offered by these instruments. This feature renders them indispensable in many micromanipulative operations in scientific laboratories and in industry, where the working tool points have to be operated between the object and the front lens of the objective.

With many of these instruments, combinations of paired eyepieces and objectives may be selected to obtain total magnifications from a few diameters to more than 200 diameters and free working distances ranging from a few millimeters at higher magnifications to several centimeters at lower magnifications. The diameters of the corresponding fields of view are in the range from a few millimeters in diameter or less, with highest powers, up to a few centimeters with lower powers, depending upon the particular microscope employed. Some instuments are equipped with rapid objective changers permitting the instant setting of a number of different magnifications while maintaining a constant working distance. In the instrument shown in Fig. 7, five different magnifications in each of the ranges from $6 \times$ to $40 \times$, $12 \times$ to $80 \times$, and $24 \times$ to $160 \times$, can be obtained by simply turning a drum. Many instruments may also be arranged for micrometry and for photography.

Natural daylight is usually sufficient for illumination of the object. For more intensive illumination, especially at higher magnifications or when the lighting conditions are not favorable, a suitable lamp may be employed, Fig. 7. In addition to examination in incident light, the microscope may be equipped with a suitable substage base with an adjustable mirror or a suitable light source for examination in transmitted light. Certain instruments, Fig. 8, provided with a stage for transmitted illumination, may also be fitted with a graduated revolving and centering object stage, as well as with accessories for examination in polarized light. These accessories comprise a filter polarizer located in the substage, an analyzer on an attachable slide, and compensators.

The versatility of these instruments is greatly extended by the numerous ways in which the stereoscopic bodies may be attached to different types of stands and the availability of stages of various designs for the examina-

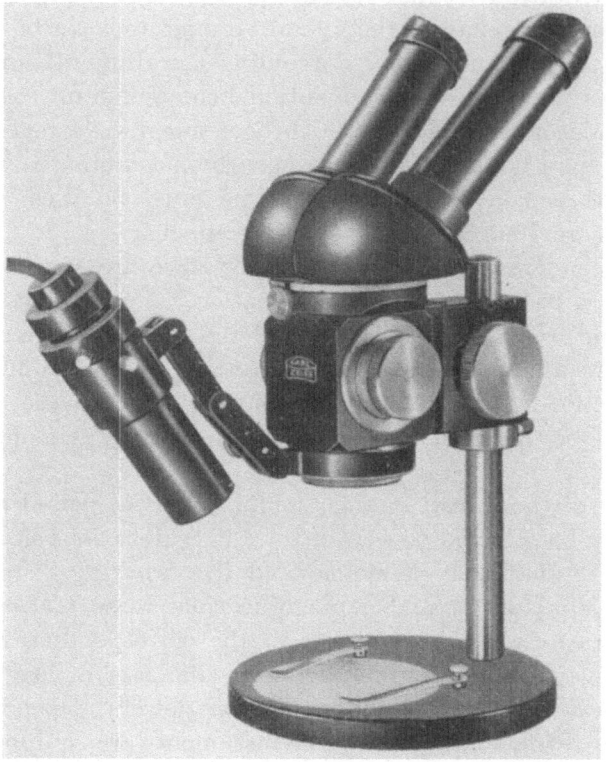

Fig. 7. Stereoscopic Microscope with Reversible Inclined Binocular Tube, Low Voltage Epi-Lamp, and Stand for Incident Illumination. Courtesy of C. ZEISS, Oberkochen/Württemberg (244).

tion in incident and transmitted light, permitting investigation of objects of the most diverse shapes and sizes. Only a few of these possibilities are illustrated in Figs.6 to 8, and also in Figs. 52, 102, 161, 176. Aside from the use in scientific and other laboratories in performing various mechanical operations under low-power microscopic observation—for instance in sampling operations, p. 187, in taking core samples of paint films, p. 196, 275, and in the opening of samples, p. 206—the three-dimensional vision, the ample working distance, the erect image, the large field of view, and the great penetrating power of the stereoscopic microscopes added to the great flexibility of mounting render these instruments invaluable also for many micromanipulative operations in precision mechanical industry. Thus the stereomicroscope is finding a widespread use in delicate and precise micromachining, microdrilling, p. 77, Fig. 52, welding, assembling, and in a diversity of other delicate mechanical operations such as those involved in making transistors, watches, electric shavers, and many other devices.

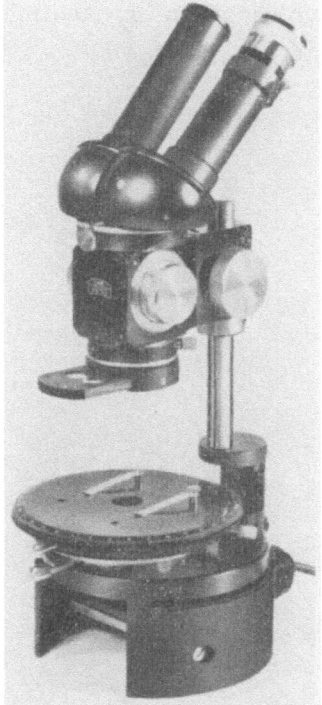

Fig. 8. Stereoscopic Polarizing Microscope with Reversible Inclined Binocular Tube. Courtesy of C. ZEISS, Oberkochen/Württemberg (244).

Stereoscopic binocular microscopes mounted on a floor stand with a many-jointed swinging arm permitting the operator to move the instrument into any desired position with great ease are successfully used in exceedingly delicate microsurgical operations of the middle and inner human ear (176, 204). The use of such operation or surgical microscopes has contributed greatly to the advance of this field of surgery. Most of the necessary operations are best performed at a magnification of about $10 \times$, though higher magnifications may be obtained without changing the working distance. Most satisfactory are a working distance of about 20 cm and a field of view of about 20 mm; both are readily offered by most modern surgical microscopes. The microscope may be equipped with a suitable demonstration eyepiece, and the field of operation may be efficiently illuminated by a built-in central illuminating device. The necessary work on the delicate bones is performed, with simultaneous suction and rinsing, by suitable tools such as frazers, drills, and diamonds. All operative tools, are manipulated by hand, however, and in human surgery, as far as the author is aware, mechanical micromanipulation proper has not been reported up to this time.

Inverted Microscopes. The essential features of these instruments are illustrated in Fig. 9, which shows an inverted microscope for the investiga-

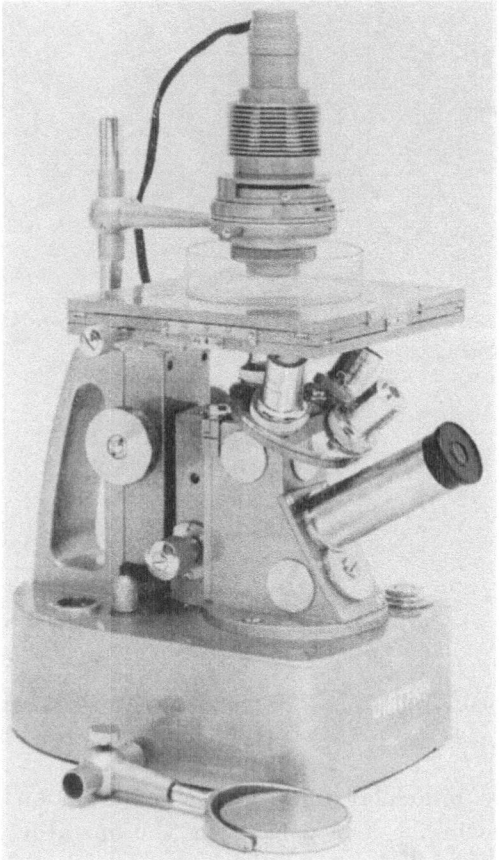

Fig. 9. Inverted Microscope for Examination of Transparent Specimens. Courtesy of UNITED SCIENTIFIC Co. (231).

tion of transparent specimens. This instrument is a Unitron inverted microscope Model MIC supplied by the UNITED SCIENTIFIC Co. (231).

The inverted microscopes are so designed that the object of investigation is placed on the microscope stage over an opening in it, while the observation system is located below the stage which therefore is free from obstructions. The body tube is inverted while the objective is directed upwards at the lower surface of the specimen located upon the stage. The field is observed through an eyepiece mounted on the observation tube. A suitable reflector is incorporated to direct the image-forming rays to the eyepiece. In several models provisions are made to permit observation and possibly photography in bright-field, dark-field, phase contrast, and polarized light. For the examination of opaque objects such as metals, ore minerals, and many other materials with inverted metallographic microscopes, the specimen is placed upside down over the opening of the stage directly above the objective. Consequently its lower surface, which becomes automatically levelled, must be illuminated from below the stage which may be done by using vertical illumination.

When using the more simple models for low-power work—for instance the E. Leitz (152) inverted chemists' microscope designed after Waldmann, which also has an image-erecting optical system—dishes, flasks, test tubes, etc. may be conveniently placed on the stage for observing through the bottom of the container such phenomena as precipitation and crystallization even if very small amounts of material may be involved. Ordinary microscope slide preparations as well as the usual mounted slides may also be examined. In comparatively large scale work with transparent objects, which does not necessitate micromanipulation, various tools such as pipets, needles, spatulas, capillaries, and glass rods may be easily manipulated by hand which is not hampered by obstructions. The tools cannot interfere with the path of rays between the specimen and the objective since the observation system is located below the stage.

In micromanipulative operations performed on transparent objects, the above mentioned features are even more valuable, and it was mainly due to them that inverted microscopes have found important applications in the micromanipulation of cells and tissues, p. 239. Delicate micrurgic operations, p. 241, may be performed in drops placed on a thin cover slip forming the floor of a moist chamber. The chamber is placed on the stage of the inverted microscope, and the necessary microtools are introduced into the moist chamber approaching the specimen from above. The condenser and a mirror for daylight illumination or a small illuminating lamp are located above the moist chamber. Aside from other advantages, p. 242, such arrangement permits micromanipulative operations to be performed under the highest magnifications. At the same time it offers an unobstructed and clear view of the whole microscopic field since the operative microtool points are not manipulated between the specimen and the observation system. In addition, this arrangement has the advantage over the hanging drop method, p. 240, that large and yet flatly spreading drops can be used more readily. Such large drops evaporate less rapidly and, being flat, restrict the movement of living organisms to be operated upon. Hanging drops can be flattened only by cleaning the cover glass extremely carefully with acids and alcohol.

The Unitron inverted microscope shown in Fig. 9 is suitable for the examination of transparent objects under various magnifications which may be over 1000 diameters. The instrument is equipped with a large built-in graduated mechanical stage and an illuminating system with transformer housed in the heavy base of the microscope. Another inverted microscope for transparent objects is shown in Fig. 13 with the volumetric submicromanipulator developed by Kopac and Harris (147). With this assembly, work can be carried out under magnifications up to 800 diameters.

Several manufacturers have designed microhardness testing devices, p. 82, to be fitted to their metallurgical microscopes. One of these devices, the Hanemann microhardness tester Model D 32, manufactured by C. Zeiss, Jena (245), is designed for the determination of Vickers hardness, p. 83. The device, p. 89, Figs. 61 and 62, essentially consists of a specially designed microscope objective with the Vickers diamond indenter cemented into the axis of its front lens so that ample free annular space is left for illumination and image formation. This high-precision microhardness testing device takes the place of a standard objective and is designed for use either with the large inverted metallographic microscope Neophot or with the small inverted metallurgical microscope Epityp. Both microscopes are produced by the manufacturer of the hardness tester. The specimen is placed in an inverted position on the mechanical stage of the microscope, and the indenter is made to act on the specimen from below by actuating the microscope adjustment. The specimen is observed through the ocular while the indentations are being made. Measurement of the area or the diagonals of the indentations are carried out by using a high-power objective. The construction of this testing device as well as its operation are described on p. 89. Microhardness indentations made with this device in the polished surfaces of different materials are shown in Figs. 155 to 159.

2. Micromanipulators.

In contrast to manipulators, such as the various remote pipetters and other varied types of large size manipulators employed in radiochemical and other radiation laboratories for handling large objects, micromanipulators are mechanical devices especially designed for holding and guiding delicate microtools used for handling minute samples in the field of the microscope at the required magnification. Thus tiny objects and very small amounts of liquids may be subjected under the microscope to a wide range of delicate operations with the aid of micropipets, microneedles, microelectrodes, micromagnets and other varied types of microtools, p. 136. These microtools are inserted in suitable tool holders which are firmly mounted on the micromanipulator by means of suitable clamps and thus are mechanically guided along the three spatial coordinates. In addition to the three-dimensional movement, which may not be enough for certain types of operations, some instruments incorporate revolving or tilting arrangements, to permit greater flexibility of motion.

Micromanipulators, originally developed for biological research, gradually became valuable tools in many fields of application. Apart from their use in several fields of biology and medicine such as the manipulation and microdissection of individual cells and small organisms and in the study of tissues, they are valuable aids in such work as solution chemistry on microgram to nanogram amounts of material, the determination of certain physical properties on very small samples, the mounting of tiny crystals and electron microscope specimens, the separation and isolation of individual specks of microscopic dimensions from metals, alloys and other material for examination. Of the other applications may be mentioned separation and study of fine fibers, assembling constituent parts of tiny devices, soldering of fine wires, the construction of quartz fiber devices, and several others. Specific examples of such applications are considered at length on p. 244.

a) General Features.

The choice of a suitable micromanipulator for a certain type of work largely depends, among other factors, upon the required magnification. A micromanipulator which acts satisfactorily under low power magnification may act under high power too rapidly or too loosely. This becomes apparent when one considers that under high magnifications microoperations are performed on exceedingly minute objects, and that the diameter of the field of view at high magnification is comparatively very small. Whereas the diameter of the microscope field offered by a low power $3.5 \times$ objective used in combination with a $6 \times$ eyepiece is about 5.6 mm, the field of view of an oil immersion $100 \times$ objective in combination with a $10 \times$ eyepiece is only about 0.14 mm in diameter.

Many micromanipulators of varied construction have been developed to fill specific needs and to satisfy various requirements of work at low to high power magnifications. A number of instruments of simple construction consisting essentially of three rack-and-pinion motions, permitting displacement of tools in the three space coordinates are commercially available. Most of these simple instruments are suitable only for types of operations requiring low magnifications usually not exceeding 200 diameters. Magnifications within this range are offered by all standard microscopes using low or medium power objectives, 16 mm and 8 mm objectives, as well as by most common stereoscopic binocular microscopes. The use of stereoscopic microscopes is recommended at low powers so as to make use of the exceptionally long free working distance, the wide field of view, in addition to the advantage of the three-dimensional vision. Some manipulators of the rack-and-pinion type, but of heavier construction and equipped with finer motions have been developed for work at low to medium magnifications up to about 500 diameters. In some of these instruments one or more additional fine motions may be provided, which act independently of the usual coarse drives. Certain massive, feed-screw operated micromanipulators such as the volumetric submicromanipulator, p. 39, are quite suitable for work at high magnifications.

Several types of high power micromanipulators also are available commercially. These are highly refined instruments equipped with exceedingly fine motions. Such fine motions are obtained by means of sensitive mechanisms based on a number of different principles which will be considered at length later. A number of these instruments are suitable for work at magnifications up to about 1000 diameters. In addition to these fine movements, high power micromanipulators are equipped with coarse motions which serve in the initial setting of the apparatus and for bringing the *assistants* (tools) into the field of the microscope prior to actual operations. Manipulation of microtools during actual operations is achieved only by means of the manipulator fine movements. This is necessary for high power manipulations requiring perfect control over the smallest changes of position of microtool points. In addition to the minuteness of the objects of investigation and the comparatively small field of view at high magnifications, the depth of focus is extremely small at high power. Consequently, the horizontal movements of the microtool points should be very accurately guided to stay nearly in the horizontal plane in order to remain in focus during delicate operations.

In contrast with low power micromanipulators, the extreme delicacy required of the operative movements of the high power instruments necessarily imposes certain other requirements and limitations of design especially concerning mobility and range of movements. Whereas in a high power micromanipulator greater emphasis is put on precision, in a low power instrument the most desirable features are convenience, mobility and speed. Thus the operative movements in the low power micromanipulators are comparatively coarse, rapid, long-range motions which, in such instruments as used in working with capillary cones, Figs. 16 and 18, may reach several centimeters.

Variable Sensitivity. One of the most important requirements in a micromanipulator to be used with high magnifications is that it should offer precise means of transmitting manual movements at greatly reduced ratios which can be varied at will to give the desired sensitivity of control. In other words, the operator should be able to adjust the ratio between the movement of his hand and the resulting motion of the microtool to correspond with the magnification employed. The magnitude of the operator's hand movements remains the same at all magnifications. This evidently results in more precise control at higher magnifications and permits the operating microtool to be confined to the extremely small field of operation, thus eliminating the possibility that the assistants be inadvertantly guided out of the field during delicate operations. If this happens under a high power objective, the only practical way may be to shift temporarily to a lower power objective in order to locate the microtool point and to get it back to the center of the field. In the B-D-H micromanipulator working under the principle of thermal expansion of electrically heated wires (3, 45), p. 65, the ratio of hand movement to microtool displacement may be varied at will from 250 : 1 to 50,000 : 1. This is achieved by means of variable transformers. In the pneumatic instrument of DE FONBRUNE (2, 76), p. 60, this ratio may be continuously varied between 50 : 1 and 2,500 : 1 by raising or lowering a sliding collar fitted around the vertical control handle. Rods attached to this collar run to each of the two pistons of the horizontal pumps. By altering the position of the sliding sleeve, the amplitude of the piston in the horizontal pumps is changed, and thus the range and sensitivity of the horizontal movements. Reduction of movement in the hydraulic micromanipulator developed by MAY (163), p. 38, is achieved in a similar way; the ratio may be varied between 30 : 1 and 300 : 1. In most instruments such variable sensitivity is confined to the two movements in the horizontal plane. For most work, however, the variation of the sensitivity of movement in the vertical plane is less important.

Under low power, working at a great reduction ratio is often inconvenient. The movements of microtools would be very slow, particularly with screw controls since it may take a number of turns of the screw for the microtool point to traverse the comparatively large field of view of low power objectives. If the instrument is not in a perfect condition, this may lead to jerky and intermitted motion. In a number of low power instruments of the rack-and-pinion type, however, one or more of the three-dimensional movements are equipped with reduction drives to enable better control of small changes in position. This is suitable for applications requiring more precision. The reduction ratio in low power instruments, however, should be comparatively very small. The BRINKMANN low power micromanipulator model MP II (35), Fig. 16, for instance, often used in electronics, has all its three movements equipped with planetary reduction drives with a reduction ratio of only 5 : 1. In the low power instrument developed by BARER and SAUNDERS-SINGER (16), p. 37, which is based on the usual principles of the pantograph and which permits accurate work up to a magnification of $200 \times$, the reduced movement is also three-dimensional with a 4 : 1 ratio, and in another low power instrument (117) the reduction ratio is about 8 : 1. In the medium power micromanipulator described by BUCHTHAL and PERSSON (42) the reduction usually employed is of the order of 50 : 1.

Erect Image. Under the microscope, the microtool should appear to move in the same direction as the operating hand. Such expected or natural motion greatly facilitates manipulation since it eliminates confusion as to the direction in which the hands should operate to produce displacement of microtools in the desired direction. In a number of micromanipulators, certain components can be arranged to give a reversal of movement so that the microtool point when viewed in the reversed field of an ordinary compound microscope appears to follow the movements of the hand; otherwise the microscope should be fitted with a suitable image-erecting eyepiece. All stereoscopic microscopes give erect images. Examples of micromanipulators that may give reversal of movement, and thus may be used with image reversing microscopes, are those of DE FONBRUNE (77), p. 60, MAY (163), p. 38, also the lever activated instrument of E. LEITZ (152), p. 48, the B-D-H micromanipulator (3, 45), p. 65, and a number of others. On the other hand with some instruments, such as the sliding micromanipulator of C. ZEISS (245), p. 55, for instance, a non-reversing microscope, or one fitted with an image-erecting eyepiece, should be used, Fig. 31.

Arrangement of Controls. In any micromanipulator, ease of operation is one of the most desirable features. For this purpose it is very helpful that the controls of the operative movements of the instrument, and also that of the injection device, be located in close proximity so as to permit selection and operation of the desired motion with the fingers of one hand without distracting the attention of the operator. Such arrangement can be seen in Figs. 23 and 26, showing the high-power micromanipulator of E. LEITZ (152).

Where two micromanipulator units are used in a single assembly, one on either side of the microscope, right- and left-hand arrangement of controls is most desirable to ensure bilateral symmetry. Each two corresponding controls in double units should have identical direction of motion. This is to permit well coordinated movements of the respective operating microtool tips.

In a number of high-power instruments all operative fine movements, in the three dimensions of space, are installed in a single control, which may remain in the hand during the whole actual operation. Examples of these instruments are the sliding micromanipulator of C. ZEISS (245), Fig. 30, the pneumatic instrument of DE FONBRUNE (2, 76), Fig. 40, and also the B-D-H micromanipulator (3), Fig. 43, all of which are described in detail, p. 48. Other examples are the hydraulic micromanipulator developed by MAY (163), p. 38, and the high power instrument described by BARER and SAUNDERS-SINGER (15) which is a rather elaborate modification of the simple instrument originally described by SCHUSTER (209), also the micromanipulators described by BUCHTHAL and PERSSON (42). The last two instruments are suitable for work at magnifications up to 400 and 500 diameters. In certain low power instruments also, all movements are carried out by means of a single control; the low power instrument of BARER and SAUNDERS-SINGER (16) may be mentioned. These single control instruments have an advantage over other micromanipulators in which three operative controls are employed, one for each space coordinate. To the latter category belong practically all the low power micromanipulators of the rack-and-pinion type (35, 168), p. 42, and also a number of high power instruments as that of CHAMBERS (58), p. 69. This in itself is not a serious disadvantage in comparatively coarse manipulations.

Freedom of Movement in One Plane. In the instruments just mentioned in which three operative controls are used, one for each space coordinate, in order to move the microtool point between any two specified points in space, the operator instead of going directly along the desired path has to do so via X, Y and Z. It would be best, however, if the micromanipulator enabled the microtool point to be moved accurately, smoothly, and directly between any two specified points in space. Due to mechanical difficulties such an ideal micromanipulator has not been developed as yet. A number of medium and high power micromanipulators, however, permit complete freedom of motion in one plane combined with a separate movement in a direction at right angles to that plane. The medium power instruments described by BUCHTHAL and PERSSON (42) and by SCHUSTER (209) referred to above, permit such free movement in a vertical plane combined with a horizontal movement at right angles. Most high power instruments used at present permit such free movement in the horizontal plane, and a separate motion in the vertical plane. This is more suitable since most of the work is usually done in the horizontal plane, the latter being the plane in focus. Examples are the instruments of E. LEITZ, p. 48, C. ZEISS, p. 55, DE FONBRUNE, p. 60, the AMERICAN OPTICAL B-D-H micromanipulator, p. 65, MAY, p. 38, and the high power instrument described by BARER and SAUNDERS-SINGER (15). Of these the first four instruments are described in detail, p. 48. In one of these instruments, that manufactured by E. LEITZ, as already mentioned, the horizontal fine motions are operated by means of a single joy stick, while the fine vertical movement is actuated by a separate control located in close proximity to the joy stick. In the other instruments the horizontal fine motions as well as the vertical motion are installed in a single control.

Stability and Freedom from Vibration. A good micromanipulator should be stable and free, as far as possible, from vibration effects. It is most desirable to eliminate the transmission of vibration to the operating microtool points. This is essential in all delicate high power manipulations. Aside from the sturdy construction of the instrument, other important factors help in eliminating vibration, such as the use of remote controls and the independent mounting of micromanipulator and the microscope on a heavy common base plate.

Massive Construction. Most micromanipulators are massively built to minimize vibration. This feature is advantageous particularly in instruments used for delicate work under high magnifications, and also in medium and low power manipulations. An exception make the instruments to be used in operations requiring great flexibility of movement and orientation.

Independent Coarse and Fine Movements. To obtain maximum stability, in practically all high power instruments, the operative fine movements are independent of the coarse controls. The latter may be put out of action after the preliminary adjustments are carried out, so that fine motions may not be disturbed during delicate operations.

Remote Controls. The use of remote control devices for operating micromanipulator fine motions, avoiding rigid connection between the tool holder carrying the microtool and the operating controls of the instruments, greatly minimizes transmission of vibration from the hand to the microtool, and thus ensures delicate and precise movements under high magnifications. It may also reduce the effects of transmitting accidental jolting of these controls to the microtool. Operation by remote control is desirable for work at high magnifications particularly with those micromanipulators having a comparatively light construction, but is not required for coarse and lower power manipulations. In the present model of CHAMBERS' micromanipulator this is achieved by means of flexible shafts with operating knobs attached to the three fine controls of the instrument, Figs. 49 and 50. The DE FONBRUNE micromanipulator, Fig. 39 is another example of an efficient remote control instrument, since the manipulator in which the control handle operating the fine motions is installed and the receiver carrying the microtool are separate units interconnected only by long flexible tubing; similar arrangement is also found in the hydraulic instrument developed by MAY (163), p. 38.

Base Plate. Micromanipulator and microscope may be combined in various ways. The commonly accepted principle is to mount them independently on a heavy common base plate. This is to obtain maximum stability and to minimize vibration. Adequate provisions are made to make it possible to return the instrument to the operating position accurately and easily after withdrawing it for changing the operation chamber or the assistants. Usually the injection apparatus also is held firmly on the same base plate. In assemblies where more than one micromanipulator is used, the base plate should be large enough to accomodate these in addition to the microscope. Provisions may be made also, if required, to permit two or more manipulators, mounted conjointly, to be arranged in a number of different positions on the same base plate while the microscope remains in an unchanged position on the base plate. It is desirable that the base plate be heavy and may rest on rubber feet to support the instrument firmly. The work bench should be rigid and free as much as possible from external disturbances.

In some micromanipulators, adapted for use for a variety of purposes, the effective part of the instrument may be detached and mounted on a sliding bar or a clamp in the required position or orientation. Some of the older instruments can even be clamped directly to the stage of the microscope as in the earlier models of BARBER's pipet holder, Fig. 10, and in one of the older models of Chambers micromanipulator (58, 60). In these instances the stability of the micromanipulator depends upon that of the microscope. This type of arrangement is still used but only in isolated instances and for rather specific purposes.

Other Features. Besides the above features, there are other requirements that a good micromanipulator would possess such as freedom from play and resistance to wear. The hand motion should be transferred to the operating instrument simultaneously, synchronically and uniformly. If rack-and-pinion movements are employed they should be very accurately machined. Even precisely made micrometer drives and rack-and-pinion mechanisms are subject to wear after prolonged use, and may show a certain amount of play.

Finally and for all instruments, robustness and absence of delicate components that are liable to be damaged, low price, and availability of spare parts at a comparatively low cost are factors to be desired when choosing between two micromanipulators of similar performance.

b) Development.
α) Early Development.

Micromanipulators, nowadays widely used in many fields of science and technology, were originally developed for biological research. Ever since the invention of the microscope, attempts have been made to carry out microdissections within the microscopic field.

A number of micromanipulators, developed mainly for this purpose, already existed in the last century. SCHMIDT (202, 203), as early as 1859, described an accurate instrument for the dissection of tissues. It consisted of a base fastened to the stage of the microscope carrying a number of clamps each of which was equipped with three screw controlled movements. The tissue was held in place by means of a lever attached to one of the clamps while the other clamps served for holding and guiding the various operating microtools. Operations were performed while the sample, the microtools, and the lower lens of the microscope objective were immersed in water or dilute alcohol. In 1887, CHABRY (53) used a spring device mounted on the stage of the microscope for performing certain operations in experimental embryology. The device was so delicate that it permitted the tip of a glass microneedle to be shot into an ovum to any desired depth.

The needle was held in a sheath in a manner which permitted the needle to be pushed into the bore of a glass capillary just wide enough to admit and hold a single ovum.

In 1907 McClendon (164) attached a vertical screw movement to an ordinary Spencer mechanical stage to obtain a three-dimensional motion, and a clamp to hold a microtool. The instrument was used for dissecting the egg cases of certain animals and enabled sucking the nucleus out of a Chaetopterus egg. In 1909 the instrument was slightly modified (165) and was used in dissecting and stimulating certain protozoa.

In earlier instruments microtools had to operate between the specimen and the microscope objective, and therefore only comparatively low power objectives could be used. Between 1904 and 1914, Barber published several papers (8, 9, 10, 11, 12) describing a mechanical pipet holder and a method for isolating microscopic organisms by manipulating micropipets in a drop hanging from the undersurface of a cover slip suspended over the moist chamber. This is a significant development since, when working in hanging drops, microtools are manipulated below the cover slip thus eliminating all obstacles between the microscope objective and the sample and permitting the use of high power and oil immersion objectives. The instrument proved very valuable also for microdissection. Needles can be manipulated in a shallow hanging drop containing the specimen. The cells to be dissected may be pressed against the undersurface of the coverslip. Barber's pipet holder has been described in detail by Chambers (55). A simple form of this instrument, clamped to one side of the microscope stage, is shown in Fig. 10, illustrating its basic construction. It consists essentially of three movements at right angles to each other. Each motion consists of a carrier pushed along a groove by means of a screw at one end. Three such carriers are built up on one another in such a way that each carrier travels in a direction perpendicular to the other two. Movements in any one of the three space coordinates may thus be imparted to the point of a microtool clamped to the top carrier. Work may be carried out in a moist chamber open at one end and held by the attachable mechanical stage of the microsocpe. The three-dimensional movement offered by the instrument, and the two movements of the moist chamber in the horizontal plane, by means of the mechanical stage of the microscope, give ample motion to carry out the necessary manipulations. The single holder is designed to carry only one microtool. A double holder has the advantage that two microtools may be used simultaneously, each microtool can be moved independently. In a later development, the holder is clamped on a stand independent of the microscope, and in a double holder the two stands are clamped together on a common base plate. For many years, from 1904 until about 1921, Barber's pipet holder was used successfully in the isolation of bacteria and in the microdissection and microinjection

of marine ova and of animal and plant cells. With it, KITE and CHAM-
BERS have been able even to drag individual chromosomes out of the
cell. Although such excellent work could be done with the BARBER
pipet holder, the instrument suffered from several disadvantages and was
replaced by finer micromanipulators of later development. The main disadvantage of BARBER's instrument is that, chiefly due to lack of adequate
contact surfaces of its sliding parts, its movements are subject to ready
wear and, after some time of use, this gives rise to a certain amount of false

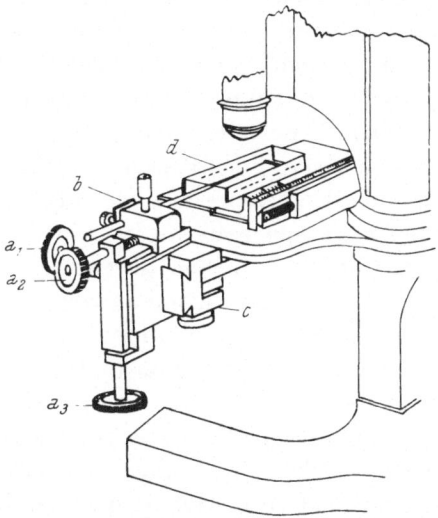

Fig. 10. Barber's Single Pipet Holder. It is clamped to microscope stage with moist chamber
in position, and with point of microtool directed upward for working in a hanging drop, a_1, a_2,
a_3, controls for horizontal and vertical movements; b, needle clamp; c, clamp to stage of
microscope; d, moist cell. After R. CHAMBERS (55).

motion rendering them unreliable for delicate high power manipulations.
This instrument, as well as practically all instruments of earlier development,
are nowadays of historical interest only.

In 1921–1922, CHAMBERS (57, 58, 60) described a micromanipulator
quite suitable for work under the highest magnifications of the compound
microscope, which soon replaced the BARBER pipet holder in various fields
of work. The operative movements of the instrument are so precisely
controlled as to make possible dissection of cell chromosomes (63). These
fine movements, Fig. 48, depend upon the spreading apart, by means of
micrometer screws, of bars of rigid metal connected at their ends by resilient
metal acting as spring hinges. Though the fine three-way movements imparted are curvilinear, the tip of the microtool practically moves in straight
lines in the tiny field of operation. The instrument has several advantages
(131) over BARBER's pipet holder. Since frictional surfaces are absent in
its fine movements, these movements can withstand long usage without

being loosened by wear and tear; thus a high degree of precision is maintained. Furthermore, backlash is avoided by the micrometer screws being opposed by the counter spring mechanism. There are other advantages in the use of various accessories to aid in manipulation of microtools. The instrument has received improvements over a period of many years. In the present commercially available models (35, 152), the precision is increased by the introduction of remote controls attached to the three fine motions, Fig. 49. This almost completely eliminates transmission of vibration from the operator's hand. The instrument is described in some detail on p. 69. In 1951, KOPAC (143) described a considerably modified model of the original instrument. This is briefly considered on p. 39.

In 1925, TAYLOR (226) devised an instrument which in principle is similar to the BARBER pipet holder. The micromanipulator, however, is far more stable mainly due to its massive construction. A right- and a left-hand manipulator unit, each consisting essentially of a massive three-dimensional movement, are mounted on a spaceous heavy cast iron base plate which also supports the microscope, by means of two clamp screws, between the two manipulator units. This arrangement largely eliminates vibration effects and offers greater stability. A modified form of the instrument is shown in Fig. 145. The massive construction of the sliding parts added to an extensive contact surface, which amounts to about 45 cm² for each pair of slides, exactly assembled, offers a high degree of precision of all three movements and makes them less subject to wear and tear. A steel gib, introduced on one side between each pair of slides and regulated by three screws each working against a stiff spring, enables wear of the sliding parts, resulting from prolonged use, to be automatically and continuously adjusted. The two manipulator units are held in position by means of clamp screws. When these screws are loosened, each manipulator unit can be rotated through 90 degrees. This facilitates exchange of the operative microtools without disturbing the microscope or the operation chamber.

In 1927, DUNN (84) discussed the use of the hydraulic pump for obtaining micromanipulation. He demonstrated that hydraulic feed can be satisfactorily adapted to obtain exceedingly fine, regulated movement. It permits large variations in fineness of movement, freedom from vibration because of the characteristics of fluids, and the flexibility of the design of apparatus. Fluid transmission of pressure, furthermore, allows simple remote control to be introduced easily and effectively. One of the widely known applications of hydraulics is the use of a combination of a small pump forcing fluid into a larger cylinder to obtain great lifting power. With a system of two cylinders and pistons connected together by a length of inexpansible tubing and filled with liquid so that all air is displaced, on moving one piston, the other piston will move an amount corresponding to the inverse ratio of the areas of the controlling and the driven pistons.

Wide variations in the fineness of movement may be obtained by varying the relative sizes of the two pistons. The controlling piston may be a suitable machine screw, and leaking may be eliminated by a packing gland. By using two suitable pistons of the proper ratio, controlled movements as small as 0.5 μm can be obtained. DUNN used paraffin oil as hydraulic fluid. To avoid lag, the oil should be free of air bubbles. Remote control can be easily obtained by using seamless or flexible metal tubing. To obtain movement in three dimensions, three plunger systems at right angles are necessary. A rotating movement may be included, if required, by using two additional plungers acting tangentially. Several instruments of varied designs were constructed and tested. With one of these designs, having a hydraulic ratio of 1:19.5, the movement was smooth and free from vibration at a magnification of 1100 diameters (84). Two micromanipulators of later development are based on the principles of hydraulics, namely the pneumatic instrument of DE FONBRUNE (77), and the hydraulic instrument described by MAY (163), p. 38.

In 1930, RICHARDS (193) developed a simple screw plunger device similar in principle to certain types of automatic lead pencils. This device can only be employed for working above the slide under low power, and its use is further limited mainly to those simple operations requiring an accurately controlled direct thrust in one plane only. The tubular, pencil-like device was held in an inclined position in a short swivel clamp supported on a suitable rack-and-pinion pillar which offers an accurate vertical movement; horizontal adjustment was crudely obtained by tapping or pushing the microscope base. RICHARDS successfully used this instrument for such operations as those involved in puncturing BOWMAN's capsule in the frog kidney and the insertion of electrodes into a renal tubule.

β) More Recent Development.

In the last three decades many forms of micromanipulators have been developed. Of these may be mentioned the instruments described by EMERSON (U.S. Patent No. 1 828 460) in about 1930, FITZ (100, 101) in 1931 and 1934, DE FONBRUNE (77) and GETTENS (106) in 1932, BUCHTHAL and PERSSON (42) in 1936, HANSEN (117) and HARDING (118) in 1938, SCHUSTER (209) in 1947, BARER and SAUNDERS-SINGER (15) in 1948, BARER and SAUNDERS-SINGER (16) in 1950, KOPAC (143) and KOPAC and HARRIS (147) in 1951, BÉKÉSY (23), BRINDLE and WILSON (34), MACKAY (157), and ORDWAY (180, 181) in 1952, BUSH, DURYEE, and HASTINGS (45), and MAY (163) and SCHOUTEN (207) in 1953, KOPAC (146) in 1955, BÉKÉSY (22) in 1956, and by ANDERSON (5) in 1958; also the lever activated micromanipulator of E. LEITZ (152) and the sliding instrument of C. ZEISS (245). Of these instruments, those of GETTENS (106) and HARDING (118), BARER and SAUNDERS-SINGER (16), ORDWAY (180,

179), MACKAY (157), and BRINDLE and WILSON (34) are low power micromanipulators; those of BUCHTHAL and PERSSON (42) and of SCHUSTER (209) are medium power instruments, while the others are suitable for work at high magnifications.

A number of commercially available low power instruments, not included in the above list, will be described on p. 42. Some commercially available high power micromanipulators will also be described in detail. These are the lever activated instrument of E. LEITZ, p. 48, the sliding micromanipulator of C. ZEISS, p. 55, the pneumatic instrument of DE FONBRUNE, p. 60, the electric B-D-H micromanipulator of the AMERICAN OPTICAL Co., p. 65, and finally that of R. CHAMBERS, p. 69. In the following, a number of the other instruments enumerated in the above list will be considered only briefly, namely those of Emerson (U.S. Patent No. 1 828 460), FITZ (100), HARDING (118), BARER and SAUNDERS-SINGER (16), BRINDLE and WILSON (34), ORDWAY (180, 181), MACKAY (157), MAY (163), KOPAC (143), KOPAC and HARRIS (147), and KOPAC (140, 146).

The EMERSON micromanipulator is an instrument of the joy stick type which has been designed by G. H. EMERSON around 1928, and is covered by U. S. Patent No. 1828460, which gives a fairly good description of the principles upon which it operates: The instrument has been produced by the J. H. EMERSON COMPANY (22 Cottage Park Avenue, Cambridge 40, Mass., U. S. A.) since about 1930. Several models are commercially available for use in biological research, in processing and assembling electronic parts, as well as for other purposes and various microscope magnifications. The instrument is designed so that as many as six units may be mounted about the microscope on a common base plate for complex operations. In most models, movements in any direction in the horizontal plane are actuated by a single control lever or joy stick with two hemispheres by which a plate, carrying the tool holder, is moved. The ratio between hand travel and the corresponding motion of the microtool point depends upon the distance between the two hemispheres of the lever, which can be readily adjusted to allow for changes from low to high power. The full travel of the control lever is about 8.7 cm, and the corresponding displacement of the microtool point can thus be varied from about 3 mm down to less than 2.5 micrometers. These models are therefore suitable also for work under the highest magnifications. In most models the manual motion is reversed so that, under an ordinary compound microscope, the microtool point appears to follow the movements of the operating hand. Certain models, however, also permit direct motion for convenient manipulation under a stereoscopic microscope. The instrument is also equipped with vertical controls, coarse adjustments, and other movements which may differ in the different models. Fig. 11, shows two manipulator units of Model A mounted on a common base plate. In this exceptionally compact model the fine vertical motion also is lever controlled and, like the joy stick controlled fine horizontal movements, has an adjustable ratio. The instrument is not provided with coarse adjustments, but the tool holder offers a wide vertical coarse adjustment and can be rotated horizontally

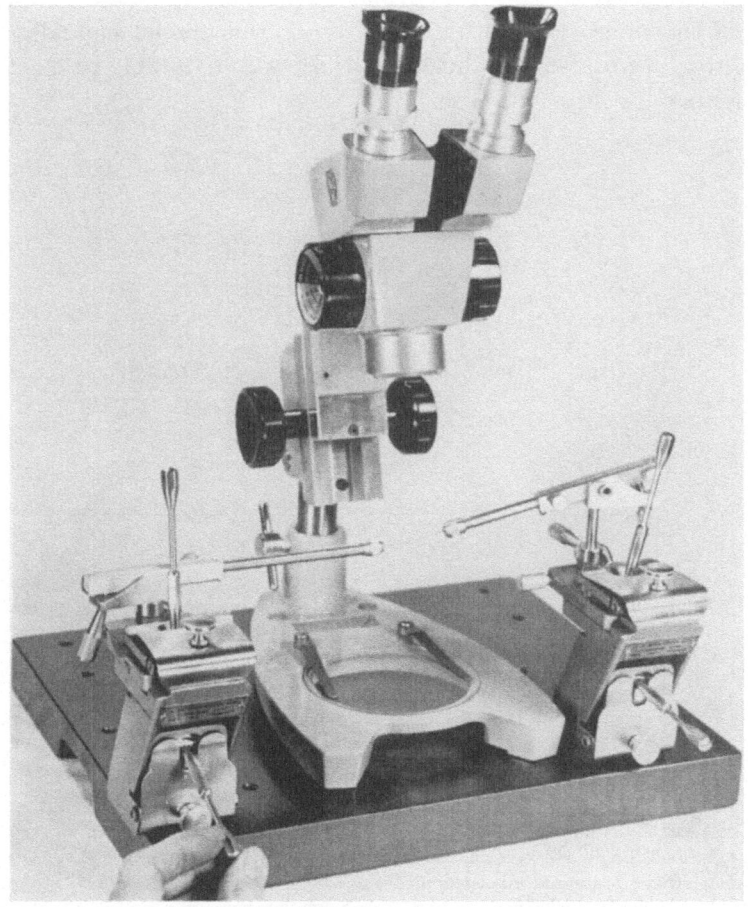

Fig. 11. EMERSON Micromanipulator Model A. Two manipulator units on a common base plate. Courtesy of J. H. EMERSON Co.

and tilted in the vertical plane. Fig. 12, shows the main construction features of Model D. In this model the instrument is equipped with several coarse adjustments. In addition to the joy stick controlled fine motion in the horizontal plane, this model is provided with fine and coarse screws for vertical adjustment. The instrument offers the choice between direct and reversed motion, according to the position of the joy stick control. When the lever is brought downward as shown in Fig. 12, direct motion is obtained. The lever, however, may be shifted into an upright position where it gives reverse motion for use with the customary compound microscope. The entire manipulator column may be adjusted vertically and may be rotated through 360°. The tool holder is clamped into a rectangular groove of a mechanical stage which may be rotated through an arc. The tool holder may be tubular or rectangular. If a tubular holder is used, it may be rotated within the rectangular groove of the stage to adjust the position of the microtool. Other features of this model can also be seen in the figure. A number of other models are also available to meet various requirements.

Fig. 12. EMERSON Micromanipulator Model D. *a*, fine vertical adjustment; *b*, coarse vertical adjustment; *c*, lock for coarse vertical movement; *d*, engaging screw for coarse rotary motion of mechanical stage; *e*, coarse adjustment for rotary motion of mechanical stage; *f*, chuck for microtool; *g*, column lock; *h*, indexing spring for rotation of column; *i*, column vertical adjustment; *j*, joy stick control for horizontal motions; *k*, lock for rotary motion of mechanical stage; *l*, horizontal motion screw of mechanical stage; *m*, clamp to retain tool holder in rectangular groove; *n*, tubular tool holder. Courtesy of J. H. EMERSON Co.

In the high power instrument described by FITZ (100), a fine thrust or sagittal motion is produced by a differential screw system. The transverse motion as well as the vertical motion are produced by a cam or wedge action on semi-sliding components. Later (101) a few changes have been introduced on the original model to make it more suitable for rapid selection of unicellular organisms for pure culture work.

HARDING's low power micromanipulator (118) is based on the usual principles of the pantograph, which is an instrument much used for mechanical copying of drawings to different scales. Similarly, in the present micromanipulator the movement of the hand is copied by the microtool point on a much smaller scale. The instrument may be used with stereoscopic microscopes. With an ordinary microscope which gives an inverted image, the instrument can be arranged to produce a reversal of movement. This instrument has been devised for dissecting under low power comparatively large animals of the size of various small Ostracoda, which are too small to be dissected by hand. Another micromanipulator based on the principle of the pantograph is the low power instrument described by BARER and SAUNDERS-SINGER (16). It consists essentially of three separate pantographs or parallel rulers permitting reduced motion in three dimensions.

BRINDLE and WILSON's low power micromanipulator (34) is a low power instrument especially designed to meet various requirements for performing chemical operations in capillary cones, p. 92. As with the instrument used by EL-BADRY and WILSON, Fig. 63, suitable clamps are provided on the top movement for direct mounting of a micrometer operated hypodermic syringe to which the micropipet is attached. The micropipet is manipulated in relation to the capillary vessels which are held on a suitable carrier on the rotating mechanical stage of the microscope. The instrument is equipped with precision screw movements in the three spatial coordinates, each permitting a 4-cm excursion, and a coarse thrust or sagittal rack-and-pinion motion permitting a fast advance over a wide range of about 15 cm. The latter movement permits the micropipet to be withdrawn quickly well out of range of the microscope or to be advanced towards the field of operation. This coarse movement is independent of the precision thrust motion, and can be put out of action in any desired position by means of locking screws. The precision thrust motion serves for manipulation of the pipet tip during actual operation. In addition, the instrument is equipped with a rotation movement of about 20° round the vertical axis. This permits adjustment of the angle of inclination of the shaft of the micropipet in relation to the capillary vessels in the microscope field. Such movement is necessary when the microscope is not equipped with a rotating stage. The syringe, directly mounted on the top movement of the micromanipulator (the precision thrust motion), is held firmly in position by means of two cork-lined, hinged clamps.

The micromanipulator devised by ORDWAY at the National Bureau of Standards (180, 181) is a low power instrument which has been used (180) in combination with a stereoscopic binocular microscope to facilitate mounting and grinding of small single-crystal specimens for x-ray diffraction studies. It consists of a heavy base plate carrying two manipulator units and a three-legged triangular platform for supporting the specimen. Each manipulator unit consists of a 45° right triangular plate carrying an ad-

justable column with ball-and-socket joints. The triangular manipulator
plate is supported in a suitable fashion on the common base plate by means
of three spring-loaded levelling screws at the corners of the triangle. The
specimen is placed on the triangular platform and is operated upon with
microtools attached by means of suitable holders to the upper ends of the
jointed columns of the two manipulators, and suspended above the center
of the platform. The various components are geometrically arranged so
that controlled two-dimensional motion of the microtool in a practically
horizontal plane is produced by turning the proper levelling screws of the
manipulator plate. It is more convenient to adjust the height before starting
actual operation since vertical adjustment needs simultaneous turning of
the three levelling screws.

MACKAY's device (157) is a low power instrument developed to facilitate
mounting small crystal specimens in x-ray goniometers. It is suitable for
the mounting of tiny crystals, down to about 15 μm in length, on glass
fibers of 15-μm diameter. The device is a simple, suitably supported
probe which can be racked up and down with the microscope tube. The
specimen is placed on a glass slide under the microscope and is operated
upon by means of a suitable microtool attached to the shorter arm of the
lever. A fine needle for selecting, cutting, or orienting the sample, or a
pip carrying a glass fiber for mounting the crystal may be employed. The
device gives a reduction ratio of 4:1, and the microtool remains stationary
in any required position. The procedure that may be used for mounting
crystals with this instrument is briefly described on p. 290.

In a simple device of ALBER (1), the adjustable tool handle is carried
by the mechanical stage so that the rack-and-pinion motions of the latter
may be utilized.

The micromanipulator developed by MAY (163) is a hydraulic instrument
employing the properties of metal bellows. The instrument possesses a
number of basic features in common with the pneumatic micromanipulator
of DE FONBRUNE, p. 60, both instruments being based on the use of fluid
transmission of pressure. The two micromanipulators are also similar in
principle and in the general construction features to those hydraulic models
devised earlier by DUNN (84), which are briefly considered on p. 32. The
present instrument consists of two essential units, a remote hand control
unit and a light compact manipulating head carrying the microtool. The
two units are set independently of one another. The hand control unit
comprises a rocking lever operating three fine mutually perpendicular
bellows or pumping devices connected hydraulically to three wider bellows
set similarly at right angles to each other on the manipulator head. Lateral
motions of the control produce a free fine motion of the microtool in the
horizontal plane. Fine vertical movement is obtained by sliding the sensitive pencil-grip control up and down. The pumping and the receiving

bellows are connected by three flexible inexpansible tubings. Each pumping bellows is connected to a receiving one and controls the motion of the microtool in one dimension. The reduction ratio is readily variable between 25:1 and 300:1. The transmitting and the receiving components can be connected to give like direction to the motions of control and image of the tool as seen in the reversed field of a compound microscope. Coarse adjustment is performed by means of an independent remote control. In the present instrument, in contrast to the DE FONBRUNE's receiver, Fig. 41, the manipulating head to which the microtool is attached can be clamped to the miscroscope tube to move with it. By means of this arrangement the point of the microtool always remains in focus. It also permits the microscope to be used at any inclination including the inverted position. Paraffin oil, free from air bubbles, was used by DUNN (84) as a hydraulic fluid. MAY, however, suggests the use of either gas-free water containing about 30% glycerol or thin vegetable or mineral oils. The use of pure water, due to its low viscosity, has the disadvantage of transmitting hand tremblings to the microtool. The other fluids above mentioned are sufficiently viscous to damp out such motions. The instrument is suitable for the manipulation and microdissection of bacteria and small organisms and also for other types of work such as manipulation of tiny crystals and separation of fine fibers.

The instrument described by KOPAC (143) is a modified form of the original CHAMBERS' micromanipulator (58), p. 69. The most important improvement is the replacement of the vertical fine and coarse control movements by a microscope body equipped with fine and coarse adjustments. Thus, in the modified instrument, the fine vertical movements imparted to the microtool points are rectilinear instead of curvilinear. This is useful in certain operations where precisely straight vertical motion is required. The range of this fine vertical movement is about 3 mm, which is sufficient for any operation.

The volumetric submicromanipulator, Fig. 13, developed by KOPAC and HARRIS (147), is designed for use in combination with an inverted microscope, giving magnification up to 800×, for preparing measured volumes of substrate mixtures for microdilatometric studies (144). Two micromanipulator units integrated with controls of the plunger device are mounted in front of the microscope. The micromanipulator movements in the two horizontal coordinates are achieved by means of spring-loaded, ball-bearing slides equipped with metric screws with large micrometric heads. The vertical movement is made of a microscope focusing fine and coarse adjustment carrying a bracket for supporting the pipet clamp. The plunger device is equipped with two steel pistons of different diameters. The fine pistons are operated by a feed screw and worm-gear drive. The plunger is connected to the pipet holder by means of a long flexible copper tubing.

The piston chamber, the copper tubing, the pipet holder, and the micro-pipet are filled with silicone oil acting as hydraulic fluid. Operation depends on the displacement of hydraulic fluid by one of the two pistons. The larger piston, the coarse volumetric control, has a diameter of about 0.4 mm, while the smaller piston, providing fine displacement, is about 0.125 mm in diameter. The linear displacement of the piston is measured with an electronic micrometer capable of measuring displacements as small as 2.5 nm corresponding to a volume of about 0.03 pl $= 3 \times 10^{-11}$ ml with the smaller piston. The electronic circuit permits pre-setting of the in-

Fig. 13. The Volumetric Submicromanipulator.   Courtesy of M. J. KOPAC (144).

strument to the required volume. This allows the worker to concentrate on the delicate operation of removing cytoplasmic material from the cells. The micropipets are siliconized inside and outside by treatment with Desicote. This treatment renders them water repellent and minimizes the sticking of cytoplasmic residues to the glass. By means of this equipment, measured volumes of cytoplasmic substances as small as 0.3 to 10 pl may be extracted from a living cell and transferred to very small amounts of substrate mixtures or implanted into other cells.

The television-micromanipulator described by KOPAC (140, 146) is a highly elaborate assembly which presents another advance in subcellular transplantation studies. The assembly, Fig. 14, includes four LEITZ lever-activated micromanipulator units, a pair on each side of the microscope. The LEITZ micromanipulator is described in detail on p. 48. The present assembly also includes a monochromator, a LEITZ Mikas attachment for the microscope coupled with a Vidicon Camera. Using essentially mono-chromatic ultraviolet illumination, the image of operation can be observed on the video screen and may be photographed with a Polaroid-Land camera. Visible monochromatic light may also be used. A shallow moist chamber is employed permitting the use of condensing systems with comparatively short working distance. By means of this equipment color producing cyto-chemical reactions can be performed and studied; the amount of light absorbed by various cellular structures can be measured. Furthermore nucleoli may be transplanted from one cell to another or removed from the nucleus to the cytoplasm of the same cell.

Fig. 14.   The Television-Micromanipulator.   Courtesy of M. J. KOPAC (146).

c) Commercially Available Micromanipulators.

A brief account of some micromanipulators of earlier development is given on p. 29. Of the micromanipulators developed during the last three decades, some twenty-four instruments are enumerated on p. 33. A short account of a number of these instruments has also been given. To the low and medium power instruments enumerated in that list, one may add the variety of models commercially available at present (35, 168). From these the prospective worker may select an instrument best suited to his requirements. In the following pages, a number of these simple low-power instruments, mostly of the rack-and-pinion type, will be described. This will be followed by a detailed description of some of the finest commercially available high power instruments, all of which are included in the list above mentioned, namely the lever activated micromanipulator of E. LEITZ, p. 48, the sliding micromanipulator of C. ZEISS, p. 55, the pneumatic instrument of DE FONBRUNE, p. 60, and the thermal expansion B-D-H micromanipulator of the AMERICAN OPTICAL CO., p. 65. Finally, one of the older high power instruments still in use, the micromanipulator of R. CHAMBERS (58), a substantially developed form of which can be obtained from E. LEITZ and from BRINKMANN INSTRUMENTS, INC., will also be described, p. 69.

Fig. 15. BRINKMANN Micromanipulator Model MP I. Instrument on small size base plate, with instrument clamp and tool holder carried on top movement. Courtesy of BRINKMANN INSTRUMENTS, INC. (35).

α) Low Power Micromanipulators.

Brinkmann Model MP I. This massively constructed instrument can be obtained from BRINKMANN INSTRUMENTS, INC. (35), and is shown in Fig. 15. It consists of three rack-and-pinion movements mounted at right angles to each other on a supporting bracket, each movement permitting a 5-cm travel. The supporting bracket may be screwed to a heavy steel base. This instrument is recommended for use at magnifications up to

75 diameters. The instrument is suitable for performing chemical operations in capillary cones using the injection apparatus described below.

In this basic model, as well as in the other BRINKMANN MP micromanipulators mentioned below, the top movement may be equipped with tilting or rotating arrangements, and scales and verniers for accurate positioning.

Large base plates may be used which accomodate in addition to the microscope one or two manipulator units, a right- and a left-hand one, and an injection apparatus. Other base plates may support two manipulator units and an insulating center post equipped with a small rotating stage with a locking screw, as shown in Fig. 16.

In the MP models, suitable tool holders carrying various microtools are fit into instrument clamps on the top movement, the thrust motion. The type of clamp commonly used, Figs. 15 and 16, is a V-grooved Bakelite bed provided with a spring clip which is suitable for mounting holders or other objects up to about 6-mm diameter. Rotation in the horizontal plane of the instrument clamp can be achieved by means of special arrangement.

One type of tool holder is equipped with a removable head suitable for holding firmly a glass capillary or a solid glass thread, up to 0.8-mm diameter, by means of a compressible rubber washer. This watertight arrangement permits the use of micropipets connected to the microsyringe by means of a flexible metal capillary tubing which is held in the tapered end of the holder. The syringe serves to control pressure in the micropipet attached to the tool holder, and may be operated directly by hand. For accurate control, a bracket with a micrometer is attached to the syringe holder. The plunger of the syringe should be loaded with a spiral spring and is operated by means of the micrometer screw. It is very desirable that the microinjection apparatus be mounted on the common base plate together with the manipulator and the microscope. Such injection apparatus is similar in general principle to that used by BENEDETTI-PICHLER for working in capillary cones, p. 112, and also that described with the high power instrument of E. LEITZ, Fig. 27. Alternatively, for mounting metal microtools, a holder having a chuck-type head with aperture which accepts rods up to 1-mm diameter may be used.

Another type of holder with two interconnected pinplug receptacles, recessed within a solid Bakelite block with tilting clamp, may be used where potentials or currents are to be measured. These are picked up by means of a probe connected to one of the plugs and transferred to a measuring instrument by an outside connection to the other plug.

For certain applications the micromanipulator may be converted into a stage with three-dimensional motion. For this purpose a rectangular platform, suitable for supporting comparatively large objects, may be placed on the top movement. The arrangement may be used for mounting various devices or small animals to be operated upon.

Brinkmann Model MP II. This instrument, Fig. 16, also supplied by BRINKMANN INSTRUMENTS, INC. (35) is similar to the MP I instrument described above, but has its three movements provided with planetary reduction drives having a ratio of 5:1. This offers greater sensitivity of movement and permits the instrument to be used at magnifications up to about 200 diameters. The micromanipulator is suitable for use in transistor work.

Fig. 16. BRINKMANN Micromanipulator Model MP II. Two manipulator units, right- and left-hand, and an insulated center post with rotating stage are all mounted on a common base plate. Courtesy of BRINKMANN INSTRUMENTS, INC. (35).

Brinkmann Model MP V. In this instrument, supplied by BRINKMANN INSTRUMENTS, INC. (35), the thrust motion is equipped with a separate micrometer screw for fine adjustment in addition to the usual coarse drive. This combination permits a rapid advance over a range of 5 cm and a change-over to the fine motion at any required point.

Brinkmann Model MP VI. Another instrument of the series, Model MP VI, consists of three combination movements similar to the thrust motion in the MP V model above described. The increased sensitivity in all three movements permits the use of the instrument at medium magnifications.

Brinkmann Model RP III. The instrument, Fig. 17, supplied also by BRINKMANN INSTRUMENTS, INC. (35), consists of two horizontal movements equipped with scales and verniers that can be read to 0.1 mm and a vertical coarse rack-and-pinion movement. The instrument is mounted on a small steel base.

In this micromanipulator and in the other BRINKMANN RP models mentioned below, the range of the two horizontal movements is 5 and 3 cm, while the vertical motion permits a 5-cm travel. The knobs operating the two horizontal movements are in close proximity. The horizontal move-

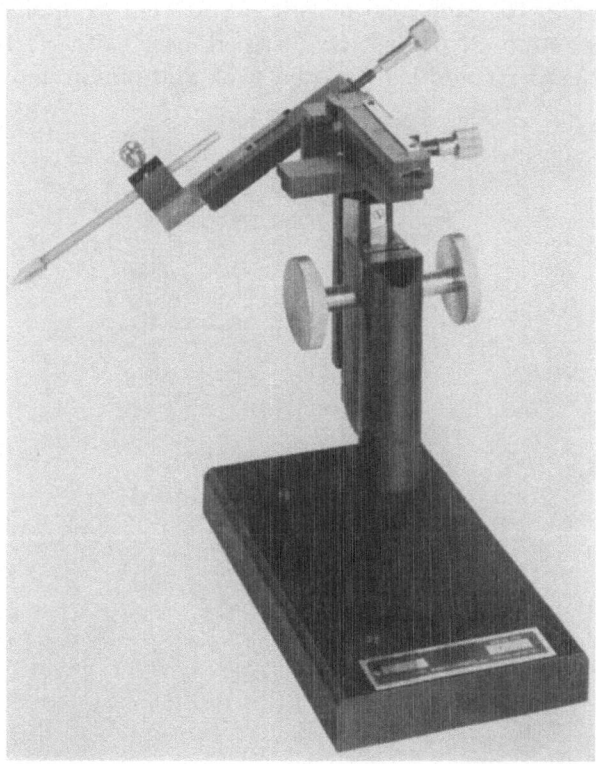

Fig. 17. BRINKMANN Micromanipulator Model RP III. Instrument clamp with tool holder on horizontal movement with tilting joint. Courtesy of BRINKMANN INSTRUMENTS, INC. (35).

ment on which the instrument clamp is mounted may be equipped with a tilting joint as shown in the figure. Common to all instruments of the RP series, the horizontal movements, as already mentioned, are provided with scales and verniers reading to 0.1 mm. The vertical support pillar may be furnished with a rotating arrangement.

The instrument holders and microinjection apparatus are the same as described under the BRINKMANN MP models.

Brinkmann Model RP IV. In addition to the features of the RP series, the vertical movement is furnished with a combination coarse and fine adjustment.

Brinkmann Model RP V. This is similar to RP III but with the vertical movement equipped with a planetary reduction drive offering an increased sensitivity of motion in this direction.

The Simple Rack-and-Pinion Type of the Microchemical Specialities Co. The simple rack-and-pinion micromanipulator No. 5020. supplied by the MICROCHEMICAL SPECIALITIES Co. (168), is shown in Fig. 18. Its three movements are similar in principle to those of the BRINKMANN MP I model, Fig. 15, each also permitting a 5-cm excursion. The present instrument is mounted on a heavy tripod base with vertical sleeve and clamp adjustment in addition to the rack-and-pinion movement.

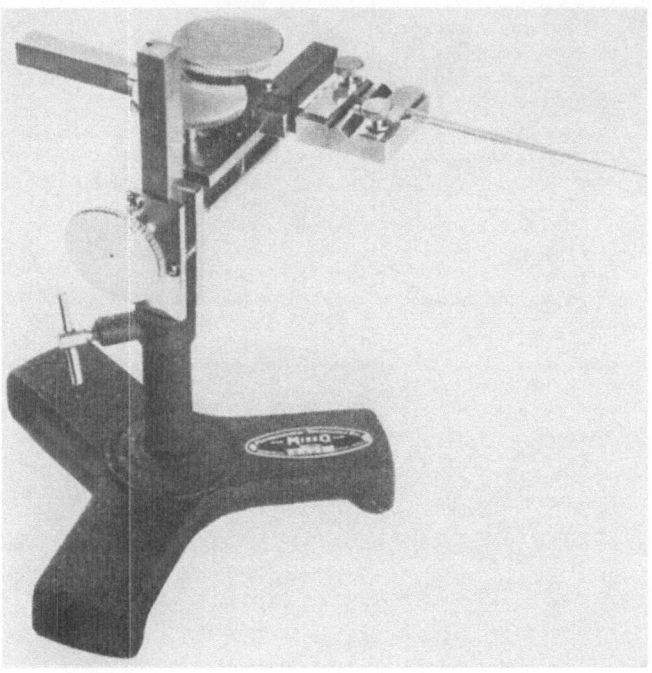

Fig. 18. Micromanipulator. Simple rack-and-pinion type, with tool holder clamp and micro-tool. Courtesy of MICROCHEMICAL SPECIALITIES Co. (168).

The Neher Type. The instrument, No. 5030, also obtainable from the MICROCHEMICAL SPECIALITIES Co. (168), has a three-jointed arm as shown in Fig. 19. It is of the type developed by NEHER (219) for use in quartz fiber work. In such work the instrument may be used in combination with a stereoscopic binocular microscope at magnifications of about 15 to 20 diameters. The instrument is useful in holding microtools or jigs. These may be manipulated in three mutually perpendicular directions by means of coarse adjustment screws.

The Kirk-Camensen Type. Finally, in the KIRK-CAMENSEN type No. 5010, Fig. 20, also supplied by the MICROCHEMICAL SPECIALITIES Co.

(168), the manipulator carrying the tool holder has three coarse mutually perpendicular movements each permitting approximately 5-cm excursion. A ball and socket swivel is provided above the vertical adjustment.

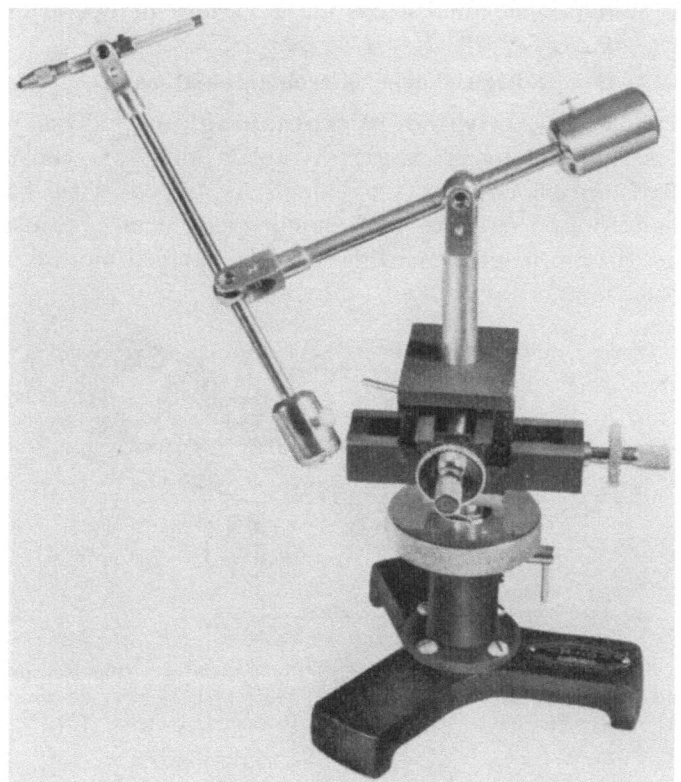

Fig. 19. Micromanipulator. NEHER type, with three-jointed arm carrying tool holder. Courtesy of MICROCHEMICAL SPECIALITIES CO. (168).

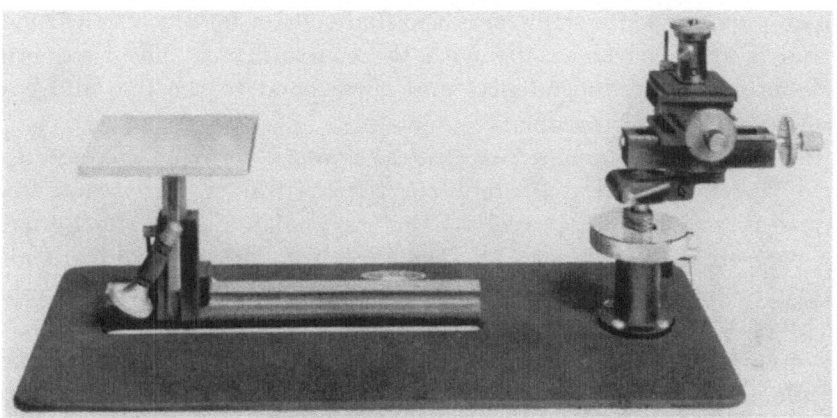

Fig. 20. Micromanipulator. KIRK-CAMENSEN type. Courtesy of MICORCHEMICAL SPECIA-LITIES CO. (168).

A porcelain table, adjustable for height by means of a special screw, is mounted on a slide for adjusting distance between table and manipulator. All adjustments are provided with locking screws. The instrument can be used with a stereoscopic microscope for a variety of operations.

β) High Power Micromanipulators.

The Leitz Lever-Activated Micromanipulator. This high power instrument is supplied by E. LEITZ, Wetzlar (152). It consists of two lever-activated fine horizontal movements equipped with ball bearing slides. In addition, there are two independent coarse adjustments for motion in the two horizontal coordinates. The vertical motion is activated by co-axial coarse and fine controls.

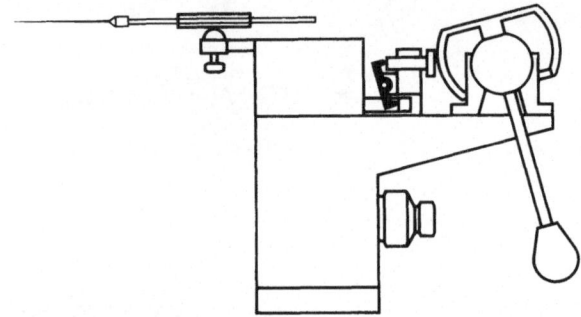

Fig. 21. General Construction of the LEITZ Micromanipulator. One manipulator column with lever control mechanism, coarse and fine controls with vertical motion, and a single instrument holder. Courtesy of E. LEITZ, Wetzlar (152).

The general construction of the instrument is shown in Fig. 21. Fig. 22, shows one manipulator column, a, for right-hand operation. The screw drive, b, provides the coarse vertical motion of the compound stage of the micromanipulator, while screw c, is the fine control for the vertical movement and is arranged co-axially with the coarse drive. The two vertical movements operate independently and correspond to the fine and coarse adjustments of the microscope.

In the horizontal plane, screw drives, d and e, provide coarse adjustments of the manipulator assembly in the sagittal or thrust and in the transverse directions, respectively. The manipulator stage can be locked in any coarsely adjusted position. The arrest of both the sagittal and the transverse movements is achieved by means of the screws, f and g, respectively.

The manipulator fine motions in the horizontal coordinates are entirely independent of the coarse adjustments and are operated in the desired direction by means of a single control lever or joy stick, h, via a segment of the sphere, i. The movement of the hand of the operator produces an excentric segment motion which is transmitted to the microtool via the compound stage of the manipulator. The ratio of transmission, that is, the ratio between travel of joy stick and travel of microtools, can be continuously varied from 16:1 to 800:1. This is done by raising or lowering the excentrically mounted segment of the sphere by turning the knurled drum, j, on the ball segment. Thus the lever movement can be adapted to any given microscope magnification and the extent of the movement by the assistant (tool) is restricted to the desired area, whereas the hand of the operator performs always the same movement regardless of the

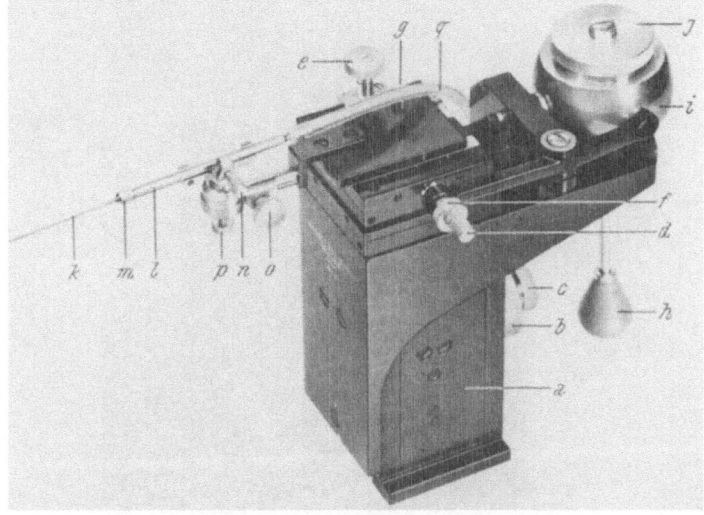

Fig. 22. The LEITZ Micromanipulator, Right Hand, with a Single Instrument Holder: a, manipulator column; b, coarse vertical adjustment; c, fine vertical adjustment; d, e, coarse horizontal adjustments; f, g, arresting screws for coarse horizontal adjustments; h, control lever for fine horizontal motions; i, segment of sphere actuating horizontal fine movements; j, knurled drum for adjusting ratio of motion; k, microtool; l, tubular holder for microtool; m, captive nut; n, clamp for tubular holder; o, screw for loosening tool holder; p, arresting screw for ball-and-socket head of microtool assembly; q, tubing from injection apparatus. Courtesy of E. LEITZ, Wetzlar (152).

magnification used, which may be up to 1000 diameters. The setting for a selected ratio can be relocated by using the calibrating marks provided on the center bolt of the segment of the sphere. The joy stick activating the fine horizontal movements and the controls for adjusting the height are so arranged that all movements of the tool within the field of the microscope can be performed without changing the position of the hand, which is shown in Fig. 23.

The available movement in the various axes are 30 mm along the X-axis, the sagittal movement, 8 mm along the Y-axis, the transverse movement, and 28 mm along the Z or vertical axis. These movements, as outlined, are available both by means of the coarse and the fine adjustments in their full range. With the full amplitude of the control lever, a circle having a diameter of 5 mm is covered; with the smallest amplitude, a circle of 0.1-mm diameter.

Each manipulator is fitted either with an instrument holder for a single microtool, Fig. 22, or with a double holder for two microtools, Fig. 24. The microtool, k, Fig. 22, whether attached to a single or a double holder, is held by a tubular holder or shaft, l, to which it is securely attached by means of a captive nut, m. The free forward end of the shaft, which takes the assistant, is conical and has an opening of 1-mm diameter. The shaft holding the microtool is secured in position on the instrument holder by

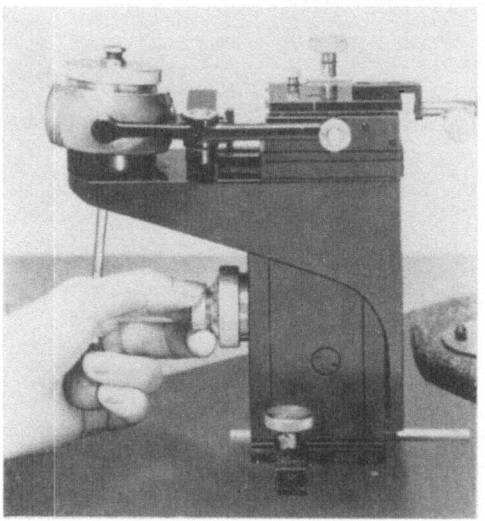

Fig. 23. Simultaneous Operation of Horizontal and Vertical Fine Movements. Courtesy of E. LEITZ, Wetzlar (152).

a spring clip, n. The instrument holder can be removed by loosening the locking screw, o. The tools can be easily interchanged. The tool holder is attached to the compound stage of the manipulator by means of a ball-and-socket joint which can be clamped by means of arresting screw, p. This joint permits the universal adjustment of the holders. In addition, the double holder, Fig. 24, has provisions permitting the coordinated movements of the two microtools independent of the actual manipulator but with the same degree of precision. One of the two holders can be moved forward and back in E-W direction in the instrument holder by means of a screw, while the other holder, by means of two screws, can be turned in the horizontal plane or inclined vertically. Thus a forceps-and-scissors motion in the required direction can be performed. The manipulator also permits to obtain erect motion in the field of an ordinary compound microscope.

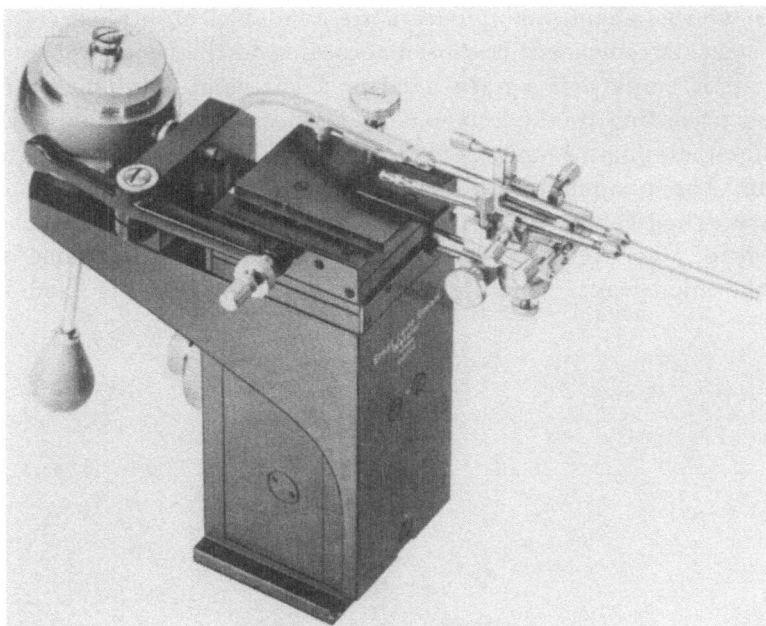

Fig. 24. Micromanipulator, Left Hand with a Double Instrument Holder. Courtesy of
E. LEITZ, Wetzlar (152).

Fig. 25. Manipulators, Right and Left Hand, on Opposite Sides of a Binocular Microscope.
Laborlux II stand with built-in illumination for transmitted light and with moist chamber
in position. Left-hand manipulator with assembly to hold two microtools. Courtesy of
E. LEITZ, Wetzlar (152).

Right- and left-hand manipulators are available, both identical in their basic designs. Arrangement is chosen according to the need. When a single instrument is employed, a right-hand or a left-hand manipulator may be employed according to the personal preference. Fig. 25, shows two complete micromanipulator units, one for left- and the other for right-hand operation. The manipulators and the microscope are mounted on a common base plate. The two micromanipulator columns are mounted with the lower grooves fitted over right and left guide rails or locating bars, one on each side of the microscope. The position of the manipulators is fixed and the

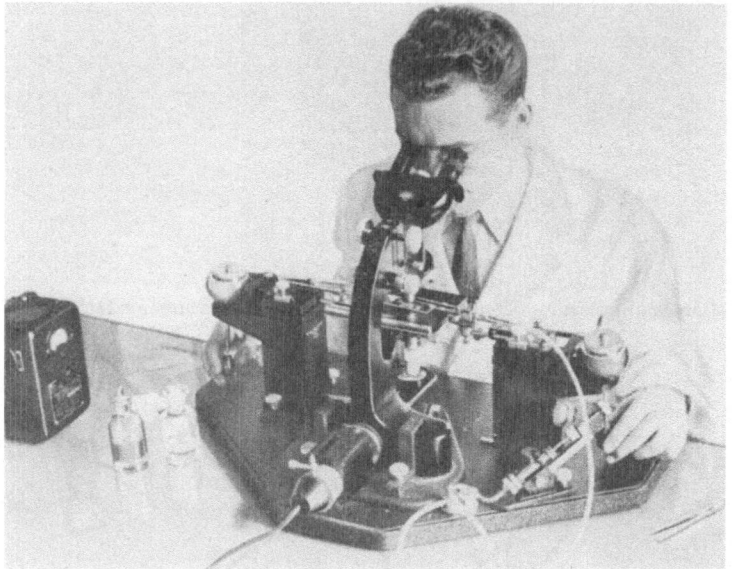

Fig. 26. Complete Assembly with Injection Apparatus and Operator in Working Position. Microscope with Laborlux II stand. Courtesy of E. LEITZ, Wetzlar (152).

necessary adjustments are carried out so that when the two manipulators are adjusted to the same height, the points of the microtools will appear in the center of the field and can, by means of the coarse adjustment of the horizontal movement, be removed and brought back into the field of operation. The manipulators are secured in position by special holding clamps.

The injection device, shown in Fig. 27, is suitable for operation under low power magnifications with micropipets having comparatively wide orifices. The device is mounted on a base support either to the left or to the right of the microscope. It consists of a holder with attached screw control for accurately operating the plunger of the syringe, a glass injection syringe of 2-ml capacity, flexible transparent plastic tubings leading to the pipet holder, and a three-way cock. The syringe and tubings are filled with distilled water to transfer pressure to the working fluid in the micropipet. A layer, a few millimeters thick, of a neutral medium such as double-distilled paraffin may be taken in before filling the pipet with the working

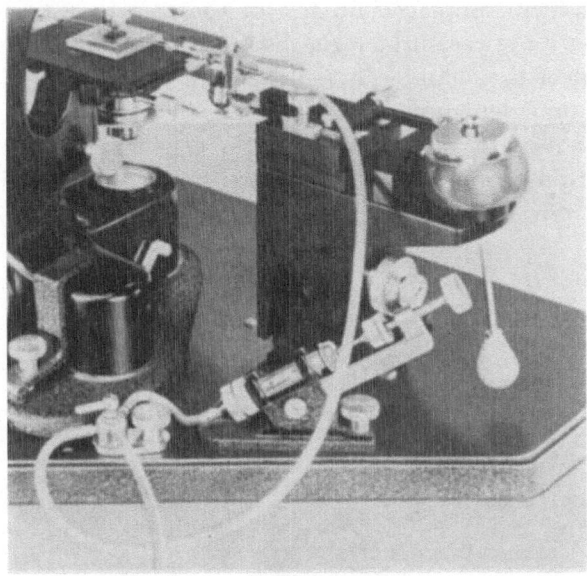

Fig. 27. Injection Device Screwed to Manipulator Base Plate. Courtesy of E. LEITZ, Wetzlar (152).

solution. This layer serves to separate the working solution in the pipet from the distilled water in the syringe. This may be necessary in working with culture fluids in order to prevent pollution of the fluid by the distilled water. With pipets having very fine orifices and consequently very high

Fig. 28. Moist Culture Chamber. Courtesy of E. LEITZ, Wetzlar (152).

capillarity at tips, it may be preferable to use special injection devices, as those based on the principle of thermal expansion.

Fig. 28 shows a simple moist culture chamber open on two sides. It is also shown in position under the microscope with the microtools introduced: Figs. 25 and 27. This chamber is built upon a rectangular plane-parallel glass plate, 4.5 cm square, forming the base of the chamber. Two square glass bars are cemented to the base plate, 3.3 cm apart. A cover glass, 3.5 cm × 3 cm, is placed on top of the two bars to form the roof which carries on its lower surface the hanging drop containing the sample. The total height of the chamber is 7 mm and its depth, from floor to ceiling, may be 3 or 5 mm depending upon the thickness of the base plate. This chamber may be used for operations on living cells under high magnification, usually in transmitted light. It can also be used as an oil chamber as described on p. 185.

Fig. 29. LEITZ Micromanipulator. Complete assembly with a Leitz binocular stereoscopic microscope (Greenough) on a Laborlux II stand. Courtesy of E. LEITZ, Wetzlar (152).

Standard microscopes, both monocular and binocular, can be used with the micromanipulators. Microscopes with vertically adjustable tubes and fixed stages are more convenient to use than those with vertically adjustable object stage. It may be preferable that the stage faces the operator, Fig. 26, which renders it easily accessible from all sides. The Laborlux II stand, p. 10, 11, microscopes Figs. 3, 25 and 26, supplied by E. LEITZ, combines these features and is equipped with a built-in source for transmitted light. The fine adjustment of the microscope may be operated by means of a remote control operated from the microscope base support, Fig. 25.

With the moist chambers described above, a BEREK bright-field condenser with two diaphragms and a long-focus condenser cap is suitable since it offers a sufficiently long focal distance to illuminate the specimen. The HEINE phase contrast condenser also, is suitable for observation in bright-field. It has a sufficiently large working distance to allow the use of moist chambers up to a total height of 7 mm. For phase contrast and dark-field examinations, the phase contrast condenser after HEINE can also be employed, but phase contrast and dark-field microscopy require a good control of the refractive conditions. Moist culture chambers cannot be used satisfactorily for phase contrast and bright-field examination. The lens effect of the hanging drop renders only a small portion of the field satisfactory for such examinations. This optically disturbing effect of the phase contrast is largely eliminated by the use of oil chambers. Immersion in an oil of somewhat similar refractive power gives a "flattening" of the hanging drop.

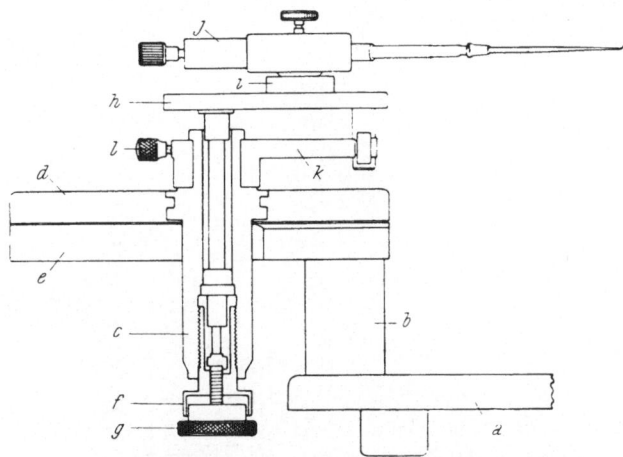

Fig. 30. Sliding Micromanipulator, schematic: *a*, base plate; *b*, upright support; *c*, control handle; *d*, *e*, sliding operating stage and lower support; *f*, coarse vertical adjustment; *g*, fine vertical adjustment; *h*, carrier plate; *i*, needle carrier; *j*, needle holder; *k*, lower support for carrier plate. Courtesy of C. ZEISS, Jena (245).

For low and medium magnifications, the use of a stereoscopic microscope, Fig. 29, is preferable since it offers three-dimensional viewing and an upright image in addition to an extraordinarily long working distance.

For work requiring sterile conditions, the complete microscope-manipulator assembly may be put under a Plexiglass protective hood with an opening for the microscope tube to protrude, and two openings with sleeves for the hands of the operator.

The Zeiss Sliding Micromanipulator. This instrument, supplied by C. ZEISS, Jena (245), has a sliding horizontal movement produced by direct transmission of pressure applied by hand to a single control handle. Finely controlled motion is obtained by the braking action of a thin film of grease or a suitable viscous fluid applied to the sliding surfaces. The vertical motion is activated by co-axial coarse and fine controls installed in the same guide handle.

The general construction features are shown schematically in Fig. 30

which shows a manipulator stand, the control handle protruding downwards with the controls for the vertical motion installed in the lower end
of the handle, the carrying plates for the needle carrier, needle holder,
and a single microtool. Fig. 31 shows a complete set-up comprising two
operating right- and left-hand micromanipulator stands, both identical in
their basic designs, one to the right and the other to the left of the microscope with moist chamber in position.

As shown in Figs. 30 and 31, each micromanipulator unit is mounted
on a heavy base plate, a, by means of an upright support, b. The micro-

Fig. 31. Sliding Micromanipulator. Complete assembly showing two manipulator stands
and a Lg microscope with moist chamber in position. a–j, same as in Fig. 30. Courtesy of
C. ZEISS, Jena (245).

scope is placed in the center of the common base plate and is fixed in the
required position by means of an adjustable stop. By means of the control
handle, c, the operating stage, d, can be moved with a sliding motion permitting a free movement in all directions in the horizontal plane. A thin
film of grease of the proper consistence is spread evenly on the contact
surfaces between the sliding stage, d, and its lower support, e. The grease
film joins the two plates firmly so that the two surfaces may slide against
each other, whereas slight accidental impulses are arrested by friction.
In this micromanipulator and in other instruments of the viscous type (23),
a constant pressure applied by hand produces a constant velocity of motion.
The resulting microtool displacement depends on the pressure applied and
its duration. By using a sliding fluid of high viscosity, controlled displace-

ment can be obtained down to very small values. In the present instrument, horizontal displacements can be controlled down to about 0.5 µm. The operator therefore can, without mechanical transformation arrangements, adjust motion in the horizontal plane to the proper dimensions of the microscopic field. Motion in the vertical plane is achieved by two co-axial screw movements installed in the control handle. The upper knurled head, f, is for coarse adjustment of height and operates a steeply pitched screw giving a rapid vertical movement. The lower knurled head, g, is for fine adjustment of height by means of a differential screw. The range of the vertical coarse movement is 13 mm, and that of the fine vertical motion is 2 mm. In the horizontal plane, the full range of the sagittal or thrust motion is 90 mm and that of the transverse motion is 15 mm.

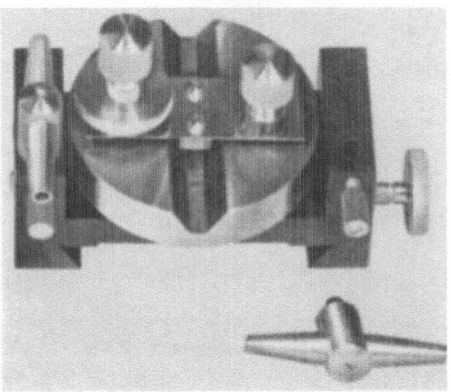

Fig. 32. Simple Needle Carrier. One of the two tubing connections detached. Courtesy of C. Zeiss, Jena (245).

The accessory implements are mounted on the operating stands and are interchangeable. A dove-tailed carrier plate, h, serves for mounting the needle carrier, i, Fig. 32, to which the instrument holder, j, is attached. The carrier plate is mounted on a lower plate, k, as shown in the figures, and the whole fixture can be turned for 360° from the working position by loosening screw l.

A simple needle carrier, Fig. 32, serves for attaching a single microtool holder as shown in Fig. 33. The needle carrier is mounted on the carrier plate of the operation stage, as shown in Fig. 31, and is fixed in position by means of a screw. The needle carrier can be tilted so that the angle of inclination of the microtool to the object may be adjusted and its position fixed at the required angle. Each needle carrier is also provided with two removable tubing connections, shown in the figures, which are attached to both sides of the carrier. These are employed when micropipets are used.

A needle holder, a needle clamp, and a pipet clamp are shown in Fig. 34. The needle holder is a piece of metal tubing which is held on the needle carrier. The needle holder can be rotated with its milled end, thus permitting the microtool to be rotated around the axis of the shank. The front end of the holder has a conical bore which receives the conical end of

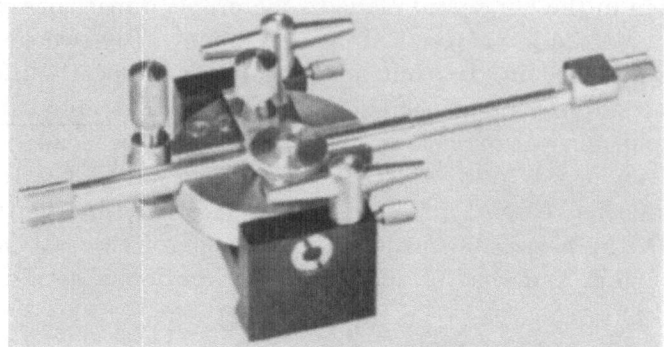

Fig. 33. Simple Needle Carrier with Needle Holder and Pipet Clamp. Courtesy of C. ZEISS, Jena (245).

either a pipet or a needle clamp, Figs. 33, 34. The needle clamp, Fig. 35, is a tubular body which takes the microtool. The clamp has a built-in notch bearing. The microtool is inserted in the clamp and is pressed by

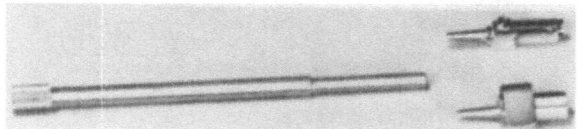

Fig. 34. Needle Holder with Needle Clamp (above) and Pipet Clamp (below). Courtesy of C. ZEISS, Jena (245).

a double flat spring against the bearing and is thus secured in position. The free ends of the flat spring are slightly bent upwards as shown in the figure, permitting the clamping jaw to be raised up by finger tip. This

Fig. 35. Needle Clamp (above) and Needle Clamp with Needle (below). Courtesy of C. ZEISS, Jena (245).

facilitates insertion and release of microtools. The micropipet is attached to the pipet clamp by means of a spring shown in Fig. 36.

A double needle carrier with two needle holders is shown in Fig. 37. It is mounted on the dove tail shaped carrier plate of the operation stage and is screwed in position in the same way as the simple needle carrier. The two needle holders stand at an acute angle to each other and co-ordinated movements of the two microtools against each other can be achieved by means of the central screw knob. Thus a pincer-like movement

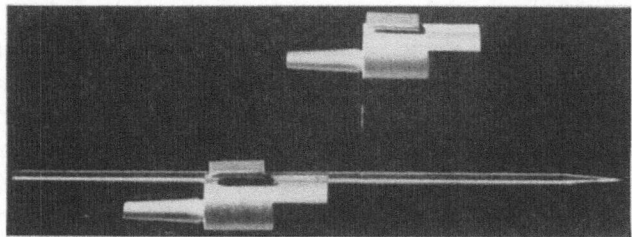

Fig. 36. Pipet Clamp (above), and Pipet Clamp with Micropipet (below). Courtesy of C. ZEISS, Jena (245).

is possible. One of the two holders can be moved forward and backward in the sagittal or E-W direction, while the other can be inclined, so that it is possible to adjust the points of the two needles exactly opposite to each other. The needle carrier can be tilted in the same way as in the simple carrier and can be locked in the desired position.

Fig. 37. Double Needle Carrier with Two Needle Holders and Needle Clamps. Courtesy of C. ZEISS, Jena (245).

For proper operation of the horizontal movements it is necessary that the sliding grease be of the proper consistency. Two kinds of grease of different consistency are supplied and these are mixed in different proportions depending on room temperature and the type of work to be performed. The sliding surface should be thoroughly cleaned with benzene or any other suitable grease solvent before the grease is applied. Grease either alone or with a little castor oil is then spread evenly with the finger tip on the sliding surfaces. Delicate work under high magnifications requires very fine movements. For this purpose the two types of grease are mixed in proportions to give high consistency producing a strong braking action. For ordinary work under comparatively low magnifications, a grease mixture of comparatively low consistency and thus weak braking action can be used. In addition, castor oil may be applied as a thin film or a little amount of this oil is added to the grease to produce a lighter sliding motion. It may be preferable to apply castor oil as drops to the evenly distributed film of grease on the sliding surfaces. Békésy, who has developed another micromanipulator of the viscous type (23), uses fly paper glue as a viscous fluid. This is prepared by mixing warm one ounce of castor oil with two ounces of rosin. The viscosity of the mixture depends on the proportion of rosin. Békésy also keeps the apparatus at a constant temperature by a thermostat so that the viscosity is not affected by changes in room temperature.

Fig. 38. Moist chamber with Connection Sleeves. Courtesy of C. Zeiss Jena (245).

The microscope should be equipped with an image-erecting tube to avoid the reversal of field given by standard compound microscopes. The microscope shown in Fig. 31, also manufactured by C. Zeiss, Jena (245), is an Lg microscope, equipped with an image-erecting inclined monocular tube L and a special bright field-dark field micrurgical condenser. This condenser permits the use of moist chambers up to 10 mm in height.

The moist chamber shown in the assembly, Fig. 31, can be closed tightly leaving two small openings on opposite sides for introducing microtools. The chamber may be provided with two connecting sleeves, Fig. 38, which serve for sealing the chamber, if required. These sleeves connect the two lateral openings of the chamber with the tool holders on opposite sides, and may be useful for work requiring particularly uniform and high humidity within the chamber.

The de Fonbrune Pneumatic Micromanipulator. The DE FONBRUNE micromanipulator (2, 76, 77, 78) is a pneumatic, lever-controlled instrument originally developed by P. DE FONBRUNE in the Pasteur Institute, Paris, in 1932. The instrument, No. 58090, can be obtained from the A. S. ALOE CO. (2).

The micromanipulator, Fig. 39, consists essentially of two independent units, the control unit (manipulator) and the receiver, interconnected by flexible rubber tubing. By moving a single control handle, air pressure from three pistons set at right angles to each other is transmitted pneumatically from the manipulator through three rubber tubings to three sensitive aneroid-type metallic membranes of the receiver, which are connected to a lever holding the microtool. The metallic membranes are sensitive to

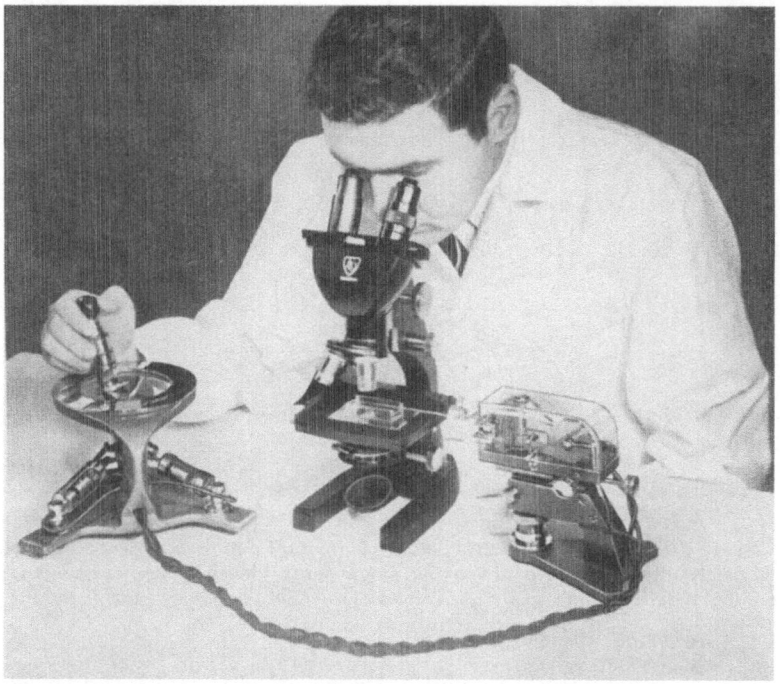

Fig. 39. The DE FONBRUNE Micromanipulator. Complete assembly with control unit and receiver on opposite sides of the microscope interconnected by flexible tubing. Courtesy of A. S. ALOE Co. (2).

the slightest movements of the control lever. Since control and receiver are built as independent units and fine movements are transmitted to the receiver through flexible tubings without any rigid connections between the control handle and the microtool, the micromanipulator is an excellent remote-control instrument. Transmission of vibration from the hand of the operator to the microtool is completely eliminated. In addition to the fine movements the receiver has knurled heads for coarse adjustments. The receiver carries two adjustable holders, a primary holder for supporting the pneumatically controlled microtool and the other for supporting an auxiliary manually controlled microtool. The primary microtool can be pneumatically manipulated over a maximum range of 3 mm in the horizontal and the vertical planes. Furthermore the micromanipulator permits erect motion under the microscope. The instrument can either be used singly or in pairs.

The control unit, Fig. 40, consists of three pneumatic pumps at right angles to each other which are mounted on a ball-and-socket type universal joint. Two of these pumps, a_1 and a_2, are horizontal and the third, a_3, is vertical. This assembly is mounted in an open support, b, having a heavy cast base with flat ring, c, at the top of which the operator can rest his hand during manipulation; it also serves to retain the upright control lever, d. Each pump consists of a glass cylinder fitted with a brass piston operating against air. Pressure applied to the pistons of the pumps by

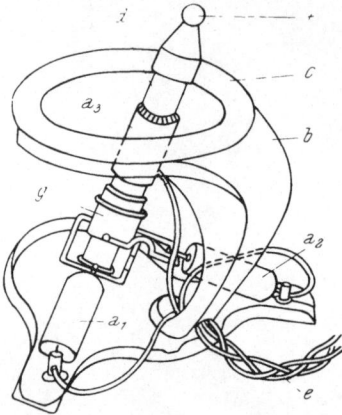

Fig. 40. Control Unit of the Manipulator: a_1, a_2, horizontal pumps; a_3, vertical pump; b, open support; c, flat ring to retain control lever; d, single control lever for fine movements; e, three connecting tubings; f, rotable knob for vertical movement; g, sleeve for adjusting ratio of movements. After A. S. ALOE Co. (2).

means of the control handle is readily transmitted through the three rubber connecting tubings, e, directly to the receiver. The glass cylinders are readily replaceable. The pneumatically controlled microtool mounted on the receiver is moved by manipulating the control lever containing the vertical pump. The two horizontal pumps, a_1 and a_2, produce motion of the microtool in the horizontal plane and are actuated by moving the upright control lever in any required direction, either at right angles to move the microinstrument in one of the two horizontal coordinates, or at any desired angle to move the microtool at a corresponding angle. The vertical pump, a_3, produces motion of the microtool in the vertical direction and is actuated by rotating the knob, f, at the top of the control lever either up or down. In the horizontal plane, the ratio of hand movement to the microtool displacement may be readily adjusted between 50:1 and 2500:1. In most cases, however, the ratio 1200:1 is satisfactory. This ratio is adjusted by means of an adjustable cylindrical sleeve, g, mounted at the base and along the axis of the control lever. Connecting rods run from this sleeve to each of the two pistons controlling movements in the horizontal plane. Raising or lowering the sliding sleeve alters the angle of movement of these piston levers. As the amplitude of the pistons in the horizontal pumps is changed, both the range and sensitivity of motion in the horizontal plane are altered. The ratio is increased when the sleeve is lowered and is decreased when the sleeve is raised. Graduation marks are provided on the upright lever to facilitate adjusting and resetting this ratio as desired. The movement in the vertical plane is unaffected, and its range is controlled directly by means of the screw knob on the control lever.

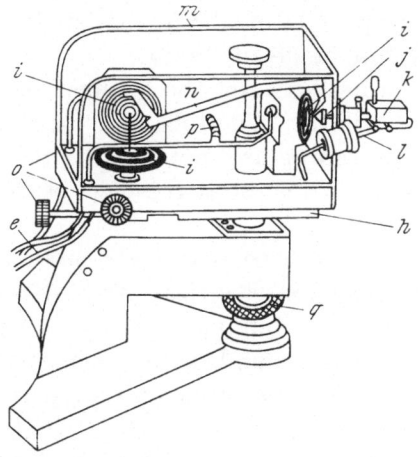

Fig. 41. The Receiver: h, adjustable plate; i, three pneumatic capsules; j, transmission rod; k, primary instrument holder; l, auxiliary instrument holder; m, plastic housing; n, connecting rod for vertical capsules; o, coarse horizontal adjustments; p, electrical binding posts; q, coarse vertical adjustments. After A. S. ALOE Co. (2).

The receiver, Fig. 41, is a pneumatic mechanism which serves to transmit motion from the pneumatic pumps to the microtool. It consists essentially of the adjustable plate, h, and a receiver mechanism, all mounted on a heavy base. The receiver mechanism is provided with three cylindrical capsules, i, fitted with sensitive metallic membranes interconnected by a transmission rod, j, which is attached to the primary instrument holder, k. The transmission rod correlates the motion of all three membranes. The pneumatic holder, used for finer manipulations, occupies a central position on the front of the receiver and supports the primary instrument. An auxiliary tool holder, l, set on the side of the receiver, serves for mounting a secondary microtool. The receiver mechanism is enclosed in a transparent plastic housing, m, to protect the capsules. The tool holders, k and l, are set at the front of the housing which is open to facilitate inserting and removing the microinstruments.

The three capsules are mounted on the adjustable plate, h, at right angles to each other corresponding with the positions of the pneumatic pumps of the manipulator. Two of these capsules are mounted vertically and are connected to the horizontal pumps while the third capsule is mounted horizontally and is connected with the vertical pump. The rod, n, connecting the two vertical capsules combines their motion and transmits it to the tool holder. The third capsule provides the vertical motion. The sensitive metallic membranes of the capsules expand or contract in exact ratio to changes in air pressure received from the pneumatic pumps. Each capsule is connected by means of a metal tube to an outlet with tip projecting from the base plate. These outlets serve for introducing and withdrawing small increments of air from each capsule. The pneumatic capsules all have an amplitude of about 1.5 mm from each side of their normal position, or a full range of 3 mm.

Three knurled screws heads, o, on the receiver plate serve for the preliminary setting of the primary microtool and for manipulating the auxiliary tool. Two of these screws are mounted on the side of the plate for sagittal motion while the third is set at the back of the plate for transverse movement.

The instrument is provided with electrical contacts so that current may be passed through the tool for electrical studies, microincineration, etc. For this purpose two electrical binding posts, p, are provided on the side of the receiver, one post is grounded and connects with the microtool while the other is electrically insulated and supports the free end when not in use.

The plate, h, of the receiver is mounted on an adjustable pillar, which may be raised or lowered by means of a vise clamp to correspond with the height of the microscope stage. More exact adjustment of the height is made by a large knurled screw, q, in the base.

The primary instrument holder, k, is mounted in a socket at the front of the pneumatic transmission rod, j. Before inserting or removing the microtool, the clamp should be removed from the socket to avoid the possibility of damaging the pneumatic mechanism. The auxiliary instrument holder, l, is provided with a universal ball-joint and is equipped with a tubular sleeve which accomodates the extension rod of the auxiliary instrument clamp. The secondary microtool can easily be adjusted in the field of the microscope and is manipulated by means of the adjustments on the receiver base plate. The two microtools can be operated simultaneously if desired.

The rubber connections between the control unit and the receiver are made in such a way as to give natural motion. The pneumatic pumps of the manipulator are numbered for easy identification. The three outlets to the pneumatic capsules located at the back of the receiver, Fig. 41, are likewise numbered for convenience in making connections to the corresponding pneumatic pumps by means of the rubber connecting tubes, *e*. The length of the rubber connecting tubes is not critical but they should be of small bore and have a wall of maximum flexibility.

The relative positions of control unit and receiver and the rubber connections may be made to give the type of motion required. Usually the corresponding numbers are connected with the receiver either placed beside or in front of the microscope stage, depending upon the type of the chamber used, to provide correlated movement between the control lever and the microtool as observed under the microscope. Before connecting the rubber tubing, certain preliminary adjustments are necessary. The adjustable sleeve on the control lever is raised as far as possible, the pistons of the

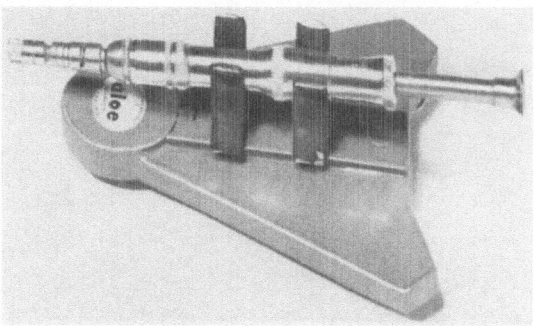

Fig. 42. Microinjector with Screw Control. Courtesy of A. S. ALOE Co. (2).

pneumatic pumps are adjusted to a central position and the control lever is set in an absolutely vertical position. In use, care should be taken that the control unit be in the correct position relative to the receiver with the axis of the horizontal pump parallel to the sides of the receiver. When not in use, the control lever should be set in a vertical position to eliminate any strain on the flexible membranes of the receiver.

For the purpose of injecting or withdrawing small amounts of fluids with micropipets, the microinjector shown in Fig. 42, also supplied by the A. S. ALOE Co., may be used. It consists of a stainless steel screw syringe of 3-ml capacity, firmly held on a heavy metal base. A long polyethylene tubing of about 0.4-mm internal diameter connects the microinjector to an angle-type micropipet holder designed for use with the micromanipulator. Capillaries inserted in the holder are firmly sealed by means of neoprene rubber gaskets. The microinjection unit may be used separately, and the holder may be mounted in any convenient support.

The B-D-H Thermal Expansion Micromanipulator. The B-D-H (BUSH-DURYEE-HASTINGS) micromanipulator (3, 45), supplied by the AMERICAN OPTICAL Co. (3), uses the thermal expansion of electrically heated fine wires to provide a system of fine controls suitable for high power manipulations. Four tool holder bars serve for mounting tool holders to which the microtools are attached. The bars are suspended under spring loading from the thermal expansion wires which are controlled by variable transformers operated by two joy sticks. Each tool holder has three sets of mechanical controls for coarse adjustment.

Fig. 43, is a general view of the instrument. It shows four rectangular manipulator heads, a, each carrying a tool holder, b, on which a microtool is mounted. Three sets of electrically heated, spring loaded thermal ex-

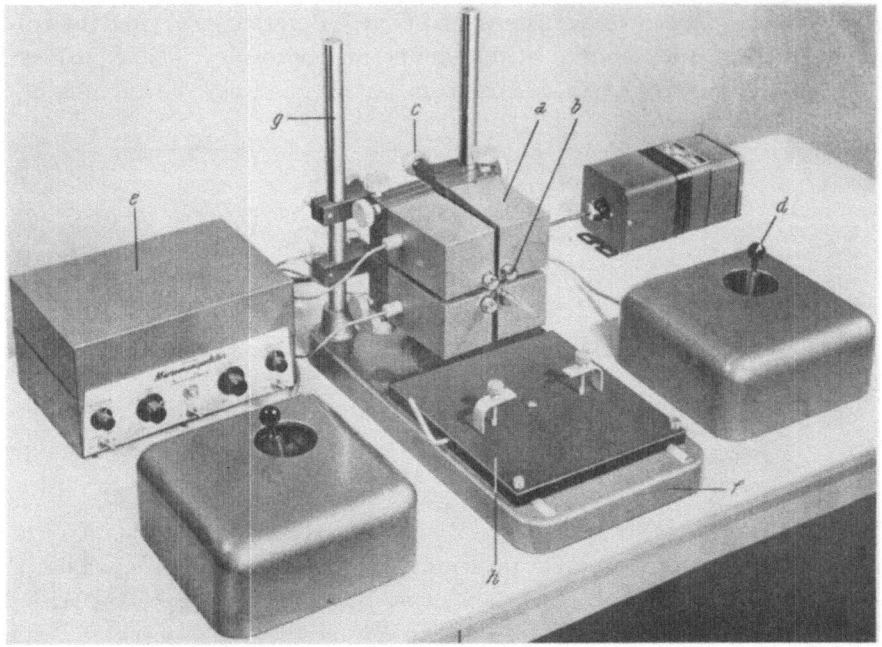

Fig. 43. The B-D-H Micromanipulator, General View: a, manipulator heads; b, tool holders; c, control knobs for coarse adjustment; d, joy sticks for fine adjustment; e, electrical control unit; f, base plate; g, pillars for supporting manipulator heads; h, sliding table for microscope. Courtesy of AMERICAN OPTICAL Co. (3).

pansion wires serve for suspending each metal tool holder bar, Fig. 45. The very fine wires are made of a heat resistant alloy. The tool holder bar carries a circular permanent magnet on its outer end for attachment of a tool holder. The tool holder has a terminal disk of soft iron which serves for mounting the tool holder to the magnetic chuck of the bar. This method of mounting allows the tool holder to be rotated freely and facilitates replacement of one holder by another without use of screws of any kind. Yet the holders are kept firmly in position.

In addition, each of the four manipulator heads, Fig. 43, accomodates three sets of mechanical controls operated by means of knobs, c, having large diameters. Operation of these knobs permits coarse adjustment of the tool holder in three directions at right angles for a range of 0.1 to 10 mm and serves to bring the microtools into the field of the microscope to within a short distance of the desired position.

Fine motion in the three spatial coordinates, from 0.5 to 400 μm, is accomplished by means of the two joy sticks, *d*. Each joy stick controls two of the manipulator heads. The instrument permits complete freedom of movement in any direction in the horizontal plane. Vertical adjustment is achieved by turning the knob of the joy stick clockwise or counter-

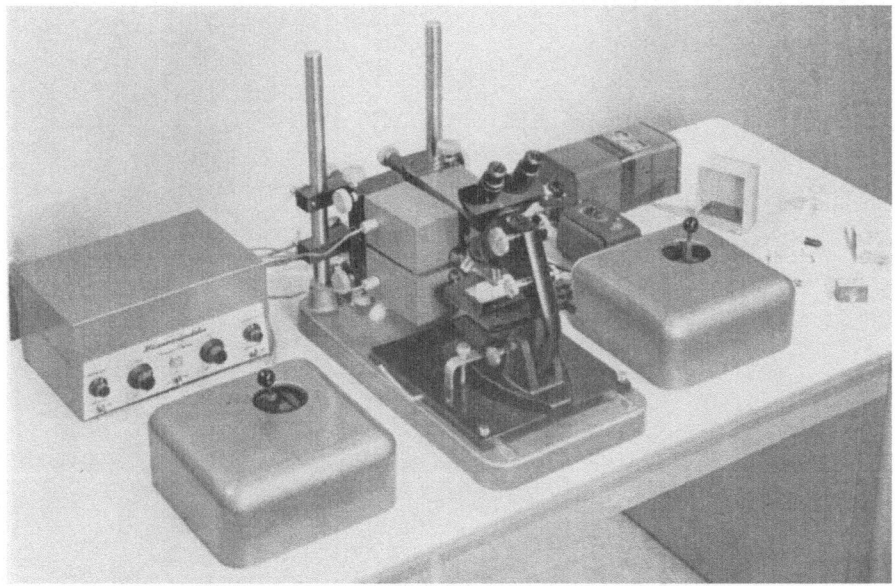

Fig. 44. Complete B-D-H Micromanipulator-Microscope Assembly with Moist Chamber in Position. Courtesy of AMERICAN OPTICAL Co. (3).

clockwise, thus causing the microtool to move down or up. According to the magnification used, the ratio of hand travel to microtool displacement may be varied at will from 250:1 to 50000:1. This is done by means of two variable transformers, one for each joy stick, in the electric control unit, *e*. The tool holders that will be influenced by each joy stick are determined by means of selector switches. A reversal of movement is also provided so as to permit erecting the image of an ordinary compound microscope. The control unit provides also for control of manipulator power supply.

The manipulator heads are mounted on a sturdy base, *f*, by means of two vertical pillars, *g*. The base carries a sliding and locking plate, *h*, on which the microscope is mounted. This plate allows the microscope to slide out of the way for rapid change of microtools without disturbing the actual set-up of the manipulator heads. It obviates the usual practice of moving the moist chamber out of the field of the microscope. The latter can be returned against a mechnical stop to within 10 μm of its original position. The sliding plate can be secured in any position by means of a safety clamp. A complete micromanipulator-microscope assembly is shown in Fig. 44.

Fig. 45 shows the three sets of thermal expansion wires and the tool holder bar with the magnetic tool holder chuck at its free end. The three

Fig. 45. Manipulator Head with Cover Removed. Visible are three sets of thermal expansion wires, the tool holder bar, and the magnetic tool holder chuck. Courtesy of AMERICAN OPTICAL Co. (3).

sets of fine wires from which the bar is suspended are so arranged that the expansion of any set produces movement in one of three directions at right angles to each other.

The metal holders, Fig. 46, are designed to accept microtools with shanks of about 0.8-mm diameter. The holder has an adjustable telescoping bar and shank which is fitted with a pierced screw cap and rubber gasket for holding glass tools. The shank of the tool holder is mounted on a flat iron button which serves for attaching the tool holder to the magnetic chuck of the tool holder bar. The length of the shank of the microtool may be adjusted by pushing it further into, or out of the tool holder. All four microtools can be placed inside the chamber at the same time, any two of which can be operated together while the other two are held ready for use as illustrated in Fig. 47.

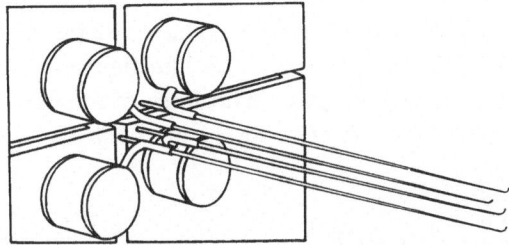

Fig. 46. The Four Tool Holders Mounted in Position with Microtools Attached. After AMERICAN OPTICAL CO. (3).

Chambers' Micromanipulator. One of the older high power micromanipulators most widely used particularly in cellular micrurgy since 1921 is based on the designs originally described by ROBERT CHAMBERS (57,

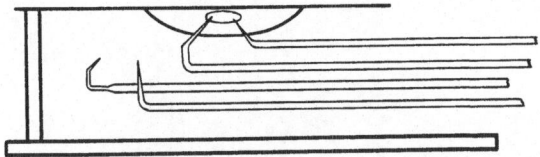

Fig. 47. Schematic Drawing of Moist Chamber with the Four Microtools Inside. After AMERICAN OPTICAL CO. (3).

58, 60). In one of these papers (58) the instrument is described in full detail. The instrument has received considerable improvements over a long period and is now supplied by E. LEITZ (152), and also by BRINK-

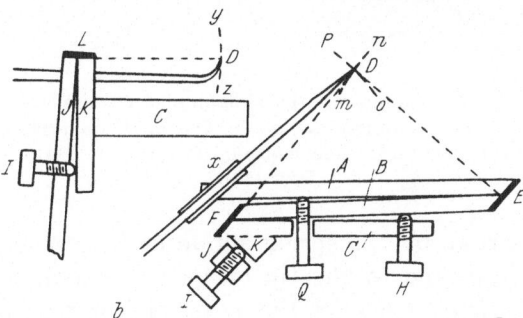

Fig. 48. CHAMBERS' Micromanipulator. Diagram illustrating the working principle. *a*, two horizontal movements, top view; *b*, vertical motion, side view; *A*, *B*, *C*, *J*, *K*, metal bars; *E*, *F*, *L*, spring hinges; *G*, *H*, *I*, micrometer screws; *mn*, *op*, *yz*, curvilinear movements. After R. CHAMBERS (58).

MANN (35). The three-dimensional fine movements depend upon the use of bars of rigid metal connected by spring hinges and separated by high grade micrometer screw feeds employing lever principles. These fine motions are independent of the coarse adjustments which serve only for placing the microtool within the field of observation.

Fig. 48, *a*, *b*, illustrates the working principle of the two horizontal fine motions and the vertical fine motion. The rigid metal bars, *A*, *B*, *C*, *J*, *K* are connected at their ends to form a Z-like figure by resilient metal strips *E*, *F*, *L*, acting as spring hinges. The bars are spread apart by means of the micrometer screws, *G*, *H*, *I*, and return back to their original position

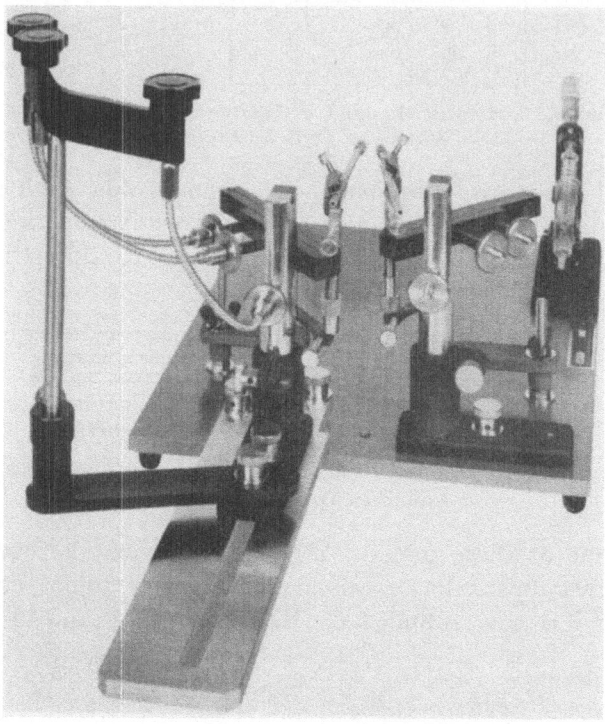

Fig. 49. CHAMBERS' Micromanipulator. Two manipulator units mounted in front of the microscope (not in place) on a common base plate. One of the manipulator units is equipped with three remote control shafts for fine motion. Courtesy of BRINKMANN INSTRUMENTS, INC. (35).

on reversing the screws by the spring action at the ends of the bars. The movements thus imparted to the tip of the microtool, which is attached in the proper position to one of the bars, are in arcs of a large circle. Fig. 48, *a*, top view, shows diagramatically how the two motions in the horizontal plane are transmitted by means of the three bars, *A*, *B*, *C*, and the two screws *G*, *H*. By turning the screws *G* and *H*, the tip of the microtool, *D*, is moved through the horizontal arcs represented by the dotted lines *mn* and *op*, respectively. Movement in the vertical plane is illustrated by Fig. 48, *b*, where the instrument is viewed from the side. By turning screw, *I*, in the stationary pillar, *J*, against the bar, *K*, the tip of the microtool moves through vertical arc yz. The full range of fine motion in each direction is about 2 mm. The fine excursions of the microtool usually required do not exceed 1 mm. Since the tip of the microtool moves in arcs of large circles, having a radius more than 6 cm, the curvature of the arcs is not appreciable and the tip of the microtool moves practically in straight lines.

In addition to the fine movements in the three spatial coordinates, the instrument is equipped with screw-controlled coarse adjustments. The tool holder clamp, Fig. 50, can be rotated and tilted and is provided with

Fig. 50. CHAMBERS' Micromanipulator. Complete assembly comprising four manipulator units with remote controls for fine motion. Courtesy of E. LEITZ, Wetzlar (152).

locking screws for each of these adjustments. In order to obtain maximum stability, the three fine movements controlling the microtool during actual operations are independent of the coarse adjustments which are put out of action by clamping them after the preliminary settings have been made. Thus the fine manipulations are not disturbed, especially under higher magnifications, by shaking, resilience, or leverage. The fine movements of the instrument operate free from backlash since the micrometer screws act against springs and are maintained in permanent contact with the metal bars which transmit motion to the microtool. Also, by making use of springs, undue wear of these movements is avoided because of the absence of frictional surfaces.

The instrument is independently mounted on a rigid pillar which is screwed into a heavy base plate on which the microscope is mounted. This offers greater steadiness and facilitates the exchange of microtools and the preparation of the apparatus for work. In one of the earlier models (58, 60) the instrument was clamped directly to the stage of the microscope and consequently its steadiness was dependent upon that of the microscope. In that model the preliminary coarse adjustment of the microtool was performed by means of sliding movements operated by hand.

The precision of the manipulator fine movements can be increased (35, 152) by using remote control. This is achieved by means of flexible shafts with operating knobs attached to the three movements as shown in Figs. 49 and 50. By using these remote controls, the transmission of vibrations from the hand to the three fine motions is almost completely eliminated. The supports for the remote controls are mounted on the base plate.

According to the type of work to be performed, one or more manipulator units may be utilized. For isolation of bacteria for example, a single manipulator unit carrying a micropipet may be sufficient, but for cell injection and for tissue cell dissection, for instance, two or more instruments may be used. Right- and left-hand instruments are available. A large, heavy cast-iron base plate may be supplied (152), permitting micromanipulator units to be attached in six different positions. This provides for up to four manipulator units in combination with the microscope which remains in a fixed position on the base plate. When a pair of tools is used and it is desired to use only one opening of the moist cell in order to minimize evaporation, both manipulators may be set up in front of the microscope, Fig. 49. Four manipulator units, a pair on each side of the microscope are shown in Fig. 50. Since the fine three-way movements imparted to the microtool points are curvilinear, this instrument—unlike other micromanipulators which can be placed at any distance from the microscope— should be mounted at a fixed distance from the microscope stage so as to maintain the relative directions of the horizontal and vertical fine movements. This imposes some limitations on the position of the instrument in its spatial relation to the microscope.

The injection apparatus shown in Fig. 50 is similar to that described under the lever activated micromanipulator of E. LEITZ, Fig. 27.

3. Microdrills.

Microdrilling machines operate drills having very small diameters or oscillating styli with exceedingly fine points. They are used when the size of the hole or the depression to be made decreases to such an extent that observation with optical aids during drilling becomes desirable or necessary. Several microdrilling machines have been developed for work under the microscope (41, 132, 138, 197). One of these machines (132) is capable of making depressions as small as 10 μm in diameter. Another machine (41, 197) is capable of drilling holes of even smaller size. Such tiny holes or depressions may be made in hard material requiring considerable force to penetrate, such as metals, alloys, minerals, and refractory materials. The depth of penetration and the size of the resulting cavities can be controlled. In metallurgical investigations, i. e., for taking samples from localized spots and for removal of inclusions of the order of one to a few cubic millimeters, KOCH *et al.* (138) use microdrills of comparatively large size. These microdrills may be made of hard steel or a hard metal fitted with diamond splinters. Such splinters are available in sizes of about 1 to 1.5 mm, while the steel tools are of about 0.5 mm in diameter. The microdrills are mounted on a micromanipulator, and are of various shapes to suit different operations such as boring, widening the drill hole, and grinding the separated sample.

Drilling under the microscope is nowadays successfully used for a variety of important scientific and technological operations. It provides a suitable mechanical means for accurately controlled microsampling of tiny metallic and nonmetallic inclusions. Such samples are usually taken from the surfaces of properly mounted and polished specimens as those normally prepared for metallography. The samples can then be removed for identification of constituents by microspectrographic analysis, x-ray diffraction studies, or by other appropriate physical or chemical methods. For emission spectrography for instance, the microsampling procedure commonly practiced (41, 197), p. 193, consists of drilling out the minute portion to be analyzed, flowing a material such as collodion over the resulting chips, and transferring the collodion together with the chips to the electrode of a standard spectrographic arc. The steps involved in this procedure are illustrated in Fig. 117. Alternatively (132), the resulting loose debris scattered in the vicinity of the sampled inclusion may be collected on the tip of a glass or silica fiber which is coated with a thin film of grease or other suitable adhesive and mounted on a simple micromanipulator, Fig. 55, p. 196.

Drilling under microscopic control is also widely applied to other operations of great importance, p. 191. Of the metallurgical applications, samples of individual layers of multiple platings can be taken for analysis by microdrilling into their cross sections. Diffusion characteristics in metals may be studied by investigating samples taken from a series of drillings. Different reaction zones in a chemically attacked metallic surface, for example when a steel tool is attacked by oven gases, may be sampled individually (138) and investigated separately.

Of the important industrial applications, p. 308, of microdrilling may be mentioned the drilling of small multiple holes in the fuel nozzle of the diesel injection system which is also used in the jet engine industry for planes and cars. The microscopic precision holes are also of vital importance in the synthetic silk and yarn industry, particularly for drilling fine holes in the spinnerettes used for the manufacture of synthetic filaments. Microdrilling plays also an important role in the fine watch and instrument industry employing multiple precision holes.

In the following, two commercially available machines of more recent design which are suitable for the removal of microgram to nanogram quantities of solid inclusions under microscopic control will be described. The two microsampling machines are based on different principles and have

Fig. 51. The NAJET Microdrilling Machine Model 7A. Main components with metallographic specimen in position. Courtesy of F. R. BRYAN and E. F. RUNGE.

different modes of operation. One of these instruments is the NAJET microdrilling machine Model 7A (173) using very fine microdrills of the pivot type which may be as small as 6.4 μm in diameter. The other machine is the ultrasonic Jack Hammer (35) which operates a 10-μm pointed stylus oscillating at ultrasonic frequencies. By means of such vibrating styli, hard and brittle inclusions may be shattered, and then removed for analysis.

a) The NAJET Microdrilling Machine Model 7A.

One of the finest machines designed for drilling proper and precise holes under microscopic control is the NAJET microdrilling machine Model 7A, which is manufactured by the NATIONAL JET COMPANY (173). This type of drilling machine was first suggested by BRYAN (39) for use in taking small metallic samples for spectrochemical analysis. The machine was subsequently used in metallurgical microspectroscopy by BRYAN and associates for taking microgram (41) to nanogram (197) amounts of metallic constituents from metallographic specimens. The instrument is capable of drilling holes as small as about 6.4 μm in diameter in relatively soft steels utilizing fine microdrills of approximately equal diameter. Drill holes up to about 0.35-mm diameter in harder material can be made with this machine.

Figs. 51 and 52 illustrate the main constructional features of the instrument and show the metallographic specimen to be sampled in position. Fig. 52, shows the complete drilling machine and microscope assembly ar-

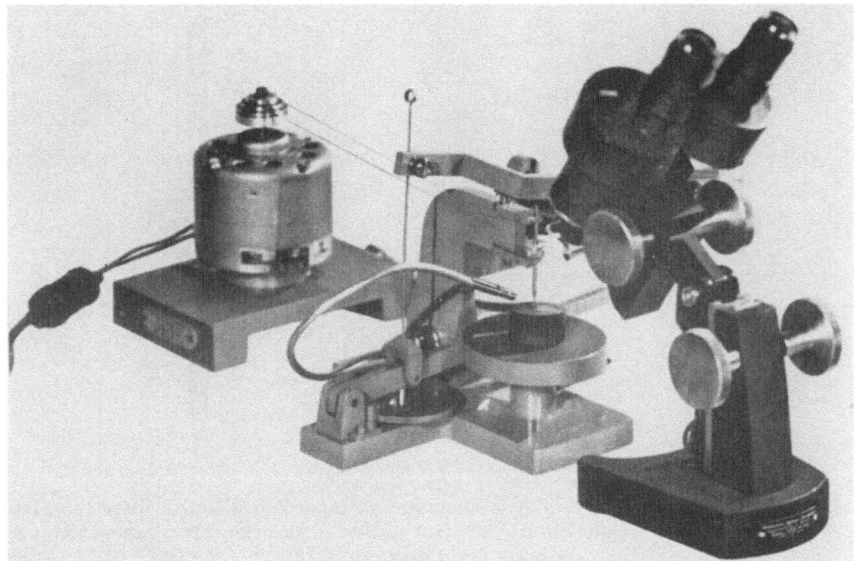

Fig. 52. Complete Microdrilling Machine and Microscope Assembly Arranged for Operation on Metallographic Specimen. Courtesy of F. R. BRYAN and E. F. RUNGE.

ranged for operation. The working dimensions of the machine accomodate a mounted and polished specimen of the size usually made for microscopic examination.

The machine is thread driven by a remote 1/40 HP electric motor providing spindle speeds from about 2500 to 5000 rpm. According to CUPLER (73), however, for satisfactory drilling, especially into very hard material approaching the hardness of the drill and when very fine drills are used, best results are obtained with speeds not exceeding about 3000 rpm. Such relatively low speeds result in holes of good quality and longer life of drill. The motor, as shown in Figs. 51 and 52, is mounted separately from the drilling machine to reduce vibration to a minimum. Rotation is imparted to the drilling machine pulley by means of a Nylon belt.

No chuck is used in this machine, but the mandrel employed is a spindle which is made as a single concentric piece with the drill and rotates in an open-jewelled V-block. Fig. 53a, shows the mandrel with the tapered drill shank fitting into a female taper in the mandrel. Fig. 53b shows the mandrel and drill with the pulley in position on the mandrel. A balanced crossarm assembly or a leverage system serves to control feed. As can be seen in Figs. 51 and 52, the pulley of the motor is at a higher level than the pulley of the mandrel so that when the motor is in operation the mandrel is kept, by the upward pull of the belt, against the crossarm of the leverage system. The depth of the hole is controlled by means of a micrometer depth stop. In using the finer microdrills, the balance weight

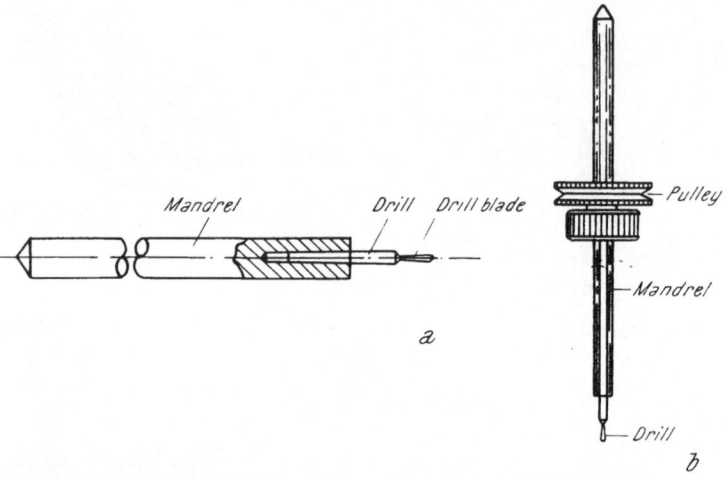

Fig. 53. Microdrill, Mandrel and Pulley. Schematic: a, tapered drill shank inserted in female taper in the mandrel; b, mandrel and drill with pulley in position on the mandrel. After NATIONAL JET COMPANY (173).

is adjusted slightly off balance so that the lever barely falls under its own weight. The micrometer depth stop is then lowered gradually to permit the microdrill to reach the required depth. This procedure (197) eliminates nerve tremor that may be introduced by the human hand.

Drills are of the pivot type which are made of tungsten carbon steel hardened to Rockwell C-65 to 68. These drills are normally used in sizes ranging from about $25\mu m$ to 0.12 mm in diameter. Larger microdrills up to about 0.35 mm are also available. The drill size should be selected according to the hardness of the material to be drilled. For hard material approaching the Rockwell C-65 hardness, when the surface area to be removed is sufficiently large, microdrills of about 125 μm and larger are most suitable. In materials of Rockwell C-30 hardness, a 50-μm drill is successfully used. For many metallurgical materials, however, it is usually easy to sample areas 25 μm in diameter or larger. The use of microdrills down to about 6.3-μm diameter may also be possible, but microdrills of appreciably smaller diameter are not only difficult to use but are fragile for most purposes. Fig. 54, illustrates the relative size of two of the finer microdrills of about 25-μm and 6.3-μm diameter in comparison with a No. 80 twist drill about 0.3 mm in diameter.

Factors affecting drill-life have been discussed by CUPLER (73). Aside from the material from which the microdrill is made, an important factor affecting the life of the drill is the trueness or concentricity with which the drill revolves. Drill diameter is another factor. Drill blades of larger

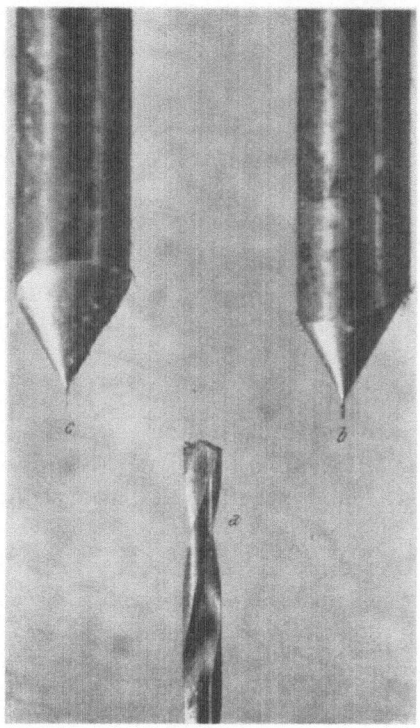

Fig. 54. Relative Sizes of Two Fine Microdrills Compared with a No. 80 Twist Drill. *a*, No. 80 twist drill of about 0.3-mm diameter; *b*, microdrill of about 25-μm diameter; *c*, microdrill of about 6.3-μm diameter. Courtesy of F. R. BRYAN and E. F. RUNGE.

diameter break less readily than finer ones. Drilling speeds also have an effect upon the drill-life. The speed of the drill is varied for the drilling of different materials. With finer microdrills, it is advisable to use lower spindle speeds especially when drilling hard material approaching the hardness of the drill. Among the other factors of importance is the use of the proper coolants or oils. Carbon tetrachloride is considered the best coolant and lubricant in spite of some undesirable qualities.

Drilling is performed under a stereoscopic binocular microscope providing magnifications of about 20 to 50 diameters. The microscope shown in Fig. 52, is an American Optical (3) Spencer stereoscopic shop microscope equipped with paired eyepieces which may be changed to obtain the required magnification. The paired objectives are protected by a glass plate which can be removed for cleaning or replacement.

As was mentioned before, p. 74, this particular machine has been used by BRYAN and co-workers for metallurgical microsampling. Microgram (41) to nanogram (197) quantities of metallic constituents were removed from metallographic specimens for identification by a microspectrographic method. The procedure used is simply to drill out the microconstituent to be analyzed and vaporizing it in an arc. Drillings are preferably transferred to the electrode by engulfing the chips in collodion, stripping the dried film, and transferring it together with the chips to the electrode. This microsampling procedure, p. 193, is illustrated in Fig. 117. The use of the machine is further illustrated in Figs. 151 and 152, in the part devoted to applications.

b) The Ultrasonic Jack Hammer.

For the removal of inclusions from metallographic specimens preliminary to identification of constituents by chemical or physical methods, KEHL and associates (132) describe a *Jack-Hammer* with a 10-μm pointed stylus oscillating at ultrasonic frequencies. By means of this vibrating stylus inclusions of about 10-μm minimum diameter can be isolated. The machine, supplied by BRINKMANN INSTRUMENTS, Inc. (35), is suitable for removing hard and brittle metallic microconstituents as well as hard glassy or loose fluffy non-metallic inclusions.

Fig. 55, shows the main components of the apparatus. The transducer and horn assembly is carried on a micromanipulator which is normally mounted on the left side of a bench-type metallurgical microscope equipped with a refracting objective. The stylus which makes contact with the inclusion to be removed is attached to this assembly. On the right side of the microscope is another micromanipulator carrying quartz fibers which serve to collect and transfer the inclusion debris. Both micromanipulators and the microscope are mounted in proper relation to one another on a common base.

A glass hood of a special design may be placed over the specimen on the stage of the microscope. This hood is provided with an opening, Fig. 55, or a slit, Fig. 56, through which the stylus enters into the hood. Another opening in the chamber permits an inert gas such as purified argon to be passed into the system during operation. This is to prevent oxidation of the material of the inclusion which may be caused by the frictional heat generated at the tip of the oscillating stylus. If the material of interest does not suffer by oxidation, the use of the hood is not necessary. In such instances a reflector is used in place of the refracting microscope objective to provide a greater working space above the specimen.

The transducer-horn assembly, also shown in position in Fig. 56, is connected to an ultrasonic driving oscillator, not shown in the figures, which is designed to match the natural frequency of the barium titanate transducer: 47 kc per second. The circuit of the generator includes a feed-back amplifier system to secure constant amplitude. The unit is capable of delivering energy sufficient for the purpose.

Fig. 55. The Ultrasonic Jack Hammer in Operating Position. One micromanipulator is carrying the transducer-horn assembly and the other is holding the pick-up fibers. The metallographic specimen and hood are in position on the microscope stage. Courtesy of BRINKMANN INSTRUMENTS, INC. (35).

Fig. 57, illustrates the transducer and horn assembly. It consists of the barium titanate transducer and the brass horn (161) which serves as a transformer to magnify the linear motion of the stylus relative to that of the transducer by a factor of about 30. The horn is especially designed to secure maximum transfer of energy to the stylus so that the stylus tip will vibrate with maximum amplitude. The throw of the tip of the stylus is approximately 25 μm each way from center. The horn is cemented to the transducer with a suitable adhesive such as Tuffbond. As can be seen in Fig. 57b, the horn is composed of two parts. The narrow part is detachable from the wide part by screw threads to facilitate the change of the stylus which is waxed to the tip of the narrow part of the horn. This allows a number of additional horn ends, fitted with styli, to be held on hand ready for use so that styli with points damaged in operation may be replaced immediately.

Styli are prepared from pieces of drill rod approximately 1 mm in diameter and 8-cm long. The rod receives a special heat treatment by brine quenching from a temperature of about 870° so as to attain maximum

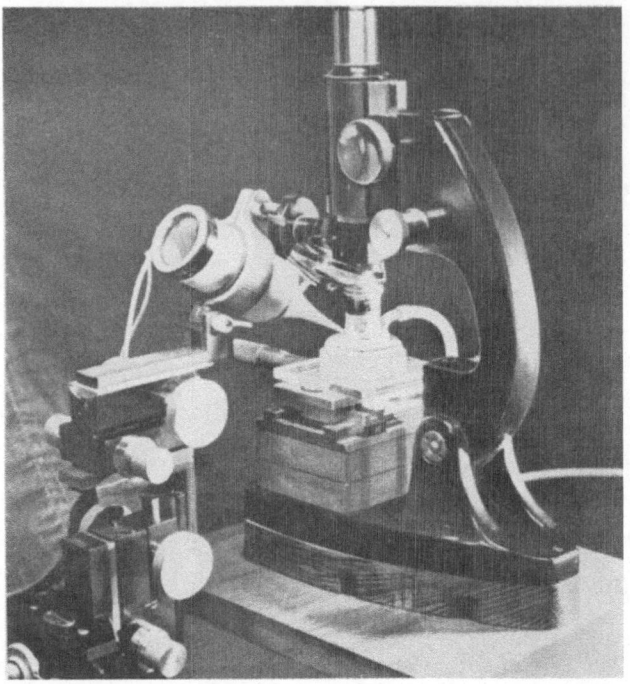

Fig. 56. The Ultrasonic Jack Hammer. Main components shown in Fig. 55, but without the micromanipulator carrying the pick-up fibers.   Courtesy of G. L. KEHL.

hardness. The heat treated wire is then pointed electrolytically by a method (132) which nearly always produces a perfect point as that shown in Fig. 58. The whole operation can be performed in about five minutes and briefly consists of making the heat treated wire one electrode, suspended in a vertical position from a micromanipulator, in a bath of equal parts of glacial acetic acid and nitric acid, which is kept cool during use by placing it in an ice bath. A strip of platinum foil is the other electrode. An emf of 8 to 10 v (ac) is applied. By means of the vertical adjustment of the micromanipulator, about 1-cm length of the lower end of the drill rod is immersed in the electrolyte and withdrawn after two seconds.   This operation is repeated 25 to 30 times to give the wire a tapered point. A sufficiently sharp point is then made by immersion into the electrolyte and withdrawal at a constant rate until the current drops to 0.2 A.   The wire is then kept immersed until the current is reduced almost to zero, whereupon it is instantly withdrawn from the solution. When the sharp point has been obtained, the drill rod is cut to a length of about 6 cm.

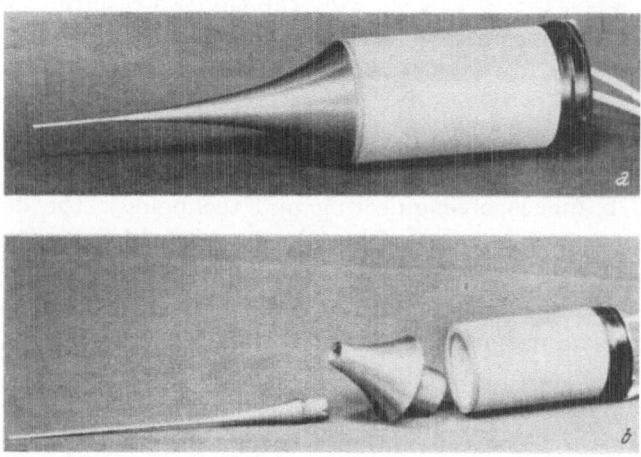

Fig. 57. The Transducer-Horn Assembly without Stylus. *a*, the barium titanate transducer and the two parts of the horn assembled; *b*, the three component parts separated. Courtesy of G. L. KEHL.

The stylus is then inserted into a hole drilled in the lower end of the horn and fixed in position by applying softened hard wax. Ready made pointed styli are available commercially (35) and may be supplied together with the rest of the apparatus.

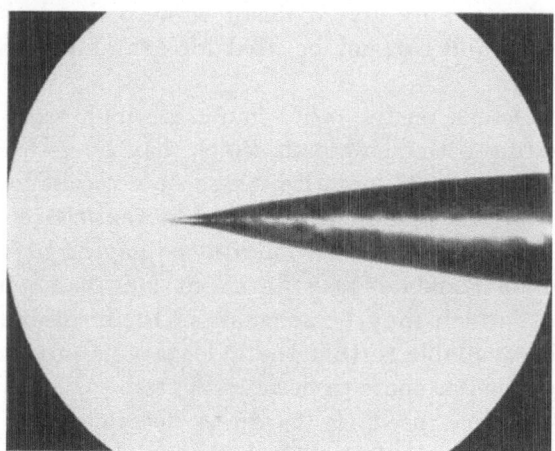

Fig. 58. Typical Shape of Electrolytically Formed Point of a Steel Stylus. Courtesy of G. L. KEHL.

The extraction procedure, using this machine, is described on p. 195. As has already been mentioned, the instrument is capable of removing inclusions of minimum diameter of about 10 µm. The micromanipulator

carrying the transducer-horn assembly must permit precise location of the
stylus point and proper control of the depth of penetration and of the size
of the resulting depression. Such precision is illustrated in the four photomicrographs, Figs. 118 to 121, showing the cavities resulting with different
materials. The inclusions are partially removed by the vibrating stylus.
The resulting debris can also be seen.

4. Microhardness Testers.

In several branches of engineering and technology, the determination
of hardness is very important. Thus the testing of hardness may provide
valuable knowledge regarding the dependable performance, workability,
and wear resistance of machines and tools. Hardness testing may also be
useful in the examination of unfinished products as well as in the identification of certain raw materials.

For testing the hardness of various materials several mechanical
devices are in use. The conventional hardness testers operate with comparatively heavy loads which generally range from about one kilogram to
several hundred kilograms depending on the dimensions of the area to be
tested and the hardness of the material. On account of the heavy loads
employed, these testers may be used only for examining materials or test
pieces which can stand such loads without danger of destruction; they
usually serve for testing hardness of semi-finished products and raw
materials by means of the BRINELL and ROCKWELL methods. Furthermore, these testers can only give a mean value of hardness over a comparatively wide area and can not be used for examining very small spots
and discrete particles.

Microhardness testers, on the other hand, are high-precision laboratory
instruments operating with small loads which may range from a few grams
to one or two kilograms. In practice, these opto-mechanical instruments
are mainly used for the examination of small textural constituents and
tiny localized spots in the field of a microscope having high magnification
which, in certain instruments, may be about 900 diameters. In testing
such minute areas, which may be as small as 10 μm in diameter, suitable
test loads must be available so that the indentations produced are smaller
in size than the particular spots to be investigated.

Microhardness testers, used for hardness determination performed on
polished surfaces of ore minerals (172, 220), may help the mineralogist in
the identification of unknown mineral species. Significant relationships,
p. 283, have also been shown (172) between hardness and the textures
encountered in the study of polished surfaces of ore minerals. In addition
to their wide use in the examination of textures of various types of steel
and other metals and alloys, these instruments also find wide application

in testing small workpieces as those encountered in the watchmaking industry, precision mechanics, and fine tool manufacture. They are useful in testing of surface layers which may be as thin as 1 μm, the surface of case-hardened work pieces, foils, varnishes, plastics, and glass. Cutting edges of tools, watch wheel spindles, and fine wires of various cross sections may be inspected.

Microhardness testers are mainly used for the determination of the VICKERS hardness, by producing indentations in the surface of the specimen by means of a pyramid-shaped diamond under a suitable load and measuring the diagonals of the indentations by means of a micrometer eyepiece. The VICKERS diamond has a four-sided base surface of equal diagonals with an apex angle of 136° between the opposed pyramid faces. The depth of penetration of the VICKERS diamond is one-seventh of the indentation diagonal. From the test loads applied and the average lengths of the indentation diagonals, the VICKERS microhardness numerals may either be found directly in special conversion tables (152) or may be calculated from:

$$HV = \frac{1854.4\,p}{d^2}$$

where HV is the VICKERS hardness in kg/mm², p is the applied load in grams, and d is the average length of the diagonals in microns. In the HANEMANN tester, p. 89, instead of measuring the lengths of the diagonals, the area of the indentation may be directly measured by means of a special centering type planimeter eyepiece.

The KNOOP method, is similar to the VICKERS method but employs a diamond pyramid with a rhombic base surface. The longitudinal diagonal is seven times the length of the transverse diagonal, while the depth of penetration is about one-thirtieth of the longer diagonal and the angles between the two opposed pairs of edges are 172° 30′ and 130°. The KNOOP hardness may either be found in special tables or is calculated from:

$$HK = \frac{14230\,p}{l^2}$$

where HK is the KNOOP hardness in kg/mm², p is the test load in grams, and l is the length of the longer diagonal in microns.

When the same material is tested under equal loads, the longer diagonal produced by the KNOOP pyramid is about three times as long as those of the VICKERS pyramid. This is advantageous in testing harder materials, since longer indentations can be measured more precisely. Under the same testing conditions the KNOOP pyramid also produces shallower indentations—about two-thirds of those of VICKERS—which are more favorable in testing very thin layers.

In addition to the VICKERS and KNOOP methods, some microhardness testers are used for the determination of scratch hardness. The tests are performed by means of special cone pointed diamonds of 90°, 120°, and 150° apex angle, or by using pyramid-shaped diamonds of 120° included angle. Normally, the scratch hardness is considered as the load producing a reference scratch width of 10 μm. The scratch hardness is of special technological importance since it gives a good idea concerning resistance of the material to wear from grinding and scoring, whereas the indentation hardness is more related to the tensile strength.

The surface area to be tested should be perfectly plane and smooth. The test area should be perfectly horizontal and the specimen should be in a stable position while indentations are being made, p. 87. Larger specimens may be held in special vises or mounting fixtures (152), p. 287, Fig. 161. Specimens too small or too thin for mounting in a clamping device, for example extremely fine wires and needles or very thin sheets of metal, are embedded in a suitable material such as Plexiglass. This also permits the cross section of the specimen to be ground and polished together with the Plexiglass. Microhardness tests can be performed on satisfactorily polished metallographic or petrographic specimens, but special care should be taken to avoid any craters.

In the following, two commercially available microhardness testers of more recent design will be described. One of these instruments is the Durimet or Miniload tester (38, 152) which is used with test loads ranging from 15 to 2000 g. The instrument is provided with exchangeable diamonds for VICKERS and KNOOP tests, as well as for scratch hardness testing. The instrument can be used for testing small as well as large objects, p. 287. Measurement of the indentation diagonals and the scratch widths is carried out by means of a precision measuring ocular and a total microscope magnification of 400 diameters. The ocular micrometer scale can be directly read with a precision of 0.5 μm, while displacements down to 0.1 to 0.2 μm may be estimated.

The other device is the HANEMANN microhardness tester Model D 32 (245) which is designed to determine the VICKERS hardness with small loads ranging from a few grams or less to about 100 g. The device essentially consists of a special microscope objective with a VICKERS indenter cemented into the axis of the front lens. In use, the device replaces the standard objective of a suitable inverted metallographic microscope, and test indentations are made by actuating the microscope adjustment while the specimen is observed under the microscope. Measurement of the area of the indentations or of the indentation diagonals is usually performed under high magnifications. These may reach 900 diameters and thus permit close examination of test areas which may be as small as 10 μm × 10 μm.

a) The Miniload (Durimet) Microhardness Tester.

One of the most versatile instruments designed for microhardness testing is the Miniload tester, manufactured by E. LEITZ, Wetzlar (152). As already mentioned, the instrument operates with test loads ranging from 15 to 2000 g and is designed for VICKERS, KNOOP, and scratch hardness testing.

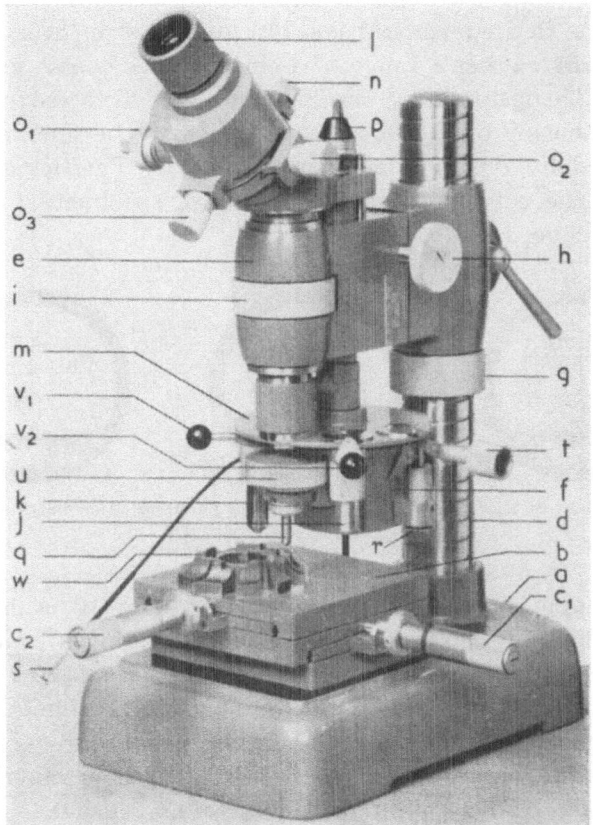

Fig. 59. The Miniload Microhardness Tester. a, base; b, compound stage; c_1 and c_2, micrometer spindles; d, upright column; e, microscope; f, indentation mechanism; g, nut for vertical adjustment of microscope and indenting mechanism; h, coarse vertical adjustment; i, fine vertical adjustment; j, low power objective; k, high power objective; l, measuring ocular; m, disk-type revolver; n, ocular locking screw; o_1, o_2 and o_3, adjusting knobs for ocular scale, transverse and longitudinal movements of ocular; p, illuminator; q, indenter; r, knurled screw for adjusting rate of lowering indenter; s, wire release; t, knob for lifting indenter; u, test weight; v_1 and v_2, handles for bringing objectives and indenter into position; w, test object. Photograph by courtesy of E. LEITZ, Wetzlar (152).

The instrument consists essentially of a stand carrying a compound stage, and a column to which the microscope and the indentation mechanism are fixed. Fig. 59 illustrates the general constructional features of the instrument. The base plate, a, carries a compound stage, b, which provides motion up to 25 mm in two directions at right angles to each other by means of two micrometer spindles, c_1 and c_2. The base plate also supports the upright column, d, carrying the microscope, e, and the indenting mechanism, f. The height of the upper part of the instrument, microscope and indentation mechanism, can be approximately adjusted in the required position along the column, d, by turning the knurled nut, g, which is able to travel 15 cm along the coarse thread of the column. The device may be locked in the desired vertical level by means of a lever. A rack-and-pinion movement having a range of 4 cm provides coarse vertical adjustment and can be operated by turning knob, h. This movement permits approximate focusing of the microscope, whereas the fine vertical adjustment is effected by turning the knurled ring, i. The device can also be swung around the column to enable testing of specimens too large to be placed on the stage, Fig. 161, p. 287.

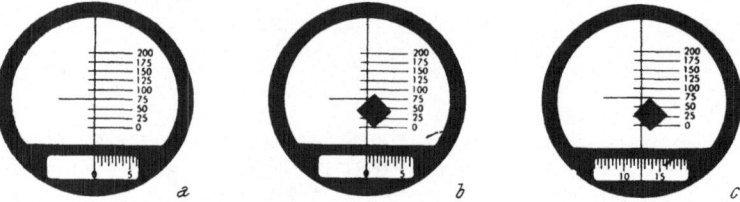

Fig. 60. Visual Field of the Measuring Ocular of the Durimet Microhardness Tester. After E. LEITZ, Wetzlar (152).

The microscope is provided with two alternatively usable objectives, j and k. The low power scanning objective, j, having a primary magnification of $10\times$, combined with the measuring eyepiece, l, $10\times$, gives a total magnification of 100. The high power measuring objective, k, has a magnification of 40. This objective in combination with the eyepiece, gives a total magnification of 400. The diameter of the field of view is 1.8 mm when using the low power scanning objective, and it is 0.45 mm when the high power measuring objective is used. The two objectives and the indentation mechanism are fixed on a disk-type revolver, m. The microscope is provided with an inclined ocular tube. The ocular can be rotated about its axis through 360° and is locked by means of locking screw, n. The graduated scale of the eyepiece as well as the transverse and longitudinal positions of the rotatable ocular can be adjusted by means of adjusting screws o_1, o_2, and o_3, respectively. Measurement of the VICKERS and the KNOOP indentation diagonals as well as the scratch widths is carried out by means of the ocular micrometer scale, using the $40\times$ objective. For this objective, each interval in the vertical or plotting scale, shown in the center of the ocular field, Fig. 60, is equivalent to 25 μm. Intermediate values are read on the horizontal or measuring scale in the lower part of the field. Each division on the latter scale is equivalent to 1 μm and is in turn divided into two divisions of 0.5 μm each. It is even possible to estimate displacements down to 0.1 μm by interpolation. Adjustable bright-field illumination is obtained by means of a removable illuminator, *p*, with an 8 v, 0.6 A incandescent bulb connected to a suitable transformer. All optical components of microscope and illuminator are coated to diminish reflection.

The indentation mechanism, *f*, as previously mentioned, is seated on a disk-type revolver, *m*. The mechanism, arranged in a housing, is provided with a slippage compensating device (38) consisting of a specially designed spring seat for the indenting body and a flexible fixation for the testing diamond, *q*. The indenting body is carried on a lever which is actuated by a spring, and the indentation speed is controlled by an oil damper. This flexible arrangement counteracts the slight lateral displacement which otherwise occurs when the indenter under load is brought into contact with the test piece. Such lateral movement, may amount to only a few microns under test loads higher than 300 g and decreases for smaller loads. By means of the compensating device, the indenter or testing diamond is kept centered on the particular spot to be tested; distorted indentations which may give rise to faulty results are thus avoided. The whole flexible mounting also protects the testing diamond from being damaged by collision with the test piece. The rate at which the indenter is lowered is set by means of the knurled screw, *r*, which controls the oil brake. It should take at least 15 seconds to lower the testing diamond. The indenter is lowered by pressing the wire release, *s*, to bring it into contact with the test piece. The indenter body is raised by turning the knurled knob, *t*, clockwise until it stops.

The active weight imposed by the indenting lever is 15 g. Other weights that may be used together with the basic weight of the lever produce loads of 25, 50, 100, 200, 300, and 500 g.

The instrument is equipped with a special device permitting the use of a Leica camera, provided with a focuser, for taking photomicrographs of indentations.

The instrument should have a vibration-free mounting, and should be placed in a perfectly horizontal position. The room temperature should be in the range of 15° C to 30° C since a larger change in temperature affects the viscosity of the lubricating oil and hence the performance of the instrument.

Testing Procedure: Directions will be given for performing indentation hardness tests, p. 83, measurement of the indentation diagonals, performance of scratch tests, and measuring scratch widths.

After setting the instrument and properly mounting the specimen, *u*, on the compound stage, *b*, Fig. 59, below the indenter, *q*, the test starts by placing the selected weight, u, in position above the indenter. Before mounting the specimen, the indenter should have been raised by turning knob, t, clockwise until it stops. It is essential that the specimen be firmly mounted, p. 286. This helps to avoid oversize indentations. The surface to be tested should be perfectly horizontal and remain horizontal during indenting. This may be checked prior to testing by focusing the high power objective sharply on the surface of the specimen and pressing on it lightly with a small rod while observing through the microscope. The specimen must remain in sharp focus when the pressure is applied.

For finding a suitable spot on the object for the indentation test, the low power scanning objective, j, is used. The specimen is brought into proper position by means of the two micrometer spindles, c_1 and c_2, of the compound stage. The high power objective, k, is then moved in position and the selected spot is brought into sharp focus and made to coincide with the cross of the ocular graticule; the eyepiece micrometer is set in the zero position. The indenter is then brought into the working position. The instrument is left for a few seconds to rest and the wire release, s, is then actuated by pressing the button. This sets the spring mechanism free to bring the indenter slowly into contact with the test object. The indenter is left in contact with the object for about 10 seconds, a time sufficient for the complete penetration of the indenter. The indenting body is then raised by turning the knurled knob, t, to the right, whereby the spring is simultaneously wound up again for the succeeding test. The high-power objective is then brought into position for measurement of the indentation. In bringing the objectives and the testing diamond into working position, both handles, v_1 and v_2, of the revolving disk, m, should always be grasped by using both hands for the purpose.

Measurement of the hardness indentation diagonals is performed by first setting the ocular scale into the zero position as shown in Fig. 60a. By rotating the ocular, the vertical line of the scale is then accurately aligned parallel to the diagonal to be measured. In this position the graduation lines of the vertical scale are perpendicular relative to the same diagonal. By manipulating the micrometer spindles of the compound stage (or when the stage is not used as when large objects are tested, by actuating the adjusting knob of the scale), the lower corner of the indentation is aligned with the edge of the zero division of the eyepiece scale, Fig. 60b, and the scale is read. By means of the adjusting knob of the eyepiece, the divisions of the eyepiece scale are moved upwards to make the top corner of the indentation coincide with the nearest division, Fig. 60c. In Fig. 60, the length of the vertical diagonal is equal to $75 + 12.5 = 87.5$ μm. In determining the VICKERS hardness, the ocular is rotated $90°$ about its axis by loosening the ocular locking screw, n, Fig. 59, and the second indentation diagonal is similarly measured. The average diagonal length is used to calculate the VICKERS units, HV, whereas the length of the longer diagonal serves to calculate the KNOOP hardness, HK. The equations are given on p. 83. Alternatively, the indentation hardness values may be found directly in special tables.

For performing scratch hardness tests, the indenter is replaced by a scratch diamond, p. 84. The indenting body with the scratch diamond is lowered to bring the diamond into contact with the specimen as described for the indentation tests. The front micrometer spindle, c_1, Fig. 59, of the compound stage is then slowly and uniformly rotated in a counterclockwise direction. This moves the specimen towards the operator and a scratch is produced. After raising the diamond and moving the high-power objective into position, the scratch width can be measured.

To measure the scratch width, the ocular is rotated until the graduation lines of the vertical or plotting scale are parallel to the edges of the scratch. By means of the knob, o_1, of the measuring scale, the two edges of the scratch are successively brought into coincidence with a graduation line of the vertical scale. The scratch width is determined as the difference between the two readings of the measuring scale. The test is repeated under different loads so that the load which produces a reference scratch width of 10 µm can be estimated. This specific load may be considered as the scratch hardness.

The use of this instrument in testing hardness of different specimens, is illustrated in Fig. 59, and also in Figs. 154 and 160 in the part on applications, p. 280.

b) The Hanemann Microhardness Tester Model D 32.

This high precision testing device, Fig. 61, is actually a specially designed microscope objective equipped with a suitable diamond indenter cemented axially into its front lens. It is manufactured by C. ZEISS, Jena (245), and it is designed for the determination of the VICKERS hardness with small test loads which range from a few grams or less to about 100 g. The tester is made for use in conjunction with the large inverted metallographic microscope Neophot or the small inverted metallurgical microscope Epityp, both produced by ZEISS. The device is simply attached in position underneath the mechanical stage to replace a standard objective; the VICKERS indentations are made by actuating the microscope adjustment while the object is observed in the microscope field. The object is mounted face down on the mechanical stage of the inverted microscope and the test diamond, which is mounted axially in the front lens of the device, approaches the specimen from below. The test load acting upon the object is indicated on a calibrated scale through an auxilliary optical system incorporated in the device. The area of indentation used to determine the VICKERS hardness is usually measured directly under a high power objective by means of a centering type planimeter eyepiece. VICKERS hardness may also be determined by measuring the length of the indentation diagonals. In the following lines, the device will be described briefly only to illustrate more clearly the principles underlying its operation. A detailed description of the tester and its operation is given in Pamphlet No. 30-G 676 b-2 of the manufacturer.

Fig. 62, illustrates the design of this testing device. A distinctive feature of this tester, as already mentioned, is the combination of the VICKERS pyramid with the front lens of the objective of the testing device to form a single unit. This feature helps in fixing accurately the point of impact

Fig. 61. The HANEMANN Microhardness Tester Model D 32. The tiny VICKERS diamond indenter is cemented into the axis of the front lens of the tester. Courtesy of C. ZEISS. Jena (245).

since the testing device itself is used as an objective while the indentations are being made. The VICKERS indenter, a, is cemented into a narrow hole drilled into the center of the front lens, b, of the objective to leave a free annular portion of the lens sufficiently large for illumination and image formation. The objective is freely suspended from two disk springs, c, so that when a test load acts on the indenter, the objective moves in the direction of its own optic axis. The extent of this motion which is a measure for the magnitude of the test load is observed with an auxiliary objective, d, which is attached to the rear element, e, of the principal objective. Both objectives receive light from the vertical microscope illuminator. By means of the mirror, f, the load indicating scale, g, is illuminated and an image of the scale is formed in the ocular. The auxiliary objective is rigidly attached to the principal objective and moves with it, whereas the scale is stationary. The magnitude of the load is read in the eyepiece as motion of the image of the load indicating scale. Calibration of this scale is carried out by using a set of calibrated weights comprising

5-, 15-, 45-, and 95-g pieces. The weight used for calibration is placed upon the front lens mounting of the testing device. This permits precise determination of the effective test loads. Sharp focusing of the image of the load indicating scale is performed by rotating ring, h, while its zero setting is adjusted by turning ring, i.

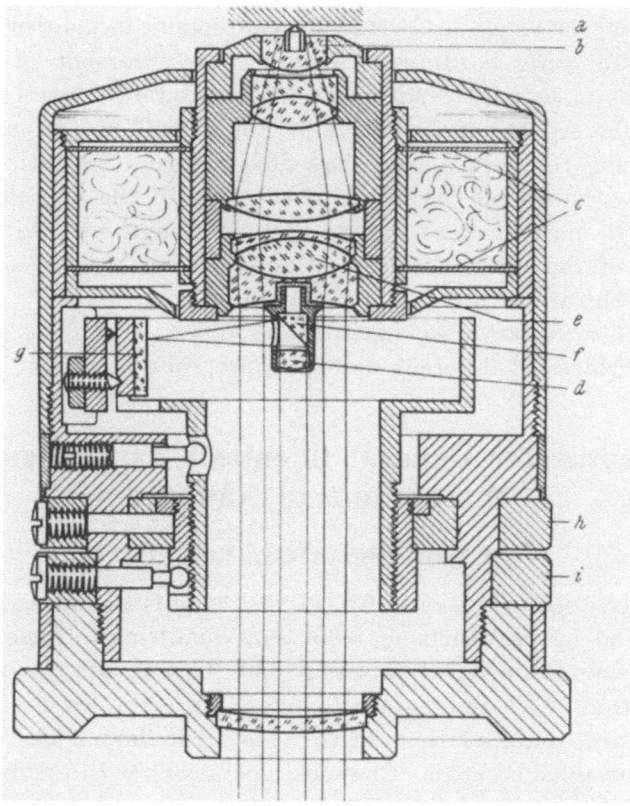

Fig. 62. The HANEMANN Microhardness Tester. Schematic sectional view, showing path of rays. a, VICKERS indenter; b, front lens tester; c, disk springs; d, auxiliary objective; e, rear element of main objective; f, mirror; g, load indicating scale; h, ring for fine focusing; i, ring for zero setting of scale. Photograph by courtesy of C. ZEISS, Jena (245).

The testing device offers a magnification of 32 diameters, and the overall numerical aperture is 0.65; the aperture of the blocked-out central area is 0.30. It is used in combination with a specially designed centerable micrometer ocular which is equipped with a $15\times$ compensating eyepiece and two reticles, one stationary and the other sliding. Each of the reticles has a right angle ruled on it so that the two angles join to form a cross when the micrometer is set in the zero position; in all other positions they form a square which serves to measure the area of the square-shaped VICKERS indentations. The length of the diagonals may also be measured

with the micrometer eyepiece, which should be calibrated for each objective
to be used in measuring. The indentations may be measured by using the
microhardness tester itself as an objective. One objection to this practice
is that the masking-out of the center of the field by the indenter gives
conditions for image formation, which may not be favourable for measurement. For this reason, the tester may only be used for measuring large
indentations, for rough and rapid measurements in control of production,
and for comparing hardness. Accurate measurement of microhardness
indentations is usually carried out with standard objectives. With the
Neophot, the Apochromats $60 \times /0.96$ and $32 \times /0.65$ are, as a rule, used
for measuring the area or the diagonals of test indentations. In testing
individual crystals, measurement of the area of the microhardness indentations gives more accurate results than the lengths of the diagonals since
the shape of the indentations may deviate from that of a perfect square
because of the anisotropy of the plastic flow.

VICKERS microhardness indentations made with this instrument in
polished surfaces of different materials are shown in Figs. 155 to 159.

5. Apparatus for General Chemical Experimentation with Micrograms to Nanograms.

a) Working in Capillary Cones.

Qualitative investigations (25, 28, 29, 30, 91) including separation and
identification by confirmatory tests, semiquantitative estimations making
use of the volumes of precipitates (25, 30, 92, 93), and gravimetric (88, 89)
determinations with microgram to nanogram samples can be carried out
by mechanical manipulations in tiny glass capillary cones, and are nowadays well established fields. Chemical operations with micrograms to nanograms in volumes of solutions of the order of one microliter to a few nanoliters are performed in tiny capillary cones of approximately one half to
a few microliters capacity. Small amounts of reagents and wash liquids
are held ready for use in reagent containers which, together with the cones,
are held ready on a glass carrier inside a moist cell to hinder evaporation
of the small volumes of solutions. Measurement of volume is carried out
in the reagent containers or in special measuring capillaries mounted on
the carrier. The moist cell is mounted on the stage of a low-power microscope and is moved by means of the revolving mechanical stage of the
microscope. The transfer of liquids is achieved by a micropipet operated
with a suitable syringe and mechanically moved by a micromanipulator
which permits displacement of the micropipet along the three spatial
coordinates. The micromanipulator together with the rotating mechanical stage of the microscope provides the means for adjusting the

relative positions of capillary vessels and micropipet. The separation of solids from liquids in the capillary cones is achieved by centrifugation and decantation. The mechanical and optical facilities permit the performance and observation of all necessary operations with ease and confidence.

For work with tenths of milligrams to micrograms of material and with relatively large volumes of solutions, up to about 50 µl and more, the capacity of the capillary cones may be proportionately increased from about 20 µl (69) to 200 µl (71, 72). This permits appreciable simplification of the apparatus. Since comparatively large volumes of solutions are usually dealt with, operations may be carried out without the use of a moist chamber. The relatively large containers employed may be mounted directly on a simple micromanipulator which serves to move the vessels relative to the micropipet. This micromanipulator takes the place of the microscope stage employed in the assemblies used for more delicate work. A calibrated capillary pipet may be directly inserted into a simple syringe control which is carried and moved by a second micromanipulator mounted in a position to face the first. These micropipets serve for measurement and transfer of solutions and reagents. No measuring capillaries or reagent containers are used as in other assemblies. A wide-field, approximately 30 power stereoscopic binocular microscope is employed to permit proper observation of the relatively large vessels and to offer the advantages of stereoscopic vision. Such simplified apparatus (69, 71, 72, 210) was successfully used by Cunningham and Werner in the last stages of isolation of the first pure samples of plutonium and in the preparation, identification, and determination of properties of several new compounds of this synthetic element. It has also been used in a similar way in the isolation of other transuranium elements (70, 160).

Detailed description of the apparatus used in qualitative microgram to nanogram analysis and the techniques of manufacture of consumable equipment such as micropipets, capillary vessels, etc. will be given, unless otherwise indicated, with reference to El-Badry and Wilson's work (90). The latter, as well as that of Cunningham and Werner (69, 71), are in fact an extension of Benedetti-Pichler and his co-workers pioneer investigations in the qualitative and semi-quantitative microgram fields (25, 28, 29, 30), that constituted the first systematic study of working in capillary cones and provided the background of the manipulative techniques involved. In the course of this description, apparatus used by Benedetti-Pichler and co-workers will also be described whenever convenient with the view of giving a practically complete idea of the requirements of the field. This will be followed by a brief description of the assembly used by Benedetti-Pichler and also that used by Cunningham and Werner which is suited for working with relatively larger samples. The latter description will be given mainly with reference to the more fundamental points of difference. It will be noted that these various sets of apparatus are fairly similar and a worker who is familiar with one set can very easily work with others.

α) Assembly of El-Badry and Wilson.

Though most consumable and other apparatus used by BENEDETTI-PICHLER and co-workers (25) are also described under this heading, greater

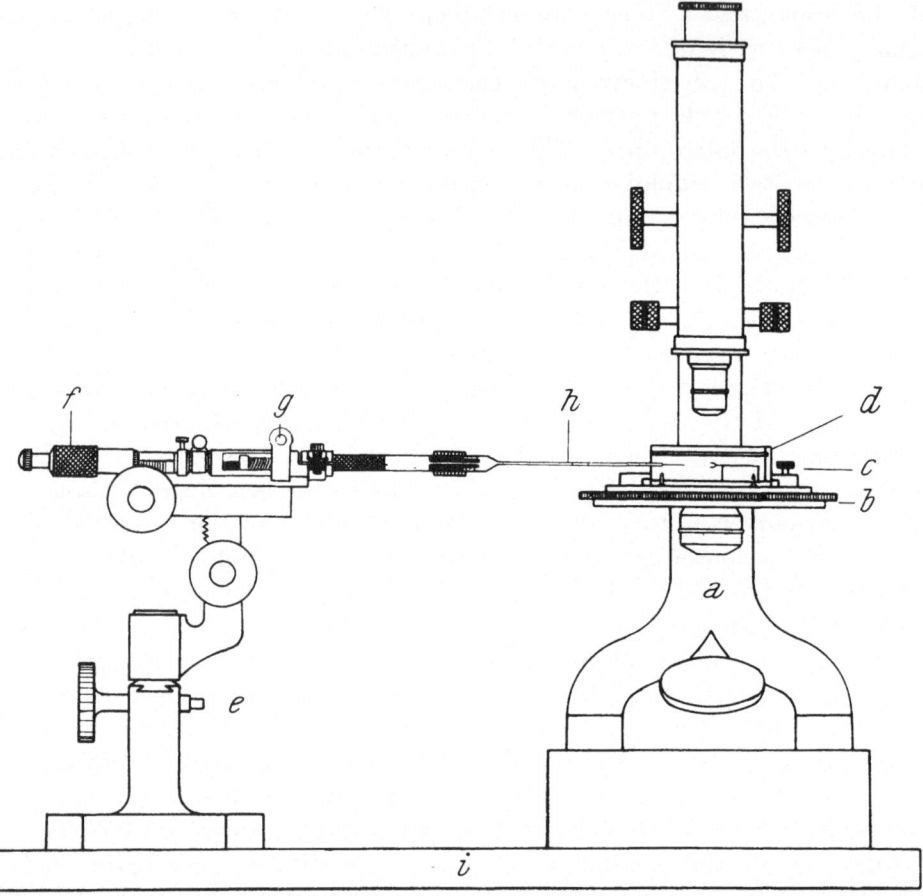

Fig. 63. Complete Set-Up Used by EL-BADRY and WILSON for Working in Capillary Cones (90). *a*, microscope; *b*, revolving stage; *c*, attachable mechanical stage holding and moving moist chamber; *d*, Perspex moist chamber; *e*, simple rack-and-pinion micromanipulator; *f*, Agla micrometer syringe; *g*, clamp for holding micrometer syringe; *h*, micropipet; *i*, wooden base.

emphasis will be placed on the apparatus used by EL-BADRY and WILSON (90). The latter apparatus, Fig. 63, is comparatively simple, notably in the replacement of the plunger device with the long copper tubing, Fig. 80, by a hypodermic syringe, in the construction of the moist chamber, and in the techniques of preparing the capillary vessels and the other consumable equipment.

Microscopes. A total magnification of approximately $40\times$ is quite suitable. It may be obtained by using a $6\times$ micrometer eyepiece in combination with a 32-mm objective. The eyepiece micrometer is calibrated for use with the 32-mm objective as mentioned on p. 7, the draw tube being fixed at a constant length. A bright-field condenser which can be focused on a plane about 1 cm above the stage may be desirable.

The microscope, a, Fig. 63, is fitted with a built-in rotating stage, b, and an attachable mechanical stage, c, which holds the moist chamber. The attachable mechanical stage provides motion in the horizontal plane in two directions at right angles to one another, thus offering added flexibility of movement. The rotating motion of the microscope stage is highly desirable in all cases to align the micropipet with the axis of the capillary vessels in the microscope field, and it is imperative if the manipulator has no rotating motion.

Micromanipulators. A micromanipulator suitable for performing chemical operations in capillary cones has to fulfil only a few basic requirements. A comparatively large sagittal or thrust motion is necessary. It should be possible during operation to bring the micropipet point to the further end of the microscope field which is usually within a few millimeters beyond the optic axis of the microscope. It should also be possible to withdraw the pipet point well out of the moist chamber and preferably out of range of the microscope altogether for washing the pipet or replacing it with another. For all this a 5-to 6-cm excursion may be sufficient, though a somewhat longer travel may be preferable. On the other hand, a relatively small transverse motion permitting about 2-to 3-cm excursion should be more than adequate. Once the pipet point is brought into the field of operation, the largest movements that may be necessary would be from one side of the field of vision to the other. Even with the low magnifications used, the diameter of the field usually does not exceed a few millimeters. The motion offered by the mechanical stage of the microscope adds to the flexibility of motion in this direction. Similarly a vertical movement of about 2 to 3 cm should be more than sufficient. Once the relative levels of the manipulator and the microscope are properly fixed, a vertical motion of a few millimeters is all that may be required. A rotating motion in the horizontal plane may be desirable, but it is necessary only when the microscope stage cannot be rotated. Even when the manipulator offers such a rotary movement, the revolving motion of the microscope stage, however, is still desirable. Furthermore, the manipulator should be reasonably stable to vibration, but this requirement is not very exacting since all operations are performed under low magnifications. A good number of low power micromanipulators easily fulfil this requirement. Finally, a requirement of all low-power instrument: mobility, speed, and convenience are among the most desirable features. All these requirements have also been discussed by BRINDLE and WILSON (34).

The simple low-power micromanipulator, e, shown as a part of the complete assembly of Fig. 63, permits motion in the three space coordinates and is suitable for work in capillary cones. It is easily constructed from three rack-and-pinion movements of discarded microscopes mounted on a microscope base, and its performance can readily be understood from a study of the figure. Each rack-and-pinion motion permits about 8-cm excursion thus providing ample movements in the three dimensions. The upper rack-and-pinion movement, the thrust motion, is provided with a clamp, g, for holding the micrometer syringe to which the micropipet is attached.

The instrument is similar in principle to the more refined simplified micromanipulator Mifyxir supplied at one time by E. LEITZ (152), and used by BENEDETTI-PICHLER, Fig. 80. It is also similar in principle to the simple instruments of the rack-and-pinion type shown in Figs. 15 and 18. These instruments may be compared with the more elaborate micromanipulator developed by BRINDLE and WILSON (34), which is briefly described on p. 37. The latter is another refined instrument which has been used for work in capillary cones (229). Though the presently used instrument is not as stable to vibration as the more refined instruments just mentioned, it gives satisfactory service when the rack-and-pinion components are in a reasonably good condition. Its use is certainly justifiable when no funds are available for aquiring a more refined instrument.

Many low-power micromanipulators manufactured by various firms may be suitable for working in capillary cones. However, for working with a micrometer syringe of the type shown in Figs. 63 and 64, the micromanipulator should be provided with suitable clamps for holding the syringe on its top movement.

Base Plate. As shown in Fig. 63, the micromanipulator is screwed permanently to a wooden board, i. The microscope is placed on a wooden platform of suitable height, fixed to the same baseboard. The microscope may be positioned by wooden blocks fastened to the platform, but it need not be permanently attached to it. No difficulty should be experienced in removing and replacing it in the working position since the motions of the manipulator and the mechanical stage of the microscope provide sufficient flexibility.

The correct relative manipulator-microscope position may be determined by the following procedure. The microscope and the manipulator carrying the micrometer syringe with a micropipet attached are placed loosely in approximate positions on the wooden board. The pinions both of the vertical and the transverse movements of the manipulator are made to engage approximately at the center of their racks, while the top pinion controlling the thrust motion is operated to advance the tip of the micropipet all the way toward the microscope. By moving the manipulator and the microscope closer to each other, adjusting the height of the microscope by means of additional boards, the tip of the micropipet can be made to take a position about one centimeter beyond the optic axis of the microscope and approximately one centimeter above the stage. Appropriate provisions are then made to maintain manipulator and microscope always in this correct working position.

The base is preferably mounted on a low table of such a height that various operations can be carried out with ease and comfort over extended periods. It is preferable to have a cover for the whole assembly which can be made to fit the baseboard to keep it clean when not in use.

Illumination. Since the moist chamber (90), shown in Fig. 66, is made of transparent material, normal daylight illumination is sufficient for most purposes. For observation of certain confirmatory tests, it is sometimes desirable to illuminate the plateau of the condenser rod, Figs. 68, 71, and 72. For this purpose a beam of light is thrown horizontally from a pencil-type electric hand torch through the base of the microcondenser to illuminate the plateau by internal reflection.

Stronger illumination of the microscope field may be required to project the image or when the sides of the moist chamber, Fig. 68, are not made of transparent material. If the microscope is not already provided with a built-in light source for observing with transmitted light, a lamp is placed in front of the microscope sending light in the direction of the arrow T, Fig. 80, to the mirror of the microscope. For observation with reflected light, another lamp equipped with a condenser lens and diaphragm capable of throwing a narrow beam of intense light is placed to the left of the microscope on a level with the stage to send light horizontally into the moist chamber in the direction R shown in Fig. 80. For rapid interchange of transmitted and reflected light, switches for both lamps should be placed in a handy location.

Projection Devices. If work is to be carried out over extended periods, it will prove a great convenience and avoid eyestrain to have the microscope fitted with a projection device giving an image on a vertical screen facing the operator. With transmitted light it is convenient by means of a simple device, Fig. 80, to project an image, 5 to 10 cm in diameter or larger, a short distance above the ocular on a screen made simply of a piece of white bristol-board mounted inside a black box (25). The screen is mounted on the microscope tube so that the projected image is always at a fixed distance from the ocular, and whenever direct observation is required, the screen is swung out of the system. Fairly sharp images can be produced by alternatingly adjusting the height of the microscope tube and the ocular until the object and the eyepiece micrometer scale simultaneously appear in sharp outline on the screen. The value of the scale divisions of the ocular micrometer will be approximately equal to their value under direct observation. Measurements can either be made with a millimeter ruler on the projected image or with the use of the eyepiece micrometer scale. For precise work, however, calibration should be performed under the exact conditions of use.

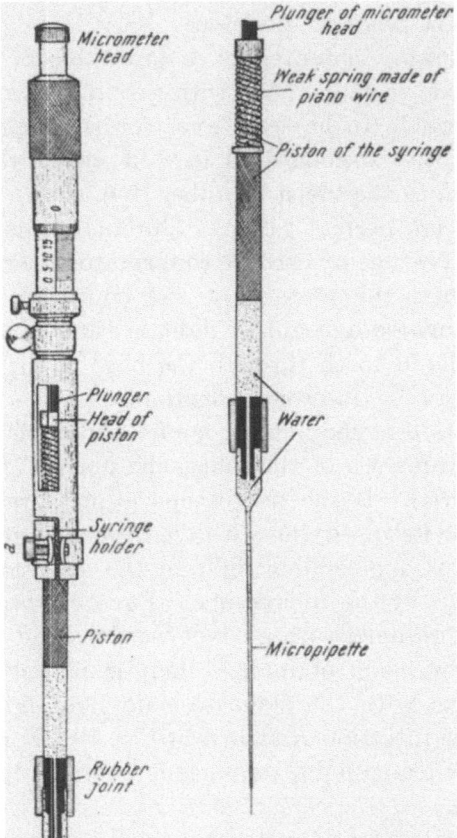

Fig. 64. Agla Micrometer Syringe and Micropipet. After EL-BADRY and WILSON (90).

Micrometer Syringe. The Agla micrometer syringe (230) is supplied by BURROUGHS WELLCOME AND Co. (44), and is illustrated in Fig. 64. It consists of an all glass hypodermic syringe of a narrow uniform bore, attached to a micrometer screw. The glass syringe is held in a rigid metal framework and is secured firmly in position by tightening the two screws of the syringe holder. The plunger of the micrometer head presses directly on the piston of the syringe. The piston is spring-loaded by placing a weak spring constructed from piano wire around its stem so that the spring is held between the head of the piston and the lip of the syringe barrel. Spring loading in this fashion ensures that the piston may be advanced or retracted without lagging behind the corresponding movement of the micrometer plunger. The syringe is held firmly in the clamp g, Fig. 63, of the micromanipulator. The micropipet is constructed in such a way as to be easily attached to the glass syringe by a simple rubber joint. In use, the glass syringe, the stem of the micropipet, and a part of the shank are completely filled with water as shown in Fig. 64. The rest of the shank contains the air cushion which serves to separate the water in the syringe from the sample or reagent solution in the shaft of the micropipet. This air cushion transfers pressure from the micrometer plunger via the piston of the syringe and the water column to the reagent solution in the shaft of the micropipet. This device

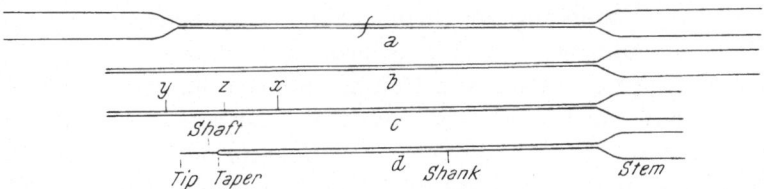

Fig. 65. Preparation of Micropipet. After EL-BADRY and WILSON (90).

for the control of liquid in the micropipet is one of the simplest devices used for microgram to nanogram solution chemistry in capillary cones. The method of attaching the pipet to the syringe for operation is described on p. 102.

Micropipet. The micropipet (90) is shown in Fig. 65d. It consists of a stem and a shank which tapers suddenly into the shaft. The stem, approximately 8 mm in external diameter, similar to the syringe barrel, 6 mm in bore, and about 2-cm long, serves to attach the micropipet to the glass syringe simply by means of a short piece of stout rubber tubing as shown in Fig. 64. The shank is about 6- to 8-cm long and approximately 1 mm in external diameter, whereas the shaft is about 8- to 10-mm long and has a 20- to 50-μm bore.

To construct the micropipets, a soft glass tube of approximately 8-mm outer diameter, 6-mm bore, and 10-cm length is heated at the center, preferably in a narrow flame such as that of a bats-wing burner. It is then drawn symmetrically, outside the flame, to produce a capillary of approximately 1-mm outer diameter, the pull being maintained until the glass cools. This produces a straight capillary with part of the original tubing attached to each end as shown in Fig. 65a. Cutting the capillary at the proper points will produce two similar pieces of the original tubing with 15 cm of capillary attached, Fig. 65b. From each of the two parts, a micropipet is prepared. Excess of the original tubing, is cut off approximately 2 cm from the capillary, preferably using a "hot point" since vibration may break the capillary if cutting is performed in the normal fashion. The 1-mm capillary is then drawn out by hand at the end furthest from the original tubing by using a microburner with a pinhead flame which may be obtained with any suitable microburner such as one of those shown in Figs. 94 and 95. To produce the shaft, the capillary is grasped between the thumb and index finger at a point x, Fig. 65c, approximately 6 to 8 cm from the stem. The thumb and index finger of the other hand grasp the capillary at a point y, approximately 4 cm from x. The hands are steadied by resting the edges of the palms on the bench. The capillary is then brought over the pinhead flame so that it is heated at a point z, midway between x and y until the glass softens. The capillary is then immediately withdrawn from the flame, and a straight horizontal pull is applied symmetrically with the edges of the palms always touching the bench while the hands are moving outwards. The amount of extension is usually from 3 to 5 cm. The fine capillary is cut approximately 8 to 10 mm from the shank and thus a micropipet as that shown in Fig. 65d, is produced. The bore at the tip is usually from 20 to 50 μm, but micropipets with finer orifices can also be used. It is not convenient to use pipets with exceedingly fine tips since they are liable to become clogged.

Elimination of Surface Tension Effects. To eliminate surface tension effects which otherwise cause poor control of liquid flow, the micropipet proper (tip, shaft and taper) is made water-repellent by treatment with either Teddol, Akard, or paraffin wax as described below. The wide shank of the pipet need not be treated since the small volumes of solutions usually employed will not fill more than the shaft and the taper.

The treated pipets operate on a principle different from that applied by BENEDETTI-PICHLER who makes use of the strong surface tension in the narrow tips. A comparison of both ways of operation of the two micropipets is given on p. 115.

Treatment with Teddol or Desicote (90). Teddol contains like Desicote a mixture of methyl chlorosilanes that is excellent for the purpose. The tip of the clean and dry pipet is dipped in the liquid and suction is applied through the stem until the liquid fills the shaft, the taper, and a short length of the shank. **Care must be taken not to inhale the vapors of the liquid.** After a few minutes, the liquid is expelled. By this treatment, the surface in contact with the liquid becomes water-repellent. The remainder of the inner surface of the shank also becomes water-repellent to some extent through contact with the vapor. Surface tension effects are almost completely eliminated, and the flow of liquid will depend almost entirely on the pressure applied by the micrometer plunger. Other advantages of siliconized micropipets will be mentioned on p. 147. It is convenient to transfer a bulk supply of Teddol to small ampules, not more than 0.5-ml capacity, and to open one of these when a stock of pipets is ready for treatment. In this way waste of the liquid is avoided.

Several preparations of chlorosilanes in suitable solvents are commercially available (110), p. 147.

Treatment with Akard (127). Akard is a solution of partially polymerized phenolformaldehyde resins. The polymerization is completed by evaporation of the solvent and heating in the stove. For use, one part of Akard is diluted with 3 to 4 parts of absolute alcohol. The shaft of a micropipet to be treated with the solution is cut 1.2 to 1.5 cm from the shank so that it is 5 to 7 mm longer than specified above. The tip is inserted into the diluted liquid, and suction is applied until the liquid fills the shaft, the taper, and a few millimeters of shank. The liquid is then expelled. The tip is left to dry in the air for a short time, whereupon it is "stoved" by heating for two hours at 140 to 150° C. The portion of the tip which was immersed in the liquid is then cut off so that the coating is retained on the inside wall only. This is necessary since some precipitates have a tendency to adhere to the coated surface.

Treatment with Paraffin Wax, first described by WIGGLESWORTH (240), is less convenient than that with Teddol, Desicote, or Akard. If the latter are not available, however, workable micropipets may be produced by coating shaft and taper with a thin layer of wax. Coating is carried out by insertion of the tip into molten wax, sucking the liquid into the pipet, and subsequently blowing it out while the tip is still immersed in the molten wax. When air bubbles begin to emerge from the tip, the pipet is withdrawn, but the stream of air is kept flowing through the pipet until the layer of paraffin has solidified on the wall. Excess of wax is then removed by dipping the coated part of the pipet into hot water for a short time while blowing air through it. When the pipet is withdrawn, the air current is maintained until the thin layer of paraffin has again solidified. After some experience, the worker should be able to determine whether a micropipet has the proper bore at the tip by noting the size and rate of air bubbles produced by blowing strongly when the tip is immersed in molten wax or in water. This avoids the necessity for measurement of the bore at the tip when preparing new pipets.

Teddol (Desicote) and Akard treatments, preferably the first, are certainly far superior to coating with paraffin wax. With the first two methods, the resulting film is hardly visible under the microscope. It does not hinder to any extent the observation of the meniscus of the liquid in the shaft. The pipets, unlike those coated with wax, are not subject to clogging so that they can be used over extended periods. Finally, a serious disadvantage of wax coating is that a portion of it occasionally becomes detached and may get into the sample. For this reason, micropipets coated with paraffin wax are absolutely unsuited for gravimetric work.

Assembly of the Pipet–Syringe Combination. A supply of micro-pipets should be available in a well covered container.

The two screws of the syringe holder, Fig. 64, are released, and the clip which they control is detached. The glass syringe with spring-loaded piston is removed from the holder. The syringe is filled with still lukewarm distilled water which has previously been boiled to drive out dissolved gases. Filling is carried out by dipping the nearly vertically held syringe, tip downwards into the water and allowing the plunger to travel gradually backwards to within 1 cm of the stop. In this way enough water, is taken up to ensure that the stem and a part of the shank of the micropipet will

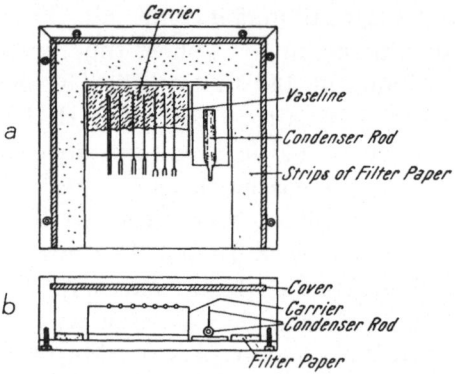

Fig. 66. Moist Chamber Made of Leucite. Capillary vessels on carrier, condenser rod, and filter paper lining are in place. *a*, top view; *b*, front view. After EL-BADRY and WILSON (90).

be filled with water when the plunger of the syringe is advanced approximately to the half-way point. Care must be taken that no air bubbles are taken up.

After filling, the syringe is reversed so that the tip points upwards. A micropipet, already fitted with a rubber joint so attached that half its length covers the stem of the pipet, is held ready for mounting. The rubber is moistened to facilitate insertion of the syringe barrel which is then pushed into place. The piston of the syringe is then advanced very gradually through the barrel until the pipet stem and part of the shank are filled with water. Care should be taken that no air bubbles are trapped. The syringe and pipet are placed approximately in position in the clamp attaching them to the micrometer, taking care to prevent the piston from sliding back and trapping air. This is easily achieved by holding the piston carefully against the spring. The plunger of the micrometer is brought up into position against the head of the piston, and is adjusted to bring the meniscus of the water about one-third way along the shank from the stem. If necessary, the position of the syringe in the clamp receives a final adjustment. The clip is replaced, and the two screws are carefully tightened.

The meniscus of the water is again adjusted using the micrometer. The pipet is now ready for use.

Loss of water in the barrel of the syringe occurs through evaporation and after a long period of use. After a micropipet has been in use for some time, it is recommended to supplement the water. For this purpose it is not necessary to detach the pipet from the syringe. The position of the piston in the syringe is advanced by turning the micrometer head until a small droplet of water forms at the tip of the pipet. The tip of the pipet is then immersed in warm, freshly boiled water in a small beaker, whereupon the piston is retracted very gradually until a sufficient amount of water has been taken up. Any excess of water taken up can be expelled through the tip.

Fig. 67. Moist Chamber. Same as shown in Fig. 66. After EL-BADRY and WILSON (88).

Moist Chambers. Fig. 66 shows the moist chamber in which various operations are carried out. The bottom, sides, and back of the chamber are made of Perspex or Leucite, while the cover is made of glass. The cover fits into the horizontal grooves of the sides and the back of the cell; it slides freely, but does not fit too loosely.

To maintain a moist atmosphere in the chamber, strips of filter paper, several layers thick, are laid on the bottom of the cell as shown in the figure and kept moistened with water. The cell is held and operated by the mechanical stage of the microscope, Fig. 63. The dimensions of the cell depend on the size which the mechanical stage of the microscope is able to accomodate. It is, however, desirable that the floor area should be as large as possible. A square floor area of 7 cm × 7 cm or 6 cm × 6 cm is convenient, but chambers of smaller dimensions may also be used. An internal vertical depth of about 1.7 to 2 cm is convenient. The Perspex plate forming the bottom of the cell may be about 3- to 4-mm thick while sides and back are approximately 5 to 6 mm in thickness. The glass cover is about 1.5- to 2-mm thick. The Leucite plates forming the bottom, sides, and back of the cell are fastened by simple screws as shown in the figure. This cell is easier to construct than that used by BENEDETTI-PICHLER (25, 29) and described below, Fig. 68. It also has the advantage of being light in weight. Since metal is not used, illumination within the chamber is better and consequently most operations can be performed in normal day light.

Fig. 68 shows the chamber designed and used by CEFOLA. A glass plate h, 60 mm × 60 mm, forms the bottom of the chamber. A brass frame b, 1-mm thick and made to fit plate h, forms a border approximately

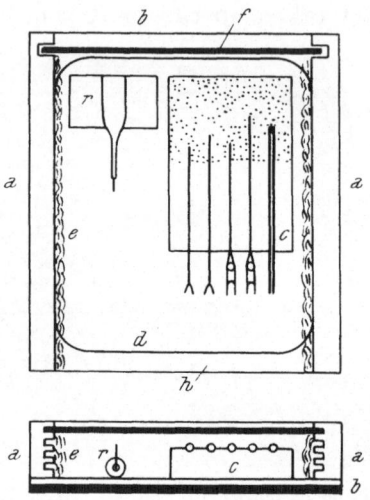

Fig. 68. Moist Chamber, top and front views. Also shown are capillary vessels on the carrier and condenser rod. After BENEDETTI-PICHLER (25).

5-mm wide all round and is mounted by means of De Khotinsky cement on the glass bottom h. The sides a consist of two pieces of brass, 67 mm long, 6 mm in thickness, and 11 mm high, and are fastened with screws to the brass frame b. The back of the chamber is formed by a narrow strip of thin glass plate f, 53 mm long and 11 mm high, which fits vertical grooves at the rear end of the sides a. The top of the cell is made of a thin glass plate d, approximately 52 mm × 63 mm, which fits into the upper-most of the horizontal grooves of the brass bars a so that it slides freely but not too loosely. Three other horizontal grooves in the brass bars are lined with cotton, e, which is moistened with water to produce a damp atmosphere in the cell.

Dry Chambers. It is useful, in certain cases, to have another chamber similar to one of the moist chambers already described to be used as a dry chamber. The dry chamber, of course, does not need the paper or cotton lining.

Carrier. It is desirable that the carrier be of such dimensions to hold comfortably as many as seven or eight capillary vessels without crowding. The carrier should also be of such height that it supports the capillary vessels in approximately the same level as that of the platform of the condenser rod. Consequently a carrier about 3 to 3.5 cm long, 2 to 2.5 cm wide, and 0.7 to 1 cm thick should be satisfactory.

Two types of carriers (90) may be used. Fig. 69 shows a carrier built up from several pieces of microscope slides of the proper dimensions. They are cemented together and stuck on a larger glass base. Canada balsam or Durofix (189) may be used as a cement. Alternatively the carrier may be cut from plate glass. The upper surface of the carrier is coated with a thin layer of vaseline to hold the assembly of capillaries. A strip of 0.5- to 1-cm width along the front, is left uncoated.

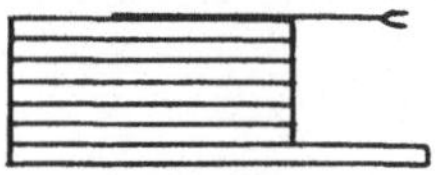

Fig. 69 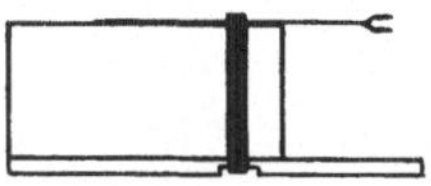   Fig. 70

Fig. 69.   Glass Carrier, side view.   After EL-BADRY and WILSON (90).
Fig. 70.   Plastic Carrier, side view.   After EL-BADRY and WILSON (90).

The other type of carrier, Fig. 70, is of the same general shape but may be constructed from a small block of Perspex cemented onto a larger base of Perspex sheet. A groove is then cut round the platform, a short distance from the front edge of the carrier, as indicated in the figure. This groove takes a thin light rubber band under which the handles of capillary vessels are inserted. This obviates the necessity of using vaseline for holding capillaries as above described.

Similarly, a thin rubber band can be employed for holding various vessels on a glass carrier shown in Fig. 69, but the lack of a groove on the base to take the band makes the carrier slightly awkward.

Condenser Rods for Slide Tests. Confirmatory tests may be performed on the circular platform of a condenser rod similar to those shown in position inside moist chambers, Fig. 66a and b. The platform has an area of about 0.03 to 0.2 mm² obtained by cutting a glass thread of 0.2- to 0.5-mm diameter. The condenser rod consists of a short piece of glass rod which tapers rapidly to a fine thread. The thread is cut off and is bent at right angles to provide the tiny platform on which small amounts of test solution and reagents may be deposited.

The condenser rod described by BENEDETTI-PICHLER and CEFOLA (25, 29) is shown in Fig. 71. A condenser rod of this type may be constructed from soft glass which should show as little color as possible. A short length of a glass rod a, 4 mm in diameter, is cut off to give a face, f, at right angles to the axis of the rod. About 2 cm from the end f, the rod is drawn out to a finer rod *b*, about 2 mm in diameter. A narrow flame is used so as to obtain a rapid taper. About 0.5 to 1 cm from the taper, the fine rod is then drawn out with a microflame to a fine thread *c*, 0.2 to 0.5 mm in diameter. The thread is then broken about 2 cm from its taper and is bent at right angles by approaching it with a microflame a few millimeters from the taper while the rod is held horizontally. The thread bends by its own weight. It receives a water repellent coating nearly up to the bend, whereafter the thread is cut approximately 0.5 to 1 cm above the bend to form platform *d*. The cutting is done by scratching with the sharp edge of broken china or a crystal of boron carbide (see also p. 108) and breaking off with forceps. It is highly desirable that the resulting platform should be flat and at right angles to the thread.

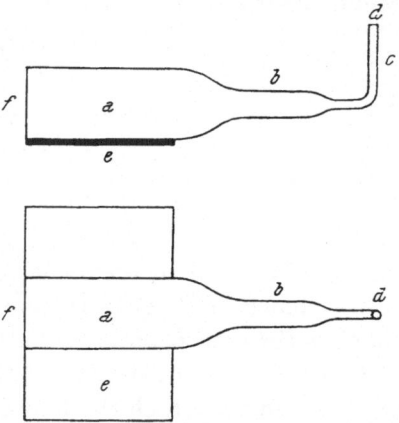

Fig. 71. Condenser Rod, side and top views. After BENEDETTI-PICHLER and CEFOLA (25, 29).

The simple condenser rod (90), shown in Fig. 72, is constructed from the same type of glass rod in a similar but somewhat easier way. After a short length of glass rod is cut off to give a face at right angles to the axis of the rod, the latter is drawn out immediately to a fine thread 0.2 to 0.5 mm in diameter. This is done about 2 cm from the end of the rod and with the use of a narrow flame. The fine thread is broken about 2 cm from its taper and is bent at right angles a few millimeters from the taper. After applying a water repellent coating to the thread, the platform is obtained in the same way as described above. With the thread in a vertical position, the rod is finally attached by Durofix or some Duco cement to a thin glass plate or to a cover slip, Fig. 71*e*, which serves as a base.

The water-repellent coating is applied to the thread by dipping it either in a suitable siliconizing fluid, p. 147, such as liquid Teddol (90) or in molten paraffin wax (25, 29). This treatment prevents test solutions from flowing down the sides of the thread so that they remain confined to the area of the platform.

The platform of the condenser rod can be efficiently illuminated by casting a beam of light through the wide end in the direction of the arrow R, Fig. 82. Most of the light is reflected internally and emerges through the platform.

Forceps. One forceps preferably made of stainless steel and having smooth parallel tips not corrugated or toothed is required. Another forceps, similar but with cork-lined tips is useful, for such type of forceps prevents the loss of capillaries which may slip the grasp of plain forceps.

Gloves. In preparing reagent containers and other capillary vessels, meticulous cleanliness should be observed. It is preferable to use thin cotton or silk gloves to handle capillaries from which capillary vessels are being prepared. Four finger cots for thumbs and index fingers may be more convenient. The finished capillary vessels are handled with forceps and

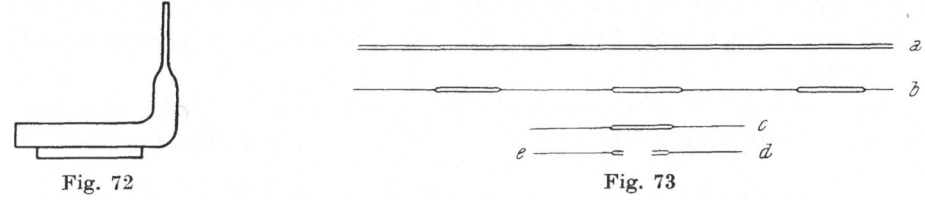

Fig. 72   Fig. 73

Fig. 72. Condenser Rod, side view. After EL-BADRY and WILSON (90).

Fig. 73. Preparation of Reagent Containers and Capillary Cones.

should not be touched with the fingers since finger prints may seriously interfere with the microscopic observation of contents.

Reagent Containers. Small volumes of solutions to be analyzed, reagents, and wash liquids are held ready for use in tiny containers placed on the carrier in the moist chamber. These containers are calibrated so as to permit the approximate measurement of volumes of reagents and wash liquids.

A large number of these containers are prepared from glass capillaries of approximately 0.5-mm uniform bore, which have been drawn out from clean narrow quill tubing and have been cut into pieces approximately 15 cm long, Fig. 73a. The bore is measured at both ends of these pieces, and the maximum deviation from the arithmetical mean should not be more than 5%.

To measure the bore, the capillary is placed in a piece of a narrow quill tubing of suitable length, fixed vertically to a movable support. The condenser and mirror of the microscope are removed, and the end of the capillary is brought from below the stage under the objective of the microscope. The evenly cut end is sharply focused and measured with the eyepiece micrometer.

A microburner is used like that shown in Fig. 94; it should have a jet opening of about 2 mm and a flame about 0.7 to 1 cm long. The capillary is fused to form an elongated bead at a point approximately 1 to 2 cm from one end. Outside the flame, the bead is then drawn out to a thin rod approximately 4 cm long, Fig. 73 *b*. This operation is repeated at 2-cm intervals until a chain-like structure consisting of 2-cm parts of the original capillary joined by solid threads, about 4 cm long, results: Fig. 73 *b*. Cutting at the proper points gives containers each consisting of 5 to 8 mm of the original capillary attached to a thread about 2 cm long, Fig. 73 *c*, *d*, *e*. These are stored in a screw-cap vial.

Cutting of capillaries can conveniently be achieved using a thin-edged carborundum plate. Probably the most satisfactory cutter, however, is a thin slip of tungsten carbide machined to provide as sharp an edge as possible. This should be reserved for capillary cutting and should not be used for ordinary glass cutting operations.

Containers of the pattern described are more or less the same as those described by BENEDETTI-PICHLER (25) but the technique of preparing them is more rapid.

The length of container which holds one nanoliter of liquid is calculated from the average diameter of the bore of the original capillary from the equation

$$L = \frac{1}{\pi \ r^2} \ \mathrm{mm}$$

where L is the length of capillary corresponding to 1-µl capacity and r is the radius of the bore in millimeters. The length of capillary which holds 1 nl is most conveniently expressed in terms of the eyepiece micrometer divisions and is recorded on the labelled vial.

Capillary Cones. Chemical operations are performed in tiny capillary cones of about 0.5-µl capacity. These are prepared from capillaries of approximately 0.5-mm uniform bore as a chain, Fig. 73 *b*, exactly in the same way as the reagent containers. From the chain, blanks are obtained by cutting the middle of each glass thread. They are pieces of capillary, about 2 cm long, with a glass thread attached at each end, Fig. 73 *c*. The sealed ends of these lengths of capillary are then examined carefully under the microscope. The tip of a capillary cone must be completely closed, and the taper should be blunt, Fig. 74. If the end is satisfactory, cutting the blank at the proper point produces a cone about 2 mm in length with a handle about 2 cm long, Fig. 73 *e*. Approximately one-third to one-half of the cone is given to the taper. Capillaries whose ends, on examination, prove to be unsuitable for cones may be used for making reagent containers.

The use of a microflame instead of the edge of the Bunsen flame (25) permits better observation and control of the process and gives wider angled cones. Such cones permit convenient agitation of a precipitate by the pipet tip, which can reach effectively to the point of the taper. Further-more, the technique of preparation is less time-consuming.

A large number of these cones are prepared and are stored in a screw-cap vial.

Measuring Capillaries. Accurate measurement of volume is carried out in measuring capillaries, whereas approximate measurements are carried out in reagent containers. Measuring capillaries are thin-walled capil-laries of approximately 0.2- to 0.3-mm uniform bore, and 2 to 3 cm long. They are drawn from thin-walled tubing 3 to 5 mm in bore. The bore of each capillary is measured at each end to ensure uniformity, and one end is then sealed in a microflame.

The length of the capillary which holds one nanoliter of liquid is calcu-lated in the same way as for the reagent containers, p. 108, and is recorded. Each measuring capillary is stored separately in a piece of wide capillary sealed at each end and labelled.

Fig. 74   Fig. 75

Fig. 74.   Good Capillary Cone with Closed Blunt Taper.

Fig. 75.   Distilling Capillary and Capillary Cone on Carrier Inside Dry Chamber.   After BENEDETTI-PICHLER (25).

Finer capillaries of 0.05-to 0.2-mm bore have been used (25) and were prepared by drawing out thin-walled capillaries of 0.8-to 1-mm inside dia-meter. These capillaries permit quite accurate determination of volumes as small as 1 to 10 nl. They may, however, make it necessary to use micro-manipulators with more precise movements. It is also necessary to use micropipets with finer shafts. Accurate measurement of volumes less than 25 to 50 nl, however, is seldom required. Such volumes are large enough to be measured with reasonable accuracy by using comparatively wide capillaries of 0.2-to 0.3-mm bore.

The inner surface of measuring capillaries may be made water-repellent by treating with Teddol or Desicote before sealing. It may even be pos-sible, by using a properly diluted solution of Teddol rather than the liquid itself, to achieve an inside surface that gives a flat meniscus.

Centrifuging Capillaries. These consist of thick capillaries approxi-mately 4-cm long and with an internal diameter of about 1 to 2 mm. These capillaries are open at both ends.

Heating Capillaries. These are capillaries approximately 5 cm in length and from 1 to 2 mm in bore containing a little water and sealed at one end. After collecting the water at the sealed end, the capillary cone is introduced handle first, whereupon the other end is sealed shut. The capillary is then inserted into a suitable opening of the multi-holed metal heating block, Fig. 79, and heated at the required temperature.

Distilling Capillaries. Liquids contained in capillary cones can be distilled into distilling capillaries placed upon the carrier inside the dry chamber. The necessary heat is supplied by a suitable heating element as that shown in Fig. 76.

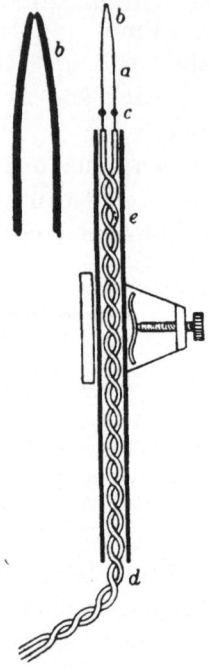

Fig. 76.   Heating Element.   After BENEDETTI-PICHLER (25).

In general appearance a distilling capillary (25) resembles a reagent container, but the taper of its completely closed tip must be blunt as with capillary cones.   The capillary part of the distilling capillary is about 10 mm long, and the bore is just wide enough so that a capillary cone fits snugly into its opening, Figs. 75, 134, and 135a. The distilling capillary is held on the carrier in the same way as other capillary vessels.

Electric Heating Element. An external heating element (25) that can be used for distillation from a capillary cone to a distilling capillary is shown in Fig. 76.   It is made of a length of No. 24 (0.5-mm diam.) copper-nickel alloy resistance wire, a, 0.7 ohm, which is bent into the shape of a V and pressed closely together by means of pliers at the point of the V. The point is then filed down carefully until the cross section of the wire is reduced to one-third of the original (33) as shown in the enlarged drawing of the point, b.

Both ends of the wire are fastened at c to an insulated copper wire d of the type used in radio work, and the copper wire is then passed through a glass tube e, 10 to 15 cm in length and of such a bore that the wire fits tightly. The wire is prevented from twisting around inside the tube by applying an insulating tape at d. The ends of the insulated copper wire

Fig. 77. Cotton Threads on Glass Holders. After EL-BADRY and WILSON (90).

are connected through a variable transformer to the ac line. Usually not more than 5 v need be supplied.

The heating element may be fastened in the clamp of the manipulator either directly by means of the glass tube or in any other convenient way.

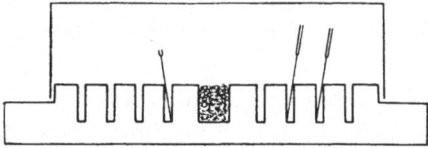

Fig. 78. Plastic Block for Storing Capillary Vessels. After EL-BADRY and WILSON (90).

Glass Rod for Thread Tests. In addition to the assembly of capillaries already described, it is often useful to add to the assembly on the carrier a straight piece of glass rod, approximately 0.5 cm in diameter and 2.5 to 3 cm long, or a bent glass thread to which a thin cotton thread

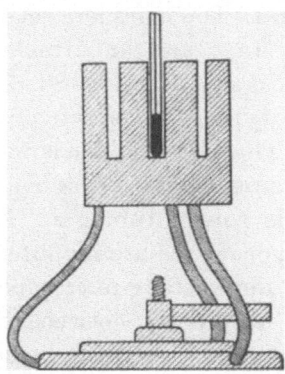

Fig. 79. Aluminium Heating Block. After EL-BADRY and WILSON (90).

can be attached (90) as shown in Fig. 77. The devices may serve for the performance of confirmatory tests.

Storage Blocks for Capillary Vessels. Capillary vessels which are in use for chemical operations may be kept in storage blocks (90) which provide clean storage containers for the vessels, Fig. 78. These are made of small Perspex or other plastic blocks which are machined to give a slight circular projection on the upper surface. Small holes that can take the handles of capillary vessels are then drilled into the face of this projection, and another hole of a larger diameter is drilled into the center of the platform. This central hole is filled with a wad of cotton which is kept moist. The lid of a Petri-dish or a crystallising dish which just fits over the circular platform provide a suitable cover.

Heating Block. This (90) is a cylindrical block of aluminium or dur-alumin, Fig. 79, provided with a central well for the thermometer and holes of various diameters for the insertion of various capillaries and micro-cones. It is heated by a small microburner having a narrow orifice, which is connected by means of a soft rubber tubing to the gas source. The base of a Bunsen burner from which the tube has been removed is suitable for the purpose. The length of the flame is adjusted by means of a screw pinch-cock on the rubber tubing to give the required temperature. A block of this type offers the simplest means for the maintainance of controlled temperatures.

Centrifuge. A small hand-driven centrifuge provided with tubes that can accomodate microcones of 1-to 2-ml capacity is suitable. The centrifuge should have provision for simple and rigid attachment to the table top. Alternatively, a motor-driven centrifuge with tubes of similar capacity is convenient.

β) Assembly Used by Benedetti-Pichler.

This assembly (25, 29) was the first to be used for performing chemical operations in capillary cones. The complete set-up including a miscroscope with a built-in revolving stage and an attachable mechanical stage, *m*, carrying the moist chamber is shown in Fig. 80. A simple micromanipulator, *a*, of the rack-and-pinion type, carries the pipet holder *b* to which the micropipet *c* is attached. The pipet holder is fastened in the clamp of the micromanipulator. The plunger device *d* is connected to the pipet holder by means of a long flexible copper tubing *e*. The microscope, the micromanipulator, and the plunger device are mounted on a common base plate. A water container *f* serves for washing micropipets.

The more fundamental differences between this set-up and EL-BADRY and WILSON's assembly already described lie in the injection apparatus and the micropipets; a description of these will be given. Other points of difference, as well as certain other equipment used by BENEDETTI-PICHLER and co-workers, that may be used also with other assemblies (moist chamber, Fig. 68, condenser rod, Fig. 71, and heating element, Fig. 76), have already been described.

Plunger Device. The injection apparatus (25), Fig. 80*d*, is made of metal and is somewhat similar in general principle to the type already described with the lever-activated micromanipulator of E. LEITZ, Fig. 27, p. 52. In the present device, Fig. 80, the flow of liquid to the micropipet is controlled by means of a plunger the motion of which is regulated by a screw with fine thread, which is operated by rotation of the milled head. The device has been adopted from one originally designed by JOHNSON and

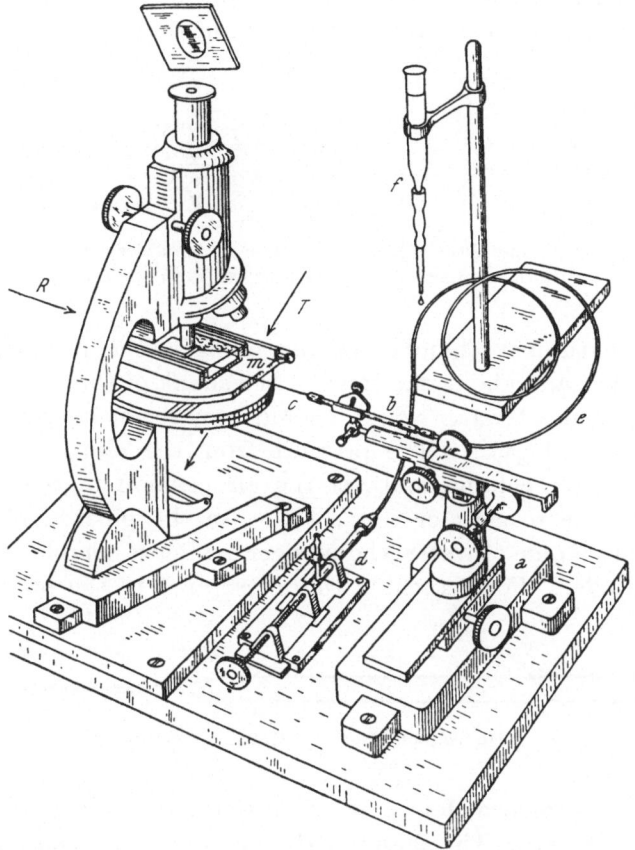

Fig. 80. BENEDETTI-PICHLER's Assembly for Working in Capillary Cones. *a*, micromanipulator; *b*, pipet holder; *c*, micropipet; *d*, plunger device; *e*, connecting tubing; *f*, water container. After BENEDETTI-PICHLER (25).

SHREWSBURY (129). A fine flexible copper tubing, at least one meter long, connects the tubular chamber of the plunger with the pipet holder in the clamp of the micromanipulator. The plunger device is fastened to the common base plate, and for convenience the copper tubing is bent to form a few helical turns.

The pipet holder, one end of which is shown in Fig. 81, consists of a metal tube *a* of 1- to 2-mm bore and approximately 12 cm in length. The micropipet is held at one end of the metal tube while the other end is connected to the fine flexible copper tubing leading to the plunger device.

Each end of the pipet holder is fitted with a rubber washer b which is a short piece of thin soft rubber tubing, a metal washer c, and a screw cap d, Fig. 81. After the shank e of the micropipet, Fig. 82, has been inserted, the screw cap is tightened causing the metal washer to advance in the well provided for the washers. The rubber washer is thus compressed between the concave face of the advancing metal washer and the bottom of the well and gives a tight seal around the shank of the pipet.

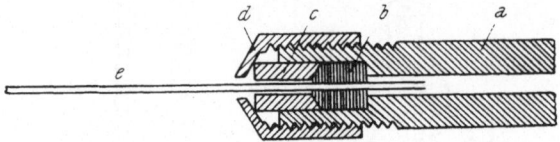

Fig. 81. Pipet Holder. a, metal tube; b, rubber washer; c, metal washer; d, screw cap; e, shank of micropipet. After Benedetti-Pichler (25).

In use, the tubular chamber of the plunger, the fine copper tubing, the pipet holder, and about one-third of the shank of the micropipet should be completely full of water which should have been boiled before use to eliminate dissolved gases. The pressure from the plunger is transmitted through the water column and via the air cushion in the shank of the micropipet to the reagent or test solution in the pipet shaft. To admit the maximum volume of water, the plunger is almost completely withdrawn from the cylinder. For attaching the micropipet to the pipet holder, the

Fig. 82. Micropipet Used by Benedetti-Pichler (25).

plunger of the pressure device is advanced until water appears at the opening of the holder. The shank of the micropipet is then inserted in the opening, and the screw cap is tightened. The plunger is advanced, if necessary, until the meniscus of water in the shank of the micropipet is approximately 3 cm from the opening of the pipet holder. The micropipet is then ready for use.

Micropipet. The micropipet (25) shown in Fig. 82 is similar to the injection pipets of the microbiologists. Different from the micropipet shown in Fig. 65d, this pipet has no stem but consists of the shank, shaft and tip. The shank of the pipet is directly inserted into the pipet holder, Fig. 81. The shank is about 10 to 12 cm in length and 0.5 to 1 mm in outside diameter. The latter is slightly less than the bore of the metal washer of the pipet holder. The shaft, having the same axis as that of the shank, is 5 to 7 mm long and should taper gradually as shown in the figure. The tip usually has an orifice of about 30 to 40 μm in diameter, but pipets with finer orifices, down to about 10 to 20 μm, may also be used.

These micropipets are constructed from capillaries 0.5 to 1 mm in outer diameter, drawn out from soft glass tubing of 6-mm bore and 8-mm outside diameter. The micropipet proper is best drawn out mechanically by means of one of the pullers described on p. 147 or drawn under the microscope, by using the microforge, p. 152. Thus identical pipets with invariably open tips can be produced rapidly and conveniently. In absence of such devices, workable micropipets may be drawn out by hand (25,123) by using a pinhead flame. A piece of capillary about 15 cm in length is grasped at its right-hand end with a pair of flat forceps, while it is seized by the thumb and index finger of the left-hand at a point 5 cm from the forceps and held horizontally in front of the operator. The hands are steadied by resting the edges of the palms on the bench. The capillary is then brought over a pinhead flame so that it is heated at a point about 1 cm from the forceps. A slight horizontal pull is applied symmetrically at once so that the drawing starts when the glass begins to soften. The steady symmetrical pull is maintained while the heated part remains above the pinhead flame. The operation is finished in less than one second when the capillary snaps at the middle of the drawn out portion. The tips thus produced are often too fine or are fused shut. A short length of the tip is snipped off, by using a pair of forceps, so as to give the required opening.

These and similar pipets used by microbiologists operate on a principle entirely different from that applying to the unwettable pipets of EL-BADRY and WILSON, p. 99. For the proper functioning of the latter pipets, the surface tension is eliminated by coating the inside of the pipet proper with a water-repellent film, p. 100. A nearly plane meniscus is thus obtained. Consequently in pipets thus treated, there is practically no resistance to the intake or delivery, and liquid flow follows every move of the micrometer plunger. As a result of eliminating the strong capillarity at the tip, it is possible to use such treated pipets with simple types of injection devices since no appreciable pressure is needed for the control of liquid flow. The way in which the pipet tapers is not critical, and satisfactory pipets are always easily obtained by drawing them out by free hand as described on p. 99.

On the other hand, the uncoated micropipets used by BENEDETTI-PICHLER and co-workers, Fig. 82, depend for operation upon the strong surface tension in the narrow tips. The pressure inside the pipet is usually slightly higher than the atmospheric pressure, but the strong surface tension of the strongly curved meniscus at the orifice prevents the outflow of the working liquid into air. Practice has shown that the proper functioning of these micropipets requires that the shank properly and rapdily tapers to the shaft, that the bore of the shaft is nearly uniform, and that the shaft again quickly tapers to form the tip. The small working droplet should reach into the taper widening to the shank so that the force on the inner meniscus would no longer be comparable to that acting at the orifice. This may be expected when taking into consideration that a volume as small as 1 nl of liquid would fill 50 mm of a capillary of 10-μm bore. Provided that the air cushion in the shank has more than atmospheric pressure, delivery is immediately obtained by immersing the tip into a liquid which eliminates the surface tension and the meniscus at the orifice. The flow is stopped by withdrawing the tip from the liquid. In this manner the meniscus at the orifice acts as a stopcock in the same way as with the measuring pipet, p. 126, used by LOSCALZO and BENEDETTI-PICHLER for titration of microgram samples (154), Fig. 87. The operation of the latter is discussed on p. 227. The same applies to the uncoated injection pipets of the microbiologists which are used also by BENEDETTI-PICHLER and RACHELE (31), Fig. 91, for performing chemical operations in hanging drops, p. 134. One advantage of the uncoated fine pipets is that, due to the strong forces at the outer meniscus, the working droplet remains at the tip without danger of outflow regardless of temperature fluctuations or of slight displacements of the plunger, which might accidentally occur. This feature, however, is not assured when nonaqueous solutions are used.

γ) Assembly Used by Cunningham and Werner.

This simplified assembly (59, 71), as has already been mentioned on p. 93, was used for the last stage in the isolation of the first samples of plutonium and other synthetic elements in the form of pure compounds (70, 71, 72, 160) and in the preparation of several compounds of these elements and the investigation of their properties and chemical behaviour at ordinary chemical concentrations. The apparatus is suitable for work with amounts of material up to hundreds of micrograms in proportionately large volumes of solutions without the use of a moist chamber.

The complete assembly is schematically shown in Fig. 83 and includes a wide-field stereoscopic binocular microscope a, and two simple low power micromanipulators, b and b_1, facing one another. One of these micromanipulators serves to carry a simple 0.5- to 1-ml syringe control c, to which the micropipet d is directly attached. The relatively large microcone e, in which chemical operations are performed, is mounted on the second micromanipulator. The calibrated pipet, used for measurement and transfer of liquids, and the microcone can thus be easily manipulated relative to one another by operating the proper controls of the two micromanipulators. Operations are conveniently observed in the wide field, f, of the stereoscopic microscope.

The more fundamental differences between this and the other microscope-micromanipulator assemblies already described, Figs. 63 and 80, lie in the use of microcones of relatively large capacity of up to 200 µl, in the dispensing with moist chambers, and in the use of a stereoscopic microscope offering a wide field of view and greater flexibility than obtainable with the microscopes used in the other assemblies. Another feature is the use of a simple syringe control and of calibrated capillary pipets. The micro-pipet is attached to the syringe control which is directly carried and moved by the micromanipulator as in the assembly shown in Fig. 63.

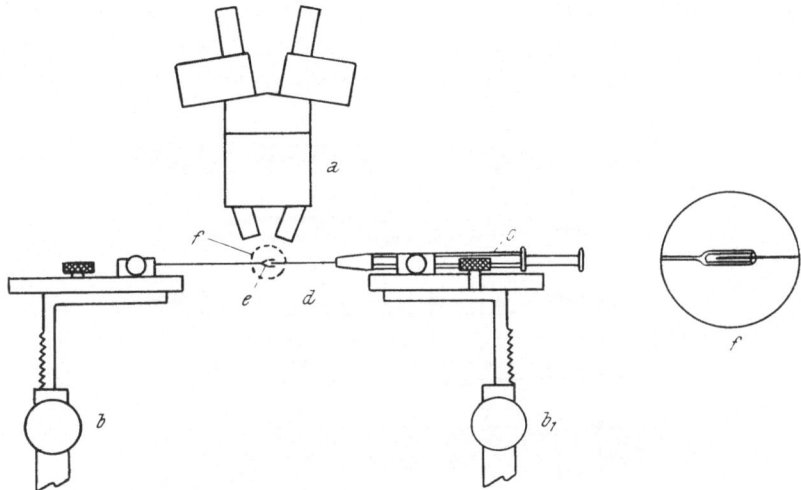

Fig. 83. CUNNINGHAM and WERNER's Assembly, schematic. *a*, stereoscopic microscope; *b*, *b₁*, micromanipulators; *c*, simple syringe control; *d*, micropipet; *e*, microcone; *f*, micro-scopic field shown at the right of figure. After CUNNINGHAM and WERNER (71).

The comparatively simple basic manipulations required by this assembly need not be described in detail. The use of the assembly is demonstrated under Applications to Nuclear Research, p. 261.

Microscope. Operations are conveniently observed through a 30-power stereoscopic binocular microscope. The microscopes used by CUNNINGHAM and WERNER (72) were either the BAUSCH & LOMB Model SKW (20) or the SPENCER Model 29 LF (8). Both have a heavy metal base.

Micromanipulators. Two simple low-power micromanipulators of the rack-and-pinion type (35), have been used (71, 72). They are mounted on separate heavy base plates and are arranged opposite to each other as shown in Fig. 83.

The micromanipulator carrying the microcone in which chemical opera-tions are performed may be replaced (69) by a simple mechanical stage mounted on a heavy metal base by means of an upright support. The microcone is then clamped to a simple metal slide which is held and moved by the mechanical stage.

Micropipets. A number of well known calibrated capillary pipets of
the transfer type, ordinarily used for milligram-scale work, can be employed.
The pipet is attached to a simple syringe control of 0.5- to 1-ml capacity
and mounted on the micromanipulator as shown in Fig. 83. The construction, calibration, and use of these pipets have been described elsewhere (25, 136, 217). They are available commercially (116, 168, 191, 235)
and are usually constructed from heavy-walled, fine-bore Pyrex capillary
tubing of about 5 to 7 mm in outside diameter. They usually range in
length from about 10 to 15 mm. For the present work, however, the drawn-
out portion of the micropipet ending into the tip should be sufficiently
fine and long to reach the bottom of the microcone used.

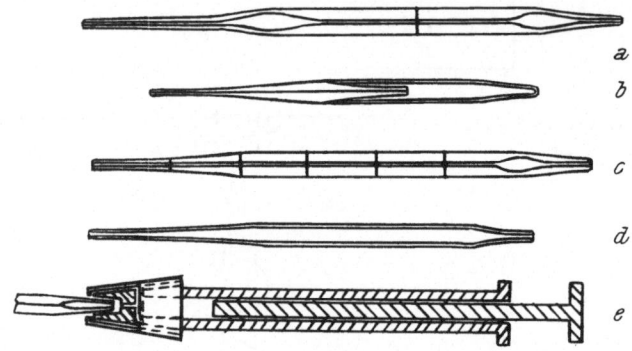

Fig. 84. Micropipets and Syringe Control. *a*, capillary transfer pipet; *b*, self-filling capillary
pipet; *c*, graduated capillary pipet; *d*, solution transfer pipet; *e*, glass syringe control.

Of these volumetric pipets, capillary transfer pipets of the style shown
in Fig. 84*a* are calibrated to hold a single volume from the tip to the mark.
Various sizes of this type are available ranging from 1-µl to more than
200-µl capacity. Pipets of this type, made of vitreous silica, are also
available (168) in various sizes ranging from 5-µl to 200-µl capacity and
larger. They can readily be calibrated for content by weighing the quantity
of mercury delivered by the pipet, and thus obtaining the total capacity.
In use, the pipets are rinsed out to recover all of the contents.

Self-filling capillary pipets, Fig. 84*b*, also calibrated to hold a single
volume, may be used for working with small volumes usually from 1 to
10 µl, whereas graduated capillary pipets, Fig. 84*c*, are usually calibrated
in five equal portions of the total volume. The latter are also available
in various capacities from 1 to 50 µl per division.

For mere transfer of liquids, a non-calibrated transfer pipet of the
qualitative type, Fig. 84*d*, may be used. These are constructed from
ordinary glass tubing of suitable outside diameter and are available in
several sizes.

The upper end of the pipets fits a suitable syringe control, 0.5 to 1 ml in capacity, which usually consists of a glass hypodermic syringe having a special metal collar attachement containing a rubber gasket, that fits tightly around the upper tapered end of the pipet, Fig. 84*e*.

Microcones. Since chemical work is carried out with comparatively large volumes of liquid, the capacity of the tiny capillary cones, p. 108, used with assemblies already described for working with smaller volumes, has been increased. Glass or vitreous silica (168) microcones of about 20- to 200-µl capacity, having the general shape shown in Fig. 85, can be employed.

δ) Semi-Quantitative Analysis.

The volume of a precipitate, if suitable precipitating forms are used, can be taken as an approximate measure of mass by comparison with the

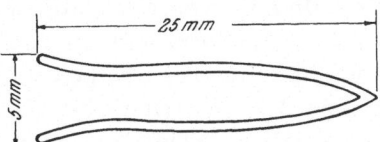

Fig. 85. Microcone of Approximately 200-µl Capacity. After CUNNINGHAM and WERNER (172).

volume of the same precipitate obtained under similar conditions from a known amount of the substance to be determined (25, 92, 93).

The apparatus is the same as that already described for qualitative microgram analysis. For the estimation of the volume of precipitates, however, special attention should be given to the capillary cones. Cones with straight-sided, symmetrical, blunt tapers must be used. The use of an electrically driven centrifuge may be preferable, but hand driven instruments are also convenient. Paraffin wax coating of micropipets is not advisable due to the separation of small particles of wax which may subsequently enter the cone, contaminate the precipitate being measured, and give high results. Treatment with Desicote, Teddol, p. 100, or Akard, p. 101, is preferable.

ε) Gravimetric Analysis.

Gravimetric microgram analysis on the classical pattern, that is, carrying out determinations involving the production, separation, washing, drying, and weighing of precipitate, is nowadays considered among the accessible techniques. Satisfactory procedures have been developed (88, 89) for the gravimetric determination of microgram amounts of inorganic ions. Needed is a suitable quartz microgram balance (88) of the torsion-restoration type in addition to the auxiliary apparatus required for performing the necessary chemical operations in capillary cones.

The Microgram Torsion Balance of Vitreous Silica.

In the following, the gravimetric technique referred to above will be considered briefly with reference to the use of EL-BADRY and WILSON's microgram torsion-restoration balance (88) which had been actually used in the development of these techniques. Since an adequate description of this instrument has been recently given by BENEDETTI-PICHLER (26), only a short account of the operation and performance of the instrument will be given in the present volume. It will be assumed that the balance is well adjusted and ready for operation.

To carry out a weighing, the balance lights are turned on, the lights of the room are switched off, and the balance is left to acclimatize for about five minutes. The beam is then released gradually while the screen is observed; the graduated dial is operated at the same time to maintain the beam nearly horizontal. Finally the beam arrest is released completely, whereafter the graduated dial is accurately adjusted so that the images of the two ends of the index fiber form a single continuous line across the screen. When the balance has thus been zeroed in, the dial is read. Two more readings are obtained by offsetting the dial by rotating it a few degrees, resetting, and rereading it. This gives a total of three zero readings. The beam and pans are arrested. The lights of the room are turned on.

The pan which is to be used is loaded and, if necessary, a counterpoise is placed on the other pan, with the pan support cradles just resting on the platforms of the pan arrests while loading is being done. The pans are released and the pan well is closed. The room lights are turned off. The beam is brought to the equilibrium position by the operation described above; when the indicator fiber images form a continuous line across the screen, the dial is read. Complete revolutions of the dial are noted by the clicking device attached at the zero mark, which is used as a revolution counter. Two offsettings, resettings, and rereadings of the dial follow to give a total of three values for the equilibrium under load.

The torsion restoration required for the load is given by the difference between the average value of the three readings under load and the average of the three zero readings. This is converted by means of the sensitivity into terms of mass.

Use of the mean of three dial settings improves the precision of weighing markedly. This procedure is therefore applied as standard practice to every weighing with the torsion-restoration balance.

The value of a scale division of the dial in terms of mass may be determined by using ten pieces of 50-gauge platinum wire, each about 2 cm long. These are weighed separately on the torsion balance and the zero reading is determined for each weighing. The ten corrected dial settings corresponding to the ten pieces of wire are then added, and the sum is equated to the weight of the same ten pieces of wire weighed collectively on a microchemical balance. From this the value of the scale of the dial is calculated in terms of nanograms or micrograms per division.

With a torsion fiber of 16-μm diameter, a sensitivity of approximately 0.137μg per degree was given by the instrument.

The standard deviation of a weighing by the triple setting technique is of the order of $\pm$ 0.04 μg for a series of ten consecutive weighings of a platinum wire.

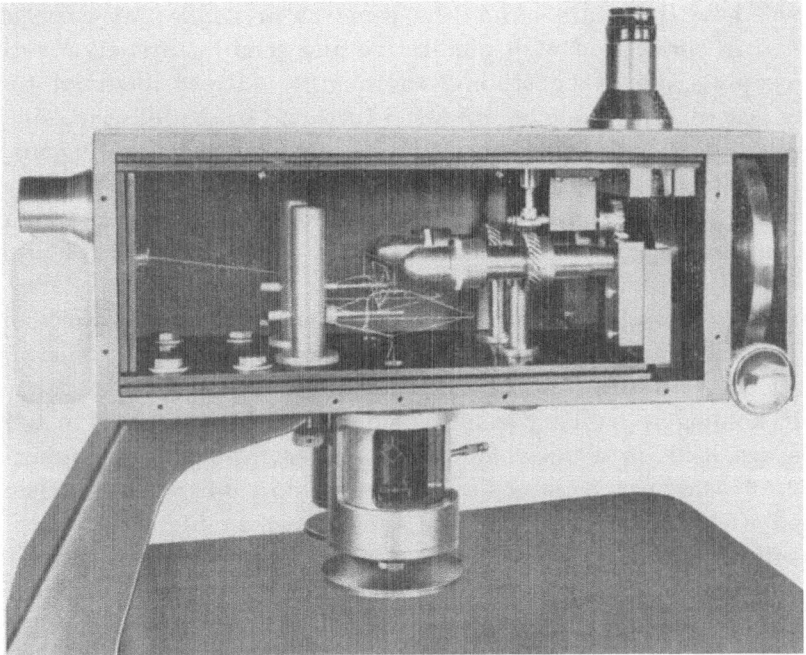

Fig. 86.   Quartz Microgram Fiber Balance of the Torsion-Restoration Type.   Balance case and a pan-well open to show internal construction.   Courtesy of MICROTECH SERVICES & CO. (169).

The maximum safe load difference on the two pans of a torsion balance can only be determined by breaking the torsion fiber. With this balance it is known that at least eight complete revolutions of the calibrated dial can be applied, corresponding to a load difference of approximately 400 μg. By the use of suitable counterpoises (empty capillary cones or lengths of platinum wire), the maximum load difference can be maintained well within the safety limit.

Other quartz microgram balances, available commercially (26, 136, 168, 169), Fig. 86, possess similar construction features and performance. These should also be suitable and their use may require only minor modifications in the weighing procedure.

Auxiliary Apparatus.

For a simple gravimetric determination, the auxiliary apparatus required comprises the following items; a microscope; a manipulator; a micrometer syringe; a micropipet; a moist chamber; a carrier; reagent containers; capillary cones; a storage block for capillary vessels; a multi-holed metal heating block; a centrifuge which may be operated by hand or electrically; gloves; a pair of flat tipped forceps; and, if possible, some means of projecting the image of the microscope field on a vertical screen so that observation can be carried out over a long period of time without strain. These items are either the same as or similar to those already described in connection with qualitative and semi-quantitative work, and these sections should be studied thoroughly. Items identical to those already described will not be discussed further. In the following lines only points of difference or points of particular importance concerning the above mentioned items will be mentioned. A few additional items used specially for gravimetric work will be described. The latter comprise a set of platinum wire counterpoises, drying capillaries, and a watch glass for reception and transference of capillary cones.

Carrier. Vessels for qualitative and semi-quantitative work may be held in position on the carrier by a thin film of vaseline. This practice, however, is obviously not permissible in gravimetric work, particularly not for holding a capillary cone which has to be weighed. Instead, the vessels are held on a suitable plastic carrier by a simple rubber band, Fig. 70. If such carrier is not available, a thin rubber band can similarly be used with a glass carrier of the type shown in Fig. 69.

Capillary Cones. These are the same as those which have already been described, but they are made from wider capillaries of approximately 1-mm internal diameter. A greater length of the original capillary, approximately 2.3 to 3 mm, is retained attached to a shorter handle about 1 cm long. The outer end of the handle must be rounded to a tiny ball in a microflame so as to leave no sharp edges which might splinter off and be lost in centrifuging. It is desirable that a set of capillary cones should be made from similar capillaries and cut to like size as far as possible. This makes it easier to counterpoise the cones. The average weight of such capillary cones is approximately 2 mg. Cones are always handled with a pair of forceps reserved specially for the purpose and kept in a stoppered tube when not in use.

Measuring Capillaries. For accurate measuring of stock solutions, comparatively wide capillaries having 0.3- to 0.4-mm bore are used, the reason being that relatively large volumes of samples are taken in gravimetric work. This will permit the whole length of the measured column of liquid to appear in the field of the microscope.

Micropipets. Coating of micropipets with paraffin wax, p. 101, should be avoided.   As has already been mentioned, small pieces of wax may become detached to contaminate precipitates and to give erratically high results.   Micropipets should be treated either with Desicote, Teddol, or Akard.   The first two, however, give a superior coating.

Drying Capillaries. These consist of capillaries approximately 4 cm in length and about 1.5 to 2 mm in bore.   They are sealed at one end. Drying operations are carried out by introducing the capillary cone, handle first, into the capillary and allowing it to slide down to the sealed end. The drying capillary containing the cone is placed in a suitable hole of the heating block at the required temperature.

Counterpoises. A set of pieces of 50-gauge platinum wire, 0.025 mm thick, is retained permanently as counterpoises.   These are lenghts of wire, approximately 2, 4, 6 and 8 cm, cut and bent into shapes equally suitable for placing on the balance pan and for recognition.   Since the wire weighs approximately 100 µg per cm, the difference between successive pieces in the series is about 200 µg.

Watch Glass. A small watch glass about 3 cm in diameter is used for reception and transference of capillary cones.   A short, thick piece of glass or vitreous silica fiber is always kept upon the watch glass; it facilitates the handling of the capillary cones which are placed on the watch glass so that their handle rests upon the fiber.

Moist Chamber. Plenty of filter paper should be placed on the floor of the chamber round the four sides of the carrier, Fig. 66.   In addition, it is advisable to place some vertical paper strips on the back and the sides of the chamber.   All linings are soaked with water before introducing the reagent and sample.   This is a provision against any evaporation of sample solution prior to measuring.

ζ) Titration of Microgram Samples.

In titrimetry of microgram and smaller samples more dilute standard solutions are frequently used (133) than are customary when working on a large scale.   In addition, higher concentrations of indicator are often taken than in corresponding macroprocedures.   This practice of using very dilute solutions for the purpose of maintaining accuracy by using larger and thus more accurately measurable volumes of liquids suffers from obvious limitations (25, 32).   As is well known, the accuracy of determining the end point is greatest when working with solutions of moderate concentrations; potentiometric redox titrations and conductimetric titrations, of course, are exceptions.   As the strength of solution is progressively decreased, the application of indicator corrections may become too inaccurate to be of any value.   It is therefore advantageous to develop techniques which permit maintaining approximately the same concentrations as are used with titrimetric determinations on the milligram and larger scales.   It is also essential for accuracy to keep within reasonable limits the amounts of indicator used.

The customary buret has a capacity of 50 ml, and the volume of the titrated solution at the end point may vary from about 50 to 500 ml. For microgram titrations, if the concentrations are to be retained as in macro-analysis, the above volumes should be divided by one million. Thus a microgram buret should have a capacity of 50 nl = 0.05 µl, and the volume of the titrated solution at the end point should be of the order of 50 to 500 nl or 0.05 to 0.5 µl. The proper handling of such volumes is possible only by means of mechanical manipulation under a low-power microscope.

In order to fulfil these requirements, LOSCALZO and BENEDETTI-PICH-LER (154) presented a technique which can be considered as an extension of the general technique for working in capillary cones. The principal apparatus is essentially the same as that used by BENEDETTI-PICHLER and co-workers, Fig. 80, and which has already been described, p. 112. There are, however, two notable additions: the measuring pipet and the stirring cone.

For the first, use has been made of the capability of the micropipet, Fig. 82, connected to a levelling device, Fig. 88, to serve as a simple buret or measuring pipet with remote control. An eyepiece micrometer provides the scale for the buret, and the micrometer readings are converted into units of volume and, finally, mass by using standard solutions of known concentration. The buret is designed for a total delivery of about 50 nl and the displacement of the meniscus is determined under the microscope.

The other feature is the addition to the assembly of the so-called titration cone, Fig. 87 B, in the form of an open capillary. The end of the capillary facing the microburet is conical and serves as a container for the solution to be titrated. The other end is connected to a plunger control device for the purpose of stirring the drop during the titration by making it move back and forth. This method of stirring by the turbulence in the drop when it is moved through a capillary proved efficient and mastered one of the major difficulties of titrating on the microgram scale. The titrated solution may have a volume of 50 to 500 nl. A relative precision of about ± 0.01 was obtained with suitable adsorption indicators.

The above technique is quite suitable only for detection of end points obtained by adsorption of the resulting coloration into the small surface area of microscopic or colloidal particles. Since the volume of the titrated solution is very small, the light path through the tiny droplet is too short to perceive a color change with dyes in the customarily given concentrations. This constitutes the principal limitation of the technique. It may be added that the same shortcoming would exist also if titration of micro-gram samples were carried out in capillary cones (25), even of relatively large size, in place of the titration cone.

This difficulty may be overcome if a practical method is found for giving the path of light through the small volume of titrated solution, 0.05 to 0.5 µl, a length of several centimeters. Four centimeter is approximately the thickness of the titrated solutions in macrotitrations with the customary concentration of the usual indicators. DORF (81), using apparatus similar to that employed by LOSCALZO and BENEDETTI-PICHLER, demonstrated that the problem can be solved by placing the titrated solution into a fine capillary of about 0.2-mm bore mounted vertically in the axis of the microscope inside the moist chamber as shown in Fig. 89 and applying the principle of the coloriscopic capillary (25, 32) for the observation of the color change. The capillary is about 2 cm long and is partly filled with mercury. The color change in the titrated solution within the capillary is observed through the microscope equipped with a vertical illuminator. Light from the illuminator passes downward through the liquid column, strikes the mercury surface, and is then reflected back through the liquid. In this way a light path equivalent to a depth of approximately 4 cm of solution is obtained. This depth is comparable to that required for proper observation of the end point in customary macrotitrations. As in the instance of the titration cone, the other end of the capillary is connected to a suitable plunger for mixing the solution during titration. A relative precision of about ± 0.01 was obtained in acid-base titrations with methyl red as indicator.

Both apparatus used for titration of microgram samples will be discussed in detail.

KOCH et al. (138) described a microburet for performing nanogram titrations under a low-power stereoscopic microscope. The buret, which is calibrated with mercury, has a total capacity of about 1 µl and can be read fairly accurately to 8 nl with the aid of the ocular micrometer. It is constructed of a narrow thermometer capillary tubing of about 0.2-mm diameter. The buret is directly attached to a microinjection apparatus which is mounted on a micromanipulator. The system works with an air cushion between a column of mercury in the syringe barrel and the working solution in the buret. Titration is performed in a tiny depression ground in a glass slide under the microscope, and efficient stirring is carried out with a simple electromagnetic stirrer. The apparatus is suitable for the determination of nanogram amounts of constituents in samples isolated from metallurgical materials.

Working in Titration Cones.

As has already been pointed out, the principal apparatus used by LOSCALZO and BENEDETTI-PICHLER (154) for titration of microgram samples under the microscope is essentially the same as that shown in Fig. 80. For volumetric work, however, there are some changes required, which need detailed description.

Moist Chamber. Calibration of the buret as well as actual titrations are performed in a moist chamber such as that used by BENEDETTI-PICHLER and RACHELE (31), Fig. 90, and schematically shown in Fig. 137 A. It is provided with two openings diametrically opposite to admit the titration vessel on one side and the buret on the other, Fig. 137 A m. The side walls of the cell are formed by a metal ring, 1 cm high and 5.6 cm in outer diameter, having two openings 2 cm wide and cut out opposite to each other. A thin round glass, 0.044 mm thick, forms the bottom of the cell, and the top is formed by a very thin, round cover glass, 5.5 cm in diameter and 0.016 mm thick. Another metal ring of slightly larger diameter fits over the first one and

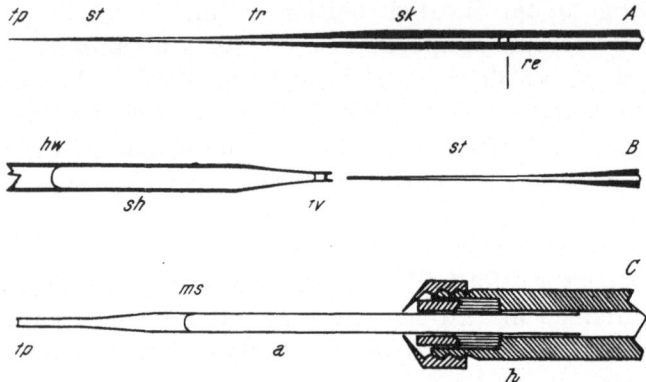

Fig. 87. Titration of Microgram Samples. *A*, buret with remote control; *tp*, tip; *st*, shaft; *tr*, taper; *sk*, shank; *re*, reference mark. *B*, open titration cone; *sh*, shank of titration cone; *tv*, titration cone proper; *st*, shaft of buret; *hw*, meniscus of hydraulic water. *C*, buret, schematic; *tp*, tip; *ms*, meniscus of standard solution; *a*, air column; *h*, pipet holder of metal with rubber washer, same as Fig. 81. After LOSCALZO and BENEDETTI-PICHLER (154).

can be rotated. This outer ring has two sections, one 2 cm and the other 4 cm wide, cut out opposite to each other in such a way that, by rotating the outer ring, the working apertures in the side walls of the cell can either be opened independently or simultaneously. Wet cotton is placed inside the cell in the channel of the inner ring to keep the atmosphere of the chamber saturated with water vapor.

The Horizontal Buret of 50-nl Capacity. A micropipet with straight tip and a shaft of uniform bore is prepared for use as a buret. Fig. 87 A shows the buret proper, several times its natural size.

The nozzle of the buret consisting of the taper, shaft and tip, is similar to that of the micropipet. The shaft is about 4 mm long and the bore at the tip is about 20 μm in diameter. Finer orifices would enable more precise control over the delivery of the standard solution, but titrations in this case would consume a longer time. The nozzle of the buret receives on the outside a very thin paraffin wax coating.

The shank of the buret used, a small length of which is shown in Fig. 87 A, is about 10 cm long, 0.1 to 0.2 mm in bore and 0.5 mm in outer diameter. Only the short length of the shank adjacent to the taper is calibrated, and a reference mark is provided 3 to 5 mm from the taper. In use, the displacement of the meniscus of the standard solution in this uniform portion of the shank is measured with the ocular micrometer. The total capacity of the "calibrated" part of the microburet is approximately 50 nl with a shank of slightly less than 0.2-mm bore. Microburets with wider shanks can also be used depending on the desired capacity of the buret. Obviously the bore of the shank must be related to the true diameter of that part of the field of the microscope which is covered by the eyepiece micrometer scale, and this length is again dependent upon the magnification of the objective. LOSCALZO used an 8× objective and could follow the movement of the meniscus over a true distance of 2 mm.

The microburets are made from pieces of heavy-walled Pyrex glass tubing of approximately 6-mm outer diameter, 2-mm bore, and about 10 cm long. The glass tubing should first be cleaned thoroughly by placing it into hot chromic-sulphuric acid for some time and then rinsing with tap water, distilled water, and alcohol. The dry piece of tubing is drawn out to produce a capillary of approximately 0.5-mm outer diameter. Such capillaries would have a bore of about 0.2 mm. The capillary is cut into lengths of 20 cm each. After the bore has been measured under the microscope, each piece of capillary is drawn out symmetrically to produce two burets having nozzles with fine tips. The nozzles of the buret are best drawn out by using a suitable mechanical pulling device, p. 147. The nozzle then receives a thin coating of paraffin wax on the outside by dipping it in smoking hot, molten paraffin while a stream of clean air is continuously forced through the fine tip. The buret is slowly withdrawn from the molten wax maintaining the air current until the film of paraffin solidifies on the wall. The fine tip of the nozzle is then snapped off by means of straight forceps to produce an orifice about 20 μm in diameter. The small length of the shank adjacent to the taper should be of a uniform bore. This part, about 5 mm long, is inspected under the microscope simply by placing the buret horizontally on a glass slide and measuring its apparent diameter at intervals. A reference mark is then made on the shank of each buret at a distance 3 to 5 mm from the taper. This may be done by heating a small amount of Canada balsam until it becomes thick. A trace of the material is then taken between the two tips of a curved forceps and spun to form a very fine thread having the thickness of a hair. The thread is then wrapped halfway around the shank of the buret at the desired place.

Leveling Device. In use, the buret is inserted in the pipet holder, Figs. 81 and 87 C, which is mounted in the clamp of the manipulator as shown in Fig. 80. The pressure may be regulated with the plunger device or, more conveniently, the pipet holder is connected by means of a rubber tubing of 2- to 3-mm bore to the bulb *a* of the leveling device shown in Fig. 88. The two bulbs of the leveling device are connected by a rubber tubing of the same kind. Both bulbs are approximately half filled with water. The bulb *b*, made of a separatory funnel of 50-ml capacity, is clamped to a mechanical stand, not shown in the figure, that permits vertical dis-

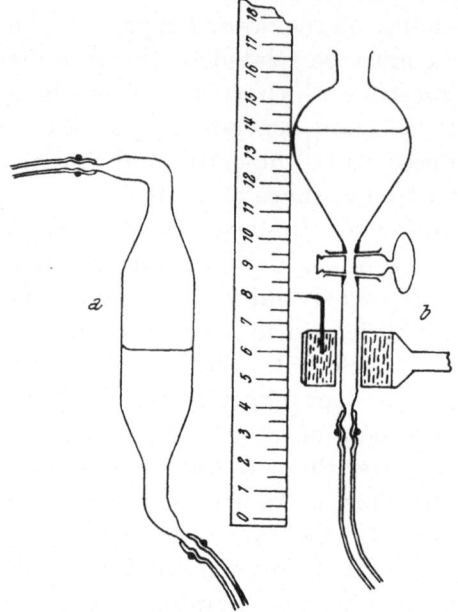

Fig. 88. Leveling Device. After Loscalzo and Benedetti-Pichler (154).

placement of the bulb in front of a vertical scale by means of a rack-and-pinion motion. The shank of the buret, the pipet holder, the connecting rubber tubing and the upper half of the bulb *a* contain the air cushion between the water in the leveling device and the standard solution in the buret. Raising or lowering of the bulb *b* will increase or decrease the pressure upon the meniscus of the standard solution in the shaft of the micro-buret. In this way the pressure necessary for proper control of the flow of the standard solution in the microburet can be regulated. As in the instance of the uncoated micropipets used by Benedetti-Pichler and co-workers for working in capillary cones, p. 114, the operation of the buret depends on the use of surface forces.

Operation and calibration of the microburet are discussed in detail on p. 227, together with the manipulations involved in titration.

Calibrating Capillaries. Since the capacity of the calibrated part of the buret is extremely small, about 50 nl or 0.05 µl, calibration is performed by measurement of the column of liquid delivered by the buret into a calibrating capillary located in the moist chamber under the microscope, Fig. 137 *A*.

The calibrating capillary has approximately 0.2-mm bore and about 0.5-mm outer diameter, as those used for making the burets. A piece of such capillary, 15 cm long, is squarely cut 1 cm from one end with a thin-edged carborundum plate or a thin slip of tungsten carbide having a sharp edge. The short piece of capillary is mounted vertically in the axis of the microscope, and the true diameter of the circular bore is accurately measured by means of the calibrated eyepiece micrometer scale. This short capillary is then discarded. The bore of the longer capillary should be uniform throughout a distance of about 5 mm starting at the freshly cut end. This length is inspected for uniformity by measuring under the microscope the apparent diameter at intervals as described for the burets.

With De Khotinsky cement or other suitable material, the other end of the capillary is then sealed into a glass tube, Fig. 137 *A*, *h*, and *D*, of approximately 4-mm bore and about 10 cm in length. The tube *h* is held in the clamp of a manipulator or a suitable stand for positioning the capillary in the moist chamber opposite to the micropipet as shown in Fig. 137 *A*.

Titration Cones. Fig. 87 *B*, left, shows the titration vessel enlarged several times its natural size. The titration vessel, made of Pyrex glass, has a shank *sh*, only a small length of which is shown in the illustration, tapering into an open cone *tv*. The shank is approximately 10 cm long, 0.5 mm in bore, and 0.7 mm in outer diameter. The titration cone proper, *tv*, in which the titration solution is placed, is 3 to 5 mm long with its bore tapering from 0.5 mm at the shank to 0.2 mm at the tip.

The titration vessels are made from pieces of Pyrex glass capillaries of approximately 0.5-mm bore, 0.7-mm outer diameter, and about 20-cm length. Each piece is drawn out at its center to a finer capillary 0.5 to 2 cm in length. The fine capillary is then cut out at the tapers to produce two open cones, *tv*, each of which is formed on a shank approximately 10 cm long.

For use, the far end of the shank is inserted into a pipet holder which is mounted in the clamp of a second micromanipulator placed opposite to the one on which the buret is mounted. If a second manipulator is not available, the holder may be mounted in the clamp of a suitable stand in such a way that the titration cone can be introduced into the moist chamber opposite the buret. The holder carrying the titration vessel is connected to a plunger device which has been completely filled with water, as that used by BENEDETTI-PICHLER and co-workers for working in capillary cones, Fig. 80, or to any suitable substitute. The plunger device and the procedure of preparing it for work has been described on p. 112. The pressure from the plunger is transferred via the column of hydraulic water to the air cushion in the shank of the titration vessel and to the titrated solution in the open cone. Before the solution to be titrated is introduced through the open tip of the cone, the meniscus *hw* of the hydraulic water should be advanced along the shank to a point about 2 cm from the titration cone proper. The drop in the titration cone can be efficiently stirred during titration by retracting and advancing a few times the plunger of the pressure device and thus moving the working drop to and fro. Displacement of the drop three to five times for about 0.5 to 1 mm after each addition of standard solution is sufficient for complete mixing.

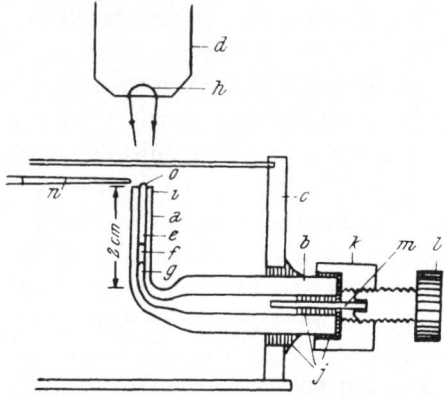

Fig. 89. Titration in Coloriscopic Capillaries. *a*, titration capillary; *b*, wide horizontal capillary; *c*, moist cell; *d*, microscope with vertical illuminator; *e*, titrated drop; *f*, air space; *g*, mercury column; *h*, microscope objective; *i*, water-repellent coating; *j*, cementing material; *k*, brass collar of plunger device; *l*, knurled screw; *m*, wire plunger; *n*, microburet; *o*, convex meniscus. After DORF (81).

Working in Coloriscopic Capillaries.

As has already been indicated, the basic apparatus used by DORF (81) for titration of microgram samples is similar to that used by LOSCALZO and PICHLER (154), which has just been described in detail. A notable development, however, lies in the titration vessel. DORF, by performing titrations in a fine capillary mounted vertically in the axis of a microscope equipped with a vertical illuminator and by using the principle of coloriscopic capillaries (25, 32) was able to obtain a light path through the column of the titrated solution, suitable for observation of the color change with the customary concentrations of indicators. Stirring of the tiny drop during titration is obtained by means of a suitable plunger device directly attached to the titration vessel. The titration vessel with attached plunger is mounted in the moist chamber, Fig. 89.

The calibrated microburet, Fig. 89*n*, is similar to that used by LOSCALZO and PICHLER, Fig. 87*A*, but is given a finer tip having an orifice a few microns in diameter.

Titration Capillary.   Fig. 89a shows the titration capillary proper, approximately 2 cm long, about 0.2 mm in bore and about 1 mm in outside diameter. This fine vertical capillary is drawn out from a wider capillary b, at a right angle to it. The wide horizontal capillary b, about 5 cm long, 6-mm outside diameter, and 1- to 2-mm bore, is connected to a plunger device. The titration capillary is completely filled with mercury and is mounted in the moist chamber c as shown in the figure.

When the chamber is placed on the stage of the microscope, the fine capillary should be vertical and its axis in line with the axis of the microscope. The drop to be titrated, e, is placed into the fine vertical capillary so that an air space f separates the titrated solution from the mercury column g occupying the lower portion of the capillary and the whole length of the wide capillary.

The color change in the titrated solution is easily observed through the microscope using a vertical illuminator. Light from the illuminator passes downwards through the solution e and is then reflected back from the mercury meniscus below the air space through the entire length of the liquid column, thus attaining a light path equivalent to about 4 cm of liquid.

The device is prepared from a piece of thermometer capillary tubing, about 10 cm long, 1 to 2 mm in bore, and approximately 6 mm in outside diameter. The capillary is carefully cleaned, dried, and then drawn out in the center to a fine capillary of approximately 1-mm outer diameter and about 0.2-mm bore. The fine capillary is cut evenly at a point about 3 cm from a taper and bent at a right angle. The free end of the fine capillary then receives a water-repellent coating on the outside to confine the titrated solution to the inside of the capillary. A rubber tube is connected to the wide end of the capillary and a stream of air is forced through it while a part of the fine capillary is dipped briefly into a suitable water-repellent liquid. Dri-Film, Desicote, Teddol, or molten paraffin may be used for the purpose.

A portion of the inner bore of the wide capillary is plugged by introducing an amount of De Khotinsky cement or any other suitable cementing material into it. The cement is also applied over the free end of the wide capillary and a part of its outer surface. Fig. 89j shows those parts where the cement is applied. A fine opening is then made in the cement plug inside the capillary by pushing a heated needle through the hardened cement and pulling the needle out quickly before it cools.

The tube is then completely filled with clean, dry mercury by holding the tip of the fine end of the capillary under mercury and applying suction from the wide end. The titration vessel is now ready to be attached to the plunger device.

Plunger Device.   The plunger device shown in Fig. 89, consists of a brass collar k, threaded internally for about 1 cm to take the knurled screw l. The remainder of the bore of the collar is unthreaded and is counter-bored to receive the wide end of the capillary tubing. The plunger m is a piece of stainless steel or Nichrome wire, about 0.5 mm in diameter and 3 cm in length, soldered into the end of the knurled screw l. The screw l is threaded to engage in the brass collar.

The titration vessel is attached to the plunger and then mounted in the moist chamber c. To this end, the brass collar is first attached to the wide capillary tubing by carefully heating the collar and the cement on the glass tubing and then pressing them together until the cement hardens. Using De Khotinsky cement or by other suitable means, one then mounts the glass tubing with the collar attached in the moist chamber so that the fine capillary, a, is vertical when the chamber is placed in position on the stage of the microscope. The wire plunger m is then heated and quickly introduced through the needle hole in the cement plug inside the wide capillary. The plunger is allowed to cool. The threads of the knurled screw, l, are then engaged in the brass collar k. Thus by turning the screw l, the plunger m can be advanced or retracted causing the mercury column in the capillary to rise or fall. In use, pressure is thus transferred from the mercury column through an air space f to the titrated solution e in the fine capillary.

b) Apparatus for Working in Hanging Drops.

BENEDETTI-PICHLER and RACHELE (31), who wished to establish the smallest scale upon which the simple confirmatory tests of customary qualitative analysis may be successfully performed, carried out experiments in which extremely small drops of test solution were treated with small droplets of reagents. For this purpose, work had to be performed in a moist chamber under comparatively high-power magnification and with the working droplets hanging from the underside of a cover slip forming the roof of the chamber. The cover slip had to be very thin to permit the use of higher magnifications.

Special condensers with long focal distance are desirable to permit efficient illumination of the tiny droplets located on the top of a chamber about 1 cm high. The performance of work on hanging drops also requires the use of micropipets with inclined shaftlets in order to deposit drops of test solution and reagent solution on the underside of the glass cover of the cell. Since, in this arrangement, micropipets are manipulated below the cover glass and below the droplets investigated, they will not interfere with the observation of the field of operation. The micropipets are carried and moved by means of suitable micromanipulators. Two micromanipulators were used, each carrying a micropipet attached to the pipet holder in the clamp of the instrument. Each pipet holder is connected by means of long fine tubing to a hypodermic syringe fastened to the common base plate. The moist chamber is held and moved by means of a mechanical stage attached to the built-in revolving stage of the microscope, which offered sufficient flexibility of movement for centering and orienting the objects of investigation in the microscopic field.

The whole arrangement is essentially similar to that commonly used by microbiologists for micrurgical investigations, Fig. 50. The apparatus has also a number of features in common with that used for performing chemical operations in capillary cones, Fig. 80.

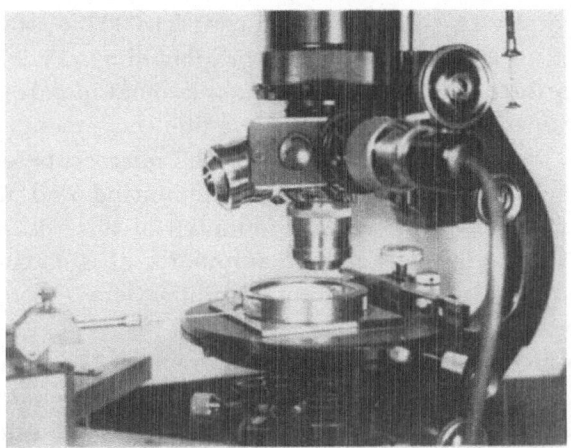

Fig. 90. Arrangement for Working with Hanging Drops. Epi-Condenser W in use (31). Courtesy of BENEDETTI-PICHLER.

The assembly is partly shown in Fig. 90. The moist chamber is in position on the stage of the microscope. The latter is equipped with an Epi-Condenser W. The figure also shows the shank of a micropipet attached to the pipet holder which is mounted in the clamp of the micromanipulator.

Microscope. The microscope, similar to those used for solution chemistry in capillary cones, Figs. 63 and 80, is fitted with a built-in rotating and centerable stage and an attachable mechanical stage for holding and moving the moist chamber in two directions at right angles to one another.

A bright-field condenser with sufficiently long focal length is used for illumination with transmitted light supplied by a 6 v microscope lamp provided with a suitable green filter. A ZEISS Epi-Condenser W, Fig. 90, equipped with objectives in a revolving nosepiece is also employed. These objectives were used with both transmitted and reflected light, and change from bright-field to dark-field was obtained by means of a light switch. In determining limits of identification, the dark-field provided by the Epi-Condenser was used for tests based upon the formation of precipitates.

This permits observation of colloidal particles while the preparation is protected by a suitable heat-absorbing glass filter. Transmitted light was used for observation of tests depending on color.

The ordinary microscope ocular was replaced by a suitable photographic eyepiece, a ZEISS Phoku eyepiece, permitting photographic recording while observing through a side tube equipped with an eyepiece having a ruled scale. This micrometer scale is used for measurement, and also serves for focusing the image on the photographic film. Direct measurement of photographic records were performed with the aid of photomicrographs of a stage micrometer with the objectives in use.

The highest total visual magnification used was $397 \times$, and a corresponding magnification of $99.2 \times$ on the photographic film.

Micromanipulators. Two CHAMBER's micromanipulators, a right-hand and a left-hand instrument, with all their fine movements provided with remote controls, were mounted in front of the microscope on the common base plate. Each instrument served for mounting and manipulating a micropipet attached to a pipet holder mounted in the clamp of the instrument, Fig. 90. The high-power micromanipulator of R. CHAMBERS, Fig. 50, suitable also for work at the medium magnifications employed, has been described in detail on pp. 69 to 72.

Injection Devices. Two hypodermic syringes are mounted on the common base plate, one on each side of the microscope. Each of these syringes is connected, by means of a fine copper tubing, to the pipet holder mounted on the micromanipulator on the opposite side of the assembly. Thus the syringe on the right-hand side operates the micropipet on the left-hand side and vice versa. The two injection devices are similar to that used by BENEDETTI-PICHLER for working in capillary cones, Fig. 80. The plunger, the copper tubing, and pipet holder, Fig. 81, as well as the operation of the device have already been described, p. 112.

Micropipets. Two types of micropipets are used. One type, Fig. 91 a, used for general work, is similar to that employed by BENEDETTI-PICHLER and co-workers, which is shown in Fig. 82. A detailed description of the latter pipet, its preparation, and operation are given on p. 114. This description applies in general to the present micropipet. For working with hanging drops, however, the shaft of the pipet has to be inclined to the axis of the shank to permit deposition of droplets on the roof of the chamber. The angle of inclination of the shank and the pipet proper is approximately 90 to 120°. Fig. 91 a shows the pipet proper consisting of tip, shaftlet, and taper. Only a small part of the shank is visible in the figure. Another important difference between this micropipet and that shown in Fig. 82 is that the opening of the tip should be less than 1 μm. This is necessary for working with droplets a few microns in diameter to obtain proper control over the volume of droplets delivered by the micropipet.

The other type of micropipet employed is the calibrated micropipet illustrated in Fig. 91 *b*. This pipet is used to estimate the delivered volumes by measuring, with a calibrated ocular micrometer, the displacement of the meniscus in the shaftlet of the micropipet. For this purpose the bore of the shaftlet should be fairly uniform. Lengths and outside diameters of the various parts are the same as for the micropipet shown in Fig. 91 *a*, but the shaftlet has a uniform bore of about 2 to 4 μm in diameter. These micropipets are prepared from fine capillaries of about 0.07-mm uniform bore and 0.7-mm outer diameter. Such capillaries are obtained by drawing out soft glass thermometer tubing of 0.5-mm bore and 6-mm outside diameter.

The diameter of the uniform bore of the shaftlet is measured, p. 233, under the microscope after immersion of the shaftlet in cedarwood oil

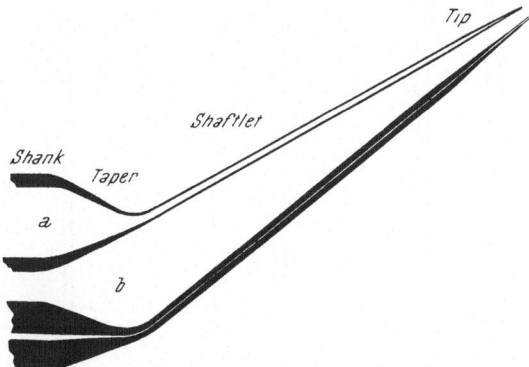

Fig. 91. Micropipets for Working with Hanging Drops. Approximately 20 times natural size. *a*, micropipet for general use; *b*, calibrated micropipet. After BENEDETTI-PICHLER and RACHELE (31).

to eliminate the lens action of the glass body of the shaftlet as a consequence of the almost equal refractive index of cedarwood oil and glass, Fig. 138 *b*.

The positions of the menisci can only be read with the calibrated part of the shaftlet located in a focal plane perpendicular to the axis of the microscope. It follows that these pipets cannot be used for measurement while droplets are delivered to the underside of the glass cover of the moist chamber. For this reason these calibrated micropipets have found only a rather limited use, as tools for measurement, in working with hanging droplets, p. 234.

Both types of micropipets shown in Fig. 91 were prepared by means of RACHELE's device, p. 151, Fig. 101.

Moist Chamber. The metal ring moist chamber of E. LEITZ has been used. This chamber, shown in position on the stage of the microscope, Fig. 90, is described in detail on p. 126.

6. Microtools.

a) Types of Microtools.

Microbiologists utilize a large diversity of microtools which are carried and guided in the field of the microscope by means of suitable micromanipulators. Such microtools, also called assistants or microinstruments, are of different types, shapes, and functions to satisfy varied requirements in different fields of study of cells and tissues. In the investigation of living cells, in single cell surgery for instance, the operator can remove the nucleus or a part of it or other cell constituents from individual cells and replace them by others. To perform these tasks the operator may use

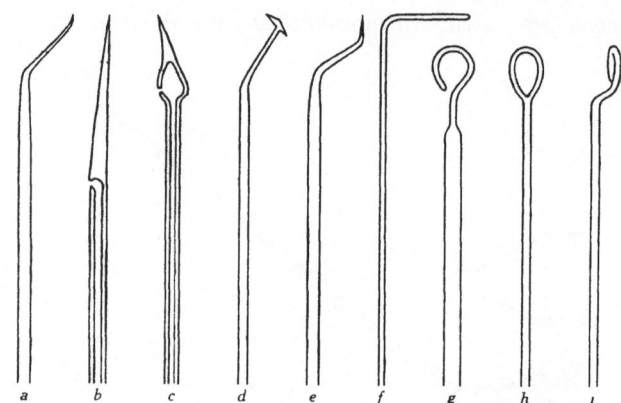

Fig. 92. Glass Microtools. *a, e*, needles; *b, c*, puncture micropipets; *d*, microscalpel; *f*, micro-rest; *g, h, i*, steadying loops. After P. DE FONBRUNE from E. LEITZ, Wetzlar (152).

microneedles, microknives, micropipets, microforceps, and a variety of other tools. Microdissection needles may be made so fine as to puncture red blood corpuscles and to tear up leucocytes. Minute quantities of fluid can be injected into a cell by means of fine micropipets connected to a suitable injection apparatus in order to study the effect of these injections on the nucleus or the protoplasm. For such operations micropipets may be made with fine orifices a few microns to less than half a micron in diameter. Micropipets may also be used in single cell isolation to deliver or to collect tiny droplets in which the bacterial cell is contained. It is also possible to measure differences in the potentials over very short distances, for example at the poles of a cell, by inserting microelectrodes. A variety of glass assistants used by microbiologists are shown in Fig. 92, and a still wider variety is illustrated in Figs. 105 to 110 in connection with the use of the DE FONBRUNE microforge in preparing microtools.

WORST (241) described two special types of microforceps. In the first type the two jaws of the forceps are opened by their own resilience and closed with the aid of a tube fitting around them. This tube can be made to slide around the pincers by means of a piston and a syringe. The closing movement can be controlled within a few microns. In the second type the action is reversed, closing is effected by resilience and opening by the force of a ball and rod. The two jaws of the forceps are fixed in a tube of suitable diameter mounted on a DE FONBRUNE micromanipulator. A number of modifications may be prepared to suit various purposes. These instruments are readily manufactured by the DE FONBRUNE micro-forge, p. 152.

Microchemists and other investigators also use various types of mechanically guided tools for performing a wide variety of operations in many fields of study. Most of these micromanipulator tools, however, are not much different from those used by the microbiologists. The variety of microtools usually needed for performing chemical operations on solutions in capillary vessels or on droplets deposited on glass slides is comparatively limited. For such work the micropipet is the most important tool. A microchemist may use micropipets with straight tips of the styles shown in Figs. 65 and 82 for performing chemical operations in capillary cones, p. 92, on the plateau of the condenser rod, p. 105, or on threads, p. 111. He may also use pipets with inclined tips, Fig. 91, for performing operations on droplets hanging from the underside of a cover glass roofing a moist chamber, p. 132. Micropipets with straight tips as those shown in Fig. 82, have been adapted (154) for use as calibrated microburets, Fig. 87, for titration of microgram samples, p. 123. Commercially available capillary transfer pipets, Fig. 84, have also been employed in micromanipulator work (59, 71). The pipet is attached to a simple syringe control, Fig. 84e, carried and moved by a low power micromanipulator, Fig. 83. Chemical reactions take place on solutions contained in glass microcones, mounted on another micromanipulator facing the first one. The calibrated pipets serve for measurement and transfer of liquids.

Another pipet, Fig. 163, of a rather special style is specifically fashioned for picking up and mounting microfossils (5), p. 293.

In chemical micrurgy (228), which is the field concerned with chemical work on particles and specks of material too small to be dealt with by application of the usual techniques of chemical microscopy (65), and in certain sampling and sorting operations for instance, each individual problem may require special adaptations. In dealing with tiny particles it may be important to manipulate them individually in order to separate them from contaminations and to identify them. It may be required to tease a particle out of the binding material or to dissect it from the material. In other instances it may be required to dissolve the material from around the particle, or it may suffice to select the particle from among other loose particles.

For manipulation of small particles, which may be only a few microns or less in diameter, fine needles preferably made of hard glass may be used. With the aid of microhooks a particle may be picked up and held. Micro-pipets with inclined tips may be employed for depositing reagent droplets on individual particles which may be placed on the undersurface of a micro-scope slide or a cover glass suitably supported under the microscope, Figs.145 to 148.

Microtools made of metal are occasionally used in certain micromani-pulative operations. Thus, in chemical micrurgic work for instance, needles made of steel or platinum may be employed in handling larger particles. One end of a fine metal wire may be ground to serve as a needle point

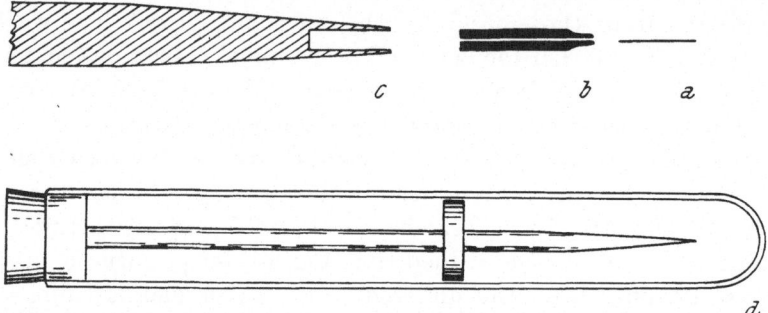

Fig. 93. Metal Microdissection Needle. *a*, tungsten wire; *b*, nickel tube; *c*, brass rod; *d*, micro-needle stored in a glass tube. After H. G. CANNON (50).

which may further be sharpened by treatment with acids. Fine wolfram needles may be sharpened by stroking the hot metal with a lump of sodium nitrite (71). Metal microtools sufficiently fine for proper manipulation of very small particles are difficult to prepare. A metal microtool a few microns in diameter at the tip or finer, is more fragile than glass instruments of the same diameter. Such fine metal points also have a tendency to bend at the slightest touch. Metal microtools have also the disadvantage of chemical reactivity which greatly limits their use. In the dissection of living cells, the use of metal needles often produces chemical injury. The reactivity of metals towards chemical reagents makes them unsuitable for use in many other fields. Chemically resistant glass does not suffer from these disadvantages and is easier to keep clean.

HARDING (118) described a metal microdissection needle for his micro-manipulator, p. 36. The needle is made by mounting a tungsten wire, about 50 μm in diameter, in a thin rod of solder so that only about 1 mm of the wire projects. The rod of solder is obtained by sucking the molten metal up a glass capillary and allowing it to solidify. The tip of the needle is sharpened either on a hone of Arkansas stone or, if the tungsten wire is pure, by holding the end of the wire in a flame to burn the tip off which leaves a very fine tapering point. Microtools may be easily made by bending such tips.

Instead of obtaining a fine point on a tungsten needle by sharpening on a hone or burning the tip in a microflame, both these methods being chancy, CANNON (50) recommends dipping the end of the wire for a very short time, depending on the thickness of the wire, into fused sodium nitrite. The salt may be fused and boiled over a spirit lamp. On dipping the end of the wire into this boiling mass, the extreme tip instantly incandesces and tapers sharply into a perfect point. With wires of about 50-μm thickness, this burning away takes place in less than one second, whereas a wire about 0.5 mm thick may require one or two seconds. Pure tungsten wires are available on the market (86). For mounting, the wire is inserted into the tapered end of a small nickel tube of about 1-mm outside diameter and 0.2-mm bore. Such nickel tubes also are commercially available (130). The tapered end of the tube is pinched with pliers to hold the wire firmly. This method of mounting the wire was more satisfactory than holding it in a rod of solder as suggested by HARDING, since molten solder will not wet tungsten. The nickel tube with mounted needle may then be inserted into the hole drilled into the end of a thin brass rod to serve as a handle or into a similar opening of a suitable tool holder mounted on the micromanipulator. It is desirable that the bore of the brass holder be just slightly larger than the diameter of the nickel tube so that the tube may be held tightly inside the bore after applying, if necessary, a slight pinch with the pliers. Alternatively the nickel tube may be soldered in position. Such needles may be kept clean and protected against damage inside a glass tube as shown in Fig. 93d. A cork guard on the brass rod prevents damage while the needle is being introduced into the glass tube or taken out for use.

Other types of metal microtools such as small pieces of fine razor blades or a very fine platinum wire heated with a weak electric current have been used, suitably mounted on a micromanipulator, for cutting out single cells from hyphae (80).

Fine pointed objects like insect mouth parts, annelidan bristles, spicules of sponges, sharp stiff hairs selected from the body of the house fly, and also fine pointed needle-like crystals have been used as tips for micro-dissection tools (53, 55). A short length of the hair or crystal was inserted into the orifice of a suitable glass micropipet, and the fine point protruding from the tip of the pipet was then used as the dissecting needle or probe. The use of such tools was limited to operating only on very delicate soft objects.

b) Preparation of Glass Microtools.

Most of the microtools or assistants used in various fields of micro-manipulation are made of glass. They are usually prepared from glass capil-

laries or solid glass threads. These may be drawn from soft glass, but hard
glass yields more durable points.

Micromanipulator tools are usually prepared by the experimenter who
uses them and makes them in accordance with specifications considering
his own special needs. It is therefore very important that the operator
masters the techniques of preparing the tools he needs, especially because
such tools have to be frequently replaced because of their delicacy.

Simple glass microtools may be prepared by hand with the use of a suitable microburner. Accurately reproducible fine micropipets and needles
may be prepared mechanically by means of needle and pipet pullers.
A great diversity of microtools of unusual shapes may be most precisely
and conveniently prepared under microscopic control by means of microforges. These methods, together with the necessary apparatus, will be considered in turn.

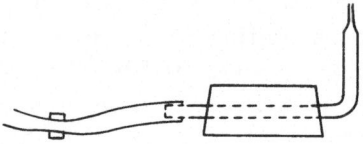

Fig. 94. Simple Glass Microburner.

α) Hand Drawing.

Simple glass micropipets and microneedles may be drawn out by hand
from a stock of glass capillaries or threads, usually of about 0.8 to 1 mm
in outer diameter. This diameter is usually determined by the opening of
the tool holder into which the microtool is to be inserted. The capillaries
and threads may be drawn out from pieces of soft glass tubing and rod
varying from about 4 to 8 mm in outer diameter (3- to 6-mm bore) in the
flame of an ordinary Bunsen burner or of a Meker burner if hard glass is
used. The large flame of the burner shown in Fig. 95, is also suitable for
drawing capillaries from glass tubings. The capillaries produced are cut
into the lengths required for preparing the microtools to be made.

Further operations for making pipet and needle points are carried out
over a pinhead flame supplied by a suitable microburner such as those
described below.

Microburners.

A simple and serviceable microburner is shown in Fig. 94. It can be
easily constructed from a length of hard glass tubing of about 5-mm internal
diameter. Over a narrow flame, one end of the tubing is drawn out into
a capillary of approximately 0.5- to 1-mm bore. The capillary is then cut
at a point approximately 7 to 10 mm from the original tubing. The tubing
is bent at right angles at a point 4 to 5 cm from the drawn-out portion.
The wide end of the original tubing is simply inserted into a large rubber

stopper which has been bored diametrically to receive it. The stopper serves as a base for the microburner as shown in the figure. The wide end of the glass tubing is connected to the gas line by means of a soft rubber tubing. The gas supply can be regulated by means of a screw pinch-cock clamped on the rubber tubing so as to give a microflame as small as a pinhead.

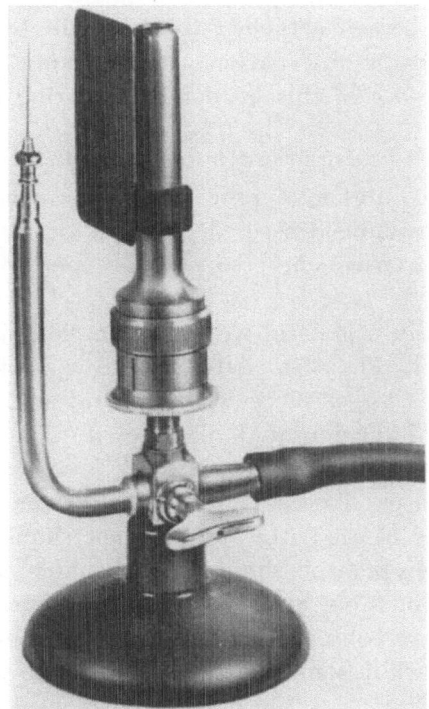

Fig. 95. Gas Burner with Pilot Flame. Courtesy of C. ZEISS, Jena (245).

A similar glass microburner, but with an orifice in the form of a narrow elongated aperture, was used by CHAMBERS (54, 55) for preparing micro-dissection needles and micropipets. The tip of the burner was made by pinching one end of the glass tubing, softened by heat, with a pair of forceps so as to close it except for the narrowest possible aperture that would retain the flame. The burner was connected with an acetylene generator or, preferably, a small compression tank which could be recharged at a small cost. CHAMBERS recommends the use of acetylene rather than ordinary gas because, with acetylene, a narrow flame can be obtained without clogging. Ordinary gas may be improved, if necessary, by passing it through alcohol or benzene.

A simple metal microburner may be made from a hypodermic needle soldered to a short length of brass tube (228). A useful, commercially available (245) burner is shown in Fig. 95. The main burner yields a large

flame suitable for drawing out glass tubing and rod into capillaries and
threads. At a distance of about 3 cm from the main tube, the burner is
equipped with a narrow pilot flame tube giving a microflame which can
be regulated to the desired size. The pilot flame tube has a conical orifice
for inserting a hollow needle similar to an injection needle. The microburner is supplied with two needle canulae of different internal diameter.
An adjustable black screen attached to the main burner tube serves as
a suitable background for observation while working over the microflame.
Fig. 96 illustrates the use of this burner in preparing microtools.

Preparing Microtools by Hand Drawing.

In the following a number of procedures for constructing simple glass
microtools by hand will be described. Such description may serve as a
useful guide for the experimenter to prepare microtools according to his
own special needs.

Fig. 96, illustrates in a general way how simple glass microinstruments
may be prepared by hand (245). After cutting glass tubing into pieces of
suitable lengths, Fig. 96b, a piece is drawn out with a large flame into
a straight capillary of the required diameter, c to e. The main flame of
the microburner shown in Figs. 95 and 96a is suitable for the purpose.
Further operations on capillaries are carried out over a microflame, f to h.

When a tiny flame is used, the microburner should be protected from
draughts of air. The operator should also be able to observe the flame
properly, and the adjustable black screen of the microburner serves as a
suitable background for observation. It may be useful, as suggested by
CHAMBERS (55), to work in semi-darkness so that the pinhead flame shows
up to best advantage.

In drawing out proper pipet and needle points over the pinhead flame,
it is important to remove the heated capillary from the flame at the correct temperature and to pull without delay and with a straight, rather
quick, horizontal motion.

The capillary may be held with both hands over the microflame. It is
of course possible to grasp the capillary at one end by means of a pair
of forceps, especially when the capillary is not sufficiently long. An ordinary
pair of forceps with accurately meeting points will be useful. It is preferable in certain instances to use a forceps with fine tips which are almost
parallel to one another for a few millimeters.

Detailed instructions for the preparation by hand of microdissection
needles and micropipets are given by CHAMBERS and KOPAC (62). CHAMBERS
(54, 55, 58) describes a method for the preparation by hand of microdissection needles, Fig. 97, and micropipets, from pieces of glass tubing
about 4 to 6 mm in outside diameter and 3 to 5 mm in bore. The diameter of the glass tubing should be selected to fit the needle holder to be

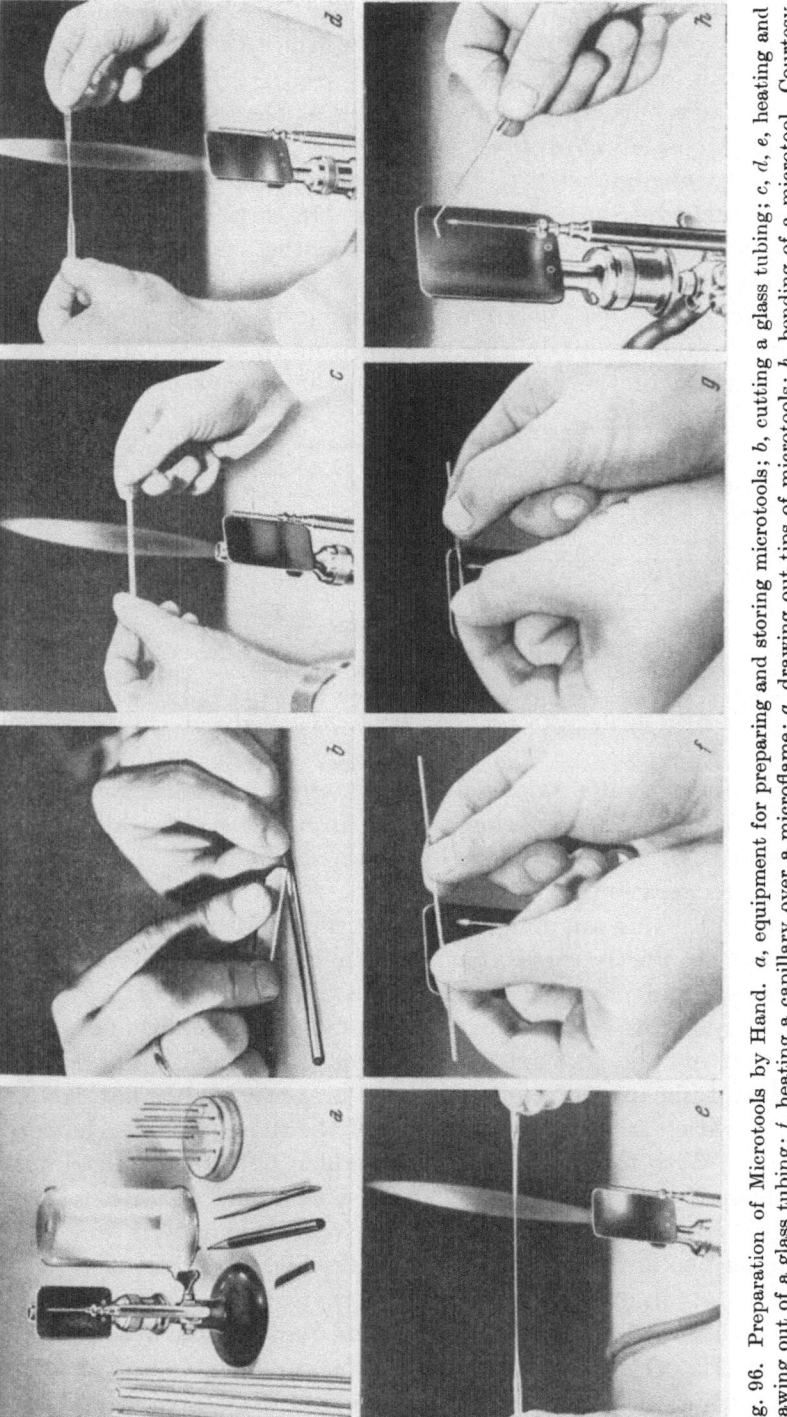

Fig. 96. Preparation of Microtools by Hand. *a*, equipment for preparing and storing microtools; *b*, cutting a glass tubing; *c*, *d*, *e*, heating and drawing out of a glass tubing; *f*, heating a capillary over a microflame; *g*, drawing out tips of microtools; *h*, bending of a microtool. Courtesy of CARL ZEISS, Jena (245).

used. The tubings may be of either hard or soft glass. Tips of needles prepared from thick-walled tubing tend to be firmer than tips of needles prepared from thin-walled tubing.

With an ordinary burner, a supply of pieces as that shown in Fig. 97*a* is prepared, each consisting of a few centimeters of the original tubing to which a capillary of about 0.3- to 0.5-mm outside diameter and at least 6 cm long is attached. Subsequent operations are carried out over a tiny flame. The free end of the original glass tubing, the shank, is held in the left hand. With the right hand, the capillary is grasped by means of a pair of forceps having flat tips coated with Canada balsam or simply with thumb and index finger at a point about 5 cm from the shank. If a forceps is used, the portion of the capillary next to the forceps is brought just

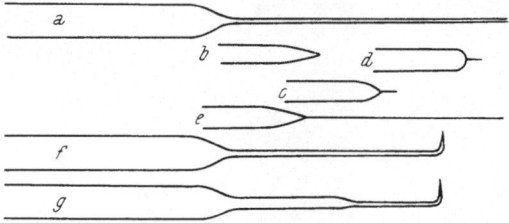

Fig. 97. Preparation of Microdissection Needles. *a*, glass tubing drawn out into a capillary; *b*, good tapering point; *c*, *d*, serviceable points; *e*, tip drawn out into a hair; *f*, microdissection needle; *g*, microdissection needle with relatively stout shank. After R. CHAMBERS (55).

over the flame and a gentle axial pull is maintained until the glass starts to soften. The capillary is then slowly lifted from the flame while a pull, slightly stronger than before, is applied with the hands resting on the table during the whole operation.

The pulling and lifting are done by turning the hands slightly outwards. If the operation is properly carried out, the capillary will part with a slight characteristic tug, and proper points as those shown in Fig. 97*b*, *c*, *d* are formed. The capillaries so formed are rigid and end in very fine points. The tips are shut but the bore extends very close to the ends. The amount of heat used and the proper timing of pull are very important factors. These must be varied slightly with the size of the flame and the diameter of the capillary. With too little heat and a sudden pull, the capillary may part with a snap and the tip break off short. Under such conditions workable micropipets may be produced. On the other hand, with too much heat, the tip may be drawn out into a long hair, Fig. 97*e*.

After making a suitable point, the end of the capillary may be turned up at right angles to produce a completed microdissection needle of the style shown in Fig. 97*f*. For this purpose the capillary is brought over a small flame just back of the point, and the end of the point is pushed up with an ordinary dissecting needle or with the tip of the forceps. The length of the needle above the bend depends upon the height of the moist chamber to be employed. If the height of the chamber is less than 8 mm, the length of the turned-up portion of the needle should not be more than 5 mm.

When stiffness is required, it is desirable to maintain a heavy shank as close to the tip as possible, Fig. 97 g. For preparing such a needle, the original tubing is first drawn out into a relatively thick capillary. The thick capillary is drawn out into a thinner capillary which is then used for making the needle point.

In the preparation of a very fine hooked needle, used by CHAMBERS for the dissection of the chromosomes of the pollen mother cells of a certain plant (63), a special twist is given while the tip of the needle is being drawn so that the resulting fine, tapering tip bends sharply at about 20 to 30 μm from its point.

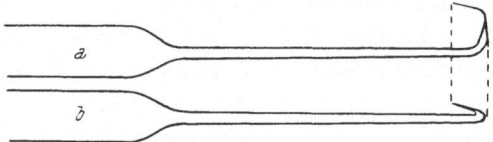

Fig. 98. Microneedle with Tip Bent Back. a lateral view; b top view. After R. CHAMBERS (55).

Fig. 98, shows a microdissection needle with tip bent back to almost a horizontal plane (55). By means of a needle of this style, a cell contained in a shallow drop hanging from the undersurface of a cover slip roofing the moist chamber may be cut cleanly into two intact pieces. This may be done either by pressing the cell against the underside of the cover slip or, with soft cells as the unfertilized sea-urchin egg cell for instance, against the lower surface of the hanging drop. This is done by means of the horizontal end of the needle. The tip of this needle being bent back seems to make it less exposed to breakage when adjusting the position of the needle in the moist chamber. For proper observation, the shank of the needle should not lie between the specimen and the light coming from below. For this purpose, the tip of the needle does not point directly back but slants off to one side as shown in Fig. 98 b. To cut the specimen properly, the angle at which the needle is bent is of importance. The end used for cutting should be neither too far from nor too near to the horizontal plane.

To give the tips of fine microdissection needles a grasping resilient surface, CHAMBERS (55) recommends dipping them into Sandarac varnish which is used by dentists.

CHAMBERS (55) describes a technique due to CHABRY (53) for making fine needle points. According to this method the tip of a capillary pipet s brought into contact with a heated mass of glass or any incandescent body to which glass will adhere (platinum) and quickly drawn away. The incandescent body, a platinum sheet for instance, is fastened to an immovable support. The glass capillary is held by the operator in a groove on a stationary support a short distance from the platinum. The platinum sheet is heated to a dull glow, and as soon as it reaches this temperature, the capillary is slid in its groove until the tip touches the hot metal, whereupon it is instantly slid back. Sliding the capillary in a groove permits producing a straight tapering point coincident with the long axis of the capillary.

Hollow needles with tips bent up, as the style shown in Fig. 97, may be converted into micropipets by breaking off their points. This is done (54) while the needle is mounted on the micromanipulator and properly attached to the injection device or connected with a rubber tubing leading to the mouth. Pressure is maintained with the plunger of the syringe or by blowing, to prevent pieces of glass from being sucked into the pipet. The tip of the needle is broken off by bringing it into a drop of water hanging from the undersurface of the cover glass of the chamber and jamming the tip against the glass surface until it breaks off.

BELKIN (24) described an instrument, the so-called microguillotine, for breaking off the tips of micropipets in a reliable and precise manner. Needles used for the preparation of micropipets should have a rapidly tapering point since a long taper is more subject to breakage. Furthermore, such needles are usually drawn out from thin-walled glass capillaries to insure that the resulting micropipets would have a reasonably large bore at the tip. Thick-walled pipets, owing to the smaller bore, are less readily breakable but are subject to rapid clogging.

The preparation by hand of some specific micropipets used by microchemists in micromanipulator work is described elsewhere in this volume. The method of preparing EL-BADRY and WILSON's micropipet (90) is described in detail on p. 99. This pipet has a wide stem fashioned to fit the syringe illustrated in Fig. 64. The stem of the pipet has approximately the same outside diameter as the syringe barrel. The pipet is attached to the syringe by means of a simple rubber joint, Fig. 64. The proper operation of this pipet depends largely upon the almost complete elimination of surface tension effects at the tip of the pipet. This is achieved by coating the shaft, taper, and a part of the shank from inside with a suitable water-repellent coating, p. 100. In such treatment the outer surface of the tip of the pipet becomes also water-repellent. The operation of such coated micropipets is different from that of the uncoated pipets as those used by BENEDETTI-PICHLER and also those usually used by microbiologists, p. 115.

A number of different substances may be used for treatment of glass surfaces to give water-repellent coatings. The most suitable of these materials for treatment of micropipets are certain preparations, such as

Desicote, Dri-Film, or Teddol, p. 100, which contain mixtures of sub-
stituted chlorosilanes in suitable volatile solvents and which can be applied
also in the vapor phase. Such preparations (82) and several silicone water-
repellents suitable for various purposes (110) are available commercially
(7, 21, 105).

Siliconized micropipets have several advantages. They are readily
cleaned and dried since adhesion of aqueous fluids to the glass is greatly
reduced by the repellent film. They also drain more completely than the
uncoated pipets, and this minimizes the need for rinsing. If water-re-
pellency is complete, micropipets deliver quantitatively the solution which
they contain. This improves the precision of delivery of measured volumes
since no residue is left which may be poorly reproducible. Complete water-
repellency also produces a flat meniscus which is more accurately adjustable. This should improve the precision of calibrated micropipets.
Siliconizing the pipet from outside hinders the sticking of solid particles
to the pipet tip which may come into contact with precipitates during
certain operations occurring in the performance of chemical work in capillary
cones. Moreover, liquids delivered with such pipets do not have the tendency
to creep up on the outside of the tip.

Various glass microtools used for working with living cells may also be
siliconized by treatment with silicone fluids of 100 to 1000 centistokes
viscosity (4). The microtool is dipped into the fluid, and the excess is
then removed by stroking the tool gently toward the tip with a small
piece of facial tissue. The silicone coating that is formed noticeably hinders
living protoplasm, such as tissue culture cells, from clinging to the working
tools. This tendency is sometimes very pronounced so that balls of material
may completely envelop the microtool point. For cytochemical work, KOPAC
(144) in 1953, used micropipets siliconized inside and outside by Desicote
(21) to reduce such tendency of cytoplasmic residue to stick to the glass.

Drawing by hand of the micropipets used by BENEDETTI-PICHLER (25),
Fig. 82, is described on p. 115. For operation of these and other wettable
micropipets, use is made of the strong surface tension in the narrow tips
as explained on p. 115. LOSCALZO and BENEDETTI-PICHLER (154) adapted
micropipets of this style to serve as calibrated microburets with remote
control, Fig. 87, for titration of microgram samples, p. 123. This type of
microburet was also used by Dorf for performing titrations in coloriscopic
capillaries, p. 130.

The preparation of a pipet, Fig. 163, designed especially for picking up
and mounting microfossils (5) is described on p. 293.

β) Automatic Needle and Pipet Pullers.

Though a great variety of microtools are used in different fields of
micromanipulator work, the greater amount of such work is carried out

with glass micropipets and needles of different sizes and shapes. Reproducible standard pipets and needles can be rapidly drawn out from glass capillaries of the required outside diameter, usually about 0.8 to 1 mm, by means of mechanical pulling devices developed especially for this purpose. The operation of these devices depends on electrically heating the central point of a glass capillary or rod of suitable length until the glass softens. The capillary is then automatically pulled apart at its hottest point, resulting in two pipets or needles with identical microtips. The electrical and mechanical settings of these machines can be adjusted to give pipet and needle tips of the required shape and size. Depending on the diameters of the glass capillaries or threads and the kind of glass used, the setting that gives the required result can only be determined by experiment. A particular setting, once determined, can be reproduced whenever required to give the same result.

By means of these machines, fine pipets and needles can be prepared and duplicated in large numbers. With one of these machines described below, the Du Bois puller (83), Fig. 99, once adjusted to give the type of tip required, the operator may be able to prepare 4 to 6 pipets or needles per minute.

These machines can also be set to give exceedingly fine, open pipet tips with the required orifice. It is possible, using Rachele's device (31), Fig. 101, to adjust the machine to give pipet tips which are invariably open, although the orifice may be so fine that it can hardly be seen under a magnification of 300 diameters. Since pipets with very fine tips, drawn out by hand, are usually closed so that the tips have to be broken open wich gives a high percentage of failures, the use of needle pullers is of great advantage.

Pipets and needles made by automatic pullers have straight tips with their axis precisely coincident with the long axis of the original capillary. If inclined tips are required, these may then be bent by hand over a pinhead flame to the desired angle.

A number of these mechanical pulling devices have been developed, but only those of Du Bois (83), Brinkmann (35), and Rachele (31) will be considered briefly.

Du Bois Needle and Pipet Puller.

This machine, supplied by E. Leitz (152), has originally been described in detail by Delafield Du Bois (83), and is shown in Fig. 99. It consists essentially of a narrow platinum heating coil in a protecting housing, *a*, located on the front center portion of the machine. Two clamping screws which can be seen at the lower part of the housing serve for adjusting the lateral and vertical positions of the coil. At the back of the housing the center portion of the machine also carries a mechanism *b* for vertical adjustment of the glass capillary. The other main component is a clamp tension mechanism for holding and stretching the glass capillary. This mechanism consists of two identical telescopic, spring-loaded devices c, at opposite sides of the platinum-coil heater.

A glass capillary of double the length required for one pipet is mounted between guides in the clamps c which are then closed and locked by taut springs. The capillary is held firmly between soft pads. In this position it passes through the platinum heating coil which surrounds a small portion of the capillary. When the current is switched on, the glass starts to melt, and when it becomes soft enough to stretch under a light pull, the capillary

Fig. 99. Du Bois Needle and Pipet Puller. a, platinum coil heater; b, mechanism for adjusting height of capillary; c, clamp tension device for holding capillary; d, tension spring. Courtesy of E. Leitz Wetzlar (152).

is automatically lifted up by the spring movement of the stretching mechanism and is immediately drawn out to give two straight, identical, fine points. When the capillary is lifted up, the force of the pull becomes stronger due to an increase of leverage, and this results in a rapid drawing out at the right instant. Thus the instrument, with the precision of a machine, copies the essential movements of the drawing of pipet and needle tips by hand.

The position of coil and capillary as well as the temperature of the coil determine the amount of heat delivered to the glass. They can be adjusted to reproduce a desired effect. The required setting, however, must be determined by experiment. With the machine adjusted at a given setting, pipets and needles can be prepared at a rate of 4 to 6 per minute.

Brinkmann's Needle and Pipet Puller.

Another commercially available instrument is the BRINKMANN needle and pipet puller (35). It consists of two units: the puller unit proper and the power control unit, both shown in Fig. 100.

In contrast to the spring loaded tension mechanism of the Du Bois puller, Fig. 99, BRINKMANN's instrument uses an electromagnetic system to pull the softened glass apart while a small portion of the capillary is being heated by a narrow Nichrome heating coil.

The puller unit is enclosed in a protective housing which is mounted on a base plate. The base carries the electrical and mechanical parts including magnets, puller arms, guide rods, and heater. The heating coil

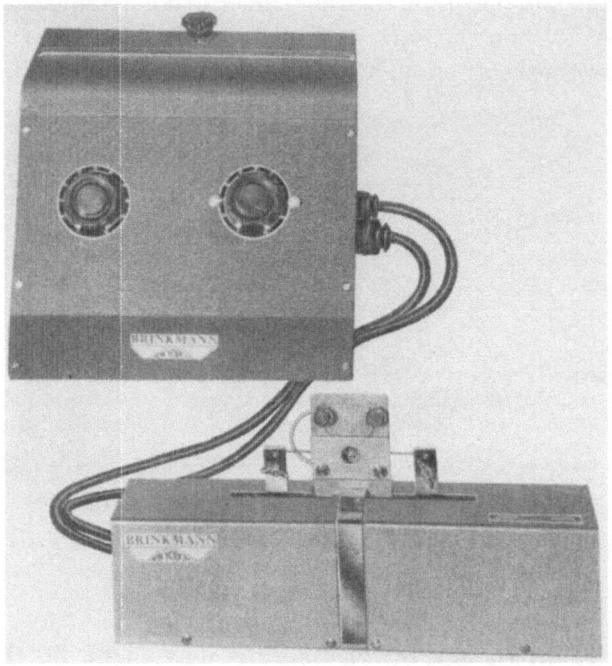

Fig. 100. BRINKMANN Needle and Pipet Puller. Puller unit (below) and the power control unit (above). Courtesy of BRINKMANN INSTRUMENTS, Inc. (35).

and the two clamps for holding the glass capillary on opposite sides of the heating coil are easily accessible as shown in the figure. The coil and capillary can be observed during the operation through a small pyrex glass window in front of the heating coil. The plate in front of the coil is removable to permit alignment of the coil or its replacement. The vertical position of the heater can be adjusted by means of a lever movement.

The power of the magnets and hence the rate of pull as well as the current supplied to the heating coil are regulated by means of control knobs on the power control unit. These controls are provided with graduated dials for accurate resetting.

For drawing out pipets or needles, a glass capillary or solid thread of the proper diameter and of the length required for two implements is mounted in the apparatus by pushing it through the coil and clamping it in place by means of the right- and left-hand clamps as shown in the figure. The glass capillary should be properly centered within the coil and the glass should not touch the wire at any point. The plate in front of the coil is removed, if necessary, and the coil is properly aligned. The height of the heater may be adjusted by the lever movement.

After the machine is properly set with the capillary properly mounted, the current to the heating coil and to the solenoid magnets is switched

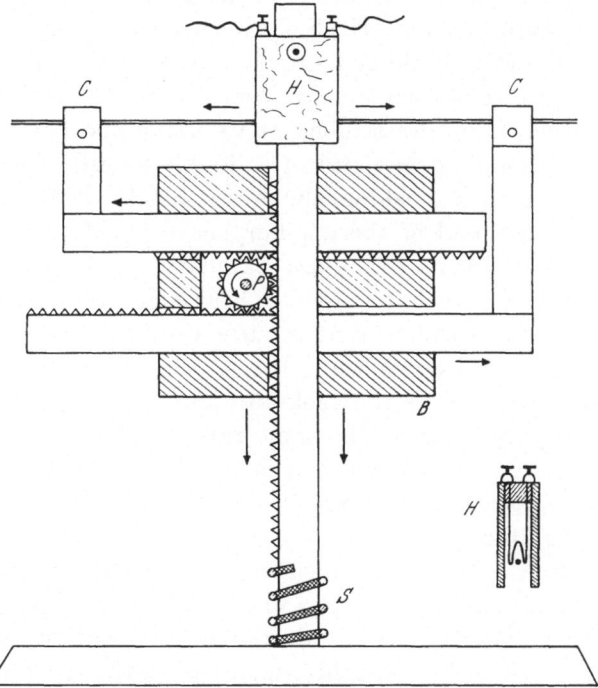

Fig. 101. RACHELE's Device for Drawing Micropipets, Schematic. C, clamps for capillary; P, pinion with axis mounted in metal block B; H, heating element; S, strong spring for absorbing the impact of the falling block. Courtesy of BENEDETTI-PICHLER.

on. When the glass becomes soft enough, it is immediately drawn apart to give two identical points. The heating current is cut off at the correct instant by means of a microswitch in the puller unit. After some experimentation, the operator will find the settings needed to obtain the desired effect depending upon the glass used and the type of tip required.

Rachele's Device.

In the device (31), shown in Fig. 101, the pulling movement required for drawing out the softened glass is achieved by using the force of gravity.

Heating is done by means of a heating coil contained in a U-shaped piece of Transite. The coil is made of a Nichrome ribbon, the ends of which are attached to two binding posts connected to a variable transformer.

Apart from the heating unit, the machine consists essentially of one vertical and two horizontal racks, all operating on one pinion. The racks are mounted on steel bars and pass through a comparatively large brass block which contains the pinion. The vertical rack serves as a supporting pillar for the brass block and also carries the heating unit which is attached to an arm sliding over the vertical pillar above the brass block. The position of the heating unit may be adjusted as desired by means of a thumbscrew. The other two racks pass horizontally through the brass block, one above and one below the steel pinion. Each of the two horizontal racks is provided with a clamp having a V-shaped groove for holding the glass capillary.

The capillary to be drawn out is fastened in the two clamps on the horizontal racks with the brass block at its maximum height. This can be achieved by first fastening one end of a double length of capillary to one of the clamps and then pushing the brass block all the way upwards, whereupon the other end of the capillary is fastened to the other clamp. In this position, Fig. 101, the glass capillary passes through the slot below the heating unit which is shaped to heat a short length of the capillary from both sides. When the current is turned on, the glass softens and the brass block starts to fall down while the two horizontal racks move apart at the same rate because of the rack-and-pinion arrangement. Thus the glass capillary is pulled out at the same rate at which it is removed from the heat. The falling brass block comes to rest on a heavy spring around the base of the supporting pillar, Fig. 101 S. Two micropipets with identical points are thus produced.

By varying the position of the heating unit and the temperature of the coil, tips of different shapes and sizes may be obtained from the same type of capillary. The heating unit and transformer can be set to give exceedingly fine pipets with invariably open tips, although the opening may be too small to be seen under a magnification of 300 diameters.

γ) The Use of the de Fonbrune Microforge.

The DE FONBRUNE microforge (2, 76), Fig. 102, is a very useful instrument for the preparation of a large diversity of micromanipulator tools under highly controlled conditions. Micropipets, microneedles, microscalpels, microhooks, microloops, microprobes, and other micro instruments can be prepared (75) to precise specifications. Microforges are especially valuable for bending micropipets near the tip and for fire polishing of their orifices. The entire operation from the initial fusion of capillaries to the final shaping of the microtools is observed under the microscope. The microforge is equipped with precise controls for temperature, air, and light as well as with means for guiding the micro instrument and the heat source relative to each other in the field of the microscope. The instrument is available commercially (2).

A simplified microforge attachement which may be fitted to a standard biological microscope was described by POWELL (187). This instrument has a number of features in common with the DE FONBRUNE microforge and was designed on similar principles. Another simplified microforge attachment was developed by HILSON (121). In this device, the hot wire is carried and moved by means of a micromanipulator. It also requires the use of a rotating mechanical stage. In the following description, however, only the DE FONBRUNE microforge and its use will be considered.

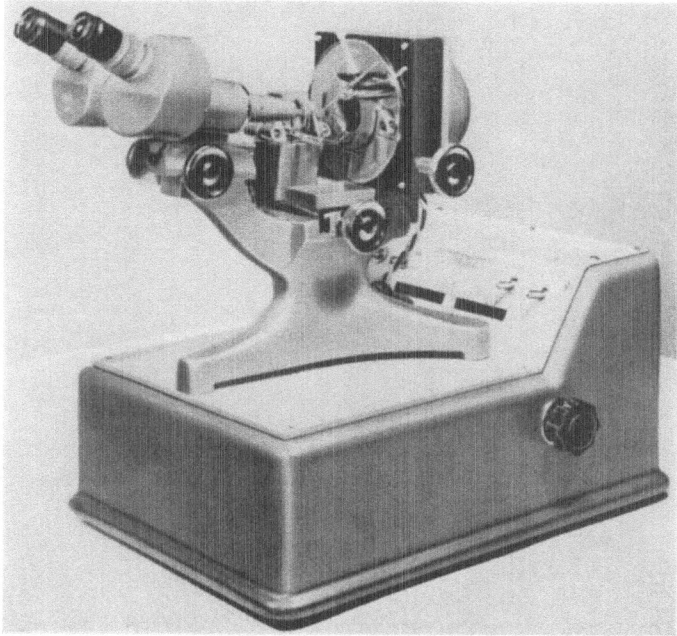

Fig. 102. The DE FONBRUNE Microforge. A general view showing the optical-electrical forge system and the electrical housing. Courtesy of A. S. ALOE Co. (2).

Main Components of the de Fonbrune Microforge.

The DE FONBRUNE microforge consists essentially of a stereoscopic binocular microscope with a built-in light source, a vise provided with controls for manipulating the microtool to be made, an electrical forge or heat source, and the base unit containing the various control equipment. These parts will be described in turn.

Microscope. The stereoscopic microscope providing three-dimensional vision and an erect image is quite suitable for the type of work required. It is equipped with an inclined binocular body with $4 \times$ paired objectives and $9 \times$ paired wide-field eyepieces giving a total magnification of 36 diameters and a sufficiently large working field. Illumination is provided by means of an incandescent light behind an opalescent glass filter and with two intensities of light available for use in working at different temperatures.

Vise. A circular disk, able to rotate about its own axis, is mounted in front of the light source, Fig. 103. An adjustable spring clamp or vise shown in the figure, which also may be rotated about its axis, is held on the periphery of this disk. This clamp or spindle serves for supporting the object before the platinum-iridium filament of the heat source within the field of the microscope. Microtools are usually prepared from fine glass

Fig. 103. The DE FONBRUNE Microforge. Relationship of heating element, air jets, micro-tool, light source and controls. Courtesy of A. S. ALOE Co. (2).

capillaries or solid glass threads about 8 cm long. These are held by means of a metal sleeve of 1-mm diameter. One end of the fine capillary or rod is inserted and cemented into the sleeve, whereafter the sleeve is inserted into the spring clamp of the microforge as shown in Fig. 103. The selected portion of the glass is brought in the desired position relative to the platinum-iridium filament by means of a lever on the vise. The entire assembly, i. e. the circular disk and the vise holding the object, can be adjusted vertically and horizontally by means of two precision screws with knurled knobs on either side of the microscope.

Heat Source. The heat source is mounted on an adjustable stage, Fig. 103, in front of the microscope objective. It consists of a 28-gauge (0.3 mm) 20% platinum-iridium filament capable of carrying an electric current of low voltage. It is supported in the form of an open **V** between two insulated cylindrical electrodes, Fig. 103. The electrodes are detachable and are mounted in a single insulated unit which fits a socket in either a vertical or a horizontal position. In practice the electrode element is usually inserted in the horizontal position. Although the 28-gauge filament is quite suitable for general work, it may be desirable to use finer wire. The filament can be exchanged by pushing in the bases of the two electrodes to expose a groove in each of the electrode tips. After the new wire has

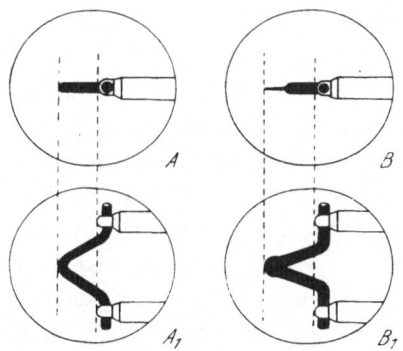

Fig. 104.   The Heat Source. A, A_1, cylindrical filament; B, B_1, cylindrical filament with flattened tip.   After A. S. Aloe Co. (2).

been inserted, a **V** is formed as shown in Fig. 104. The round filament may be used as it is, Fig. 104 A and A_1, or the tip of the filament may be flattened as shown in Fig. 104 B and B_1. The filament with flattened tip offers a minimum hot surface area, and its use is sometimes advantageous in more delicate work as in the bending of fine threads by unequal remote fusion, Fig. 109.

Two air jets, Fig. 103, are mounted on the adjustable stage which supports the heat source. These jets converge on the filament and serve to control the temperature of the hot tool during the forging process and to allow spot cooling as desired. They are equipped with an adjustable valve, located on the left side of the microscope. The switch for the air pump is located on the instrument panel, Fig. 102.

The assembly including the electrodes and the air jets can be rotated as one unit horizontally through a short arc by using one of the air jets as a lever. The unit can also be moved in a vertical or in a lateral direction by means of two knurled knobs.

The optical-electrical forge assembly rests securely on three pins provided on the base unit.

Base Unit. The base unit supporting the microscope assembly is essentially a housing containing a transformer and a rheostat for controlling the temperature of the platinum-iridium filament, a rheostat for controlling the intensity of the light bulb, an electrical air pump with motor, and switches conveniently placed for regulating each of these parts from outside.

On the instrument panel, Fig. 102, are four switches for controlling the electric supply, the air pump, the 2-way illumination, and the heating element. A linear scale with a pointer is also found on the panel below the switches. The pointer is mounted on a resistance slide wire for controlling the temperature of the filament. This rheostat is controlled by means of a knob on the right hand side of the base. Also situated on the base are the electrical inlets and the outlet for the air hose. The latter connects to the air line on the microscope.

The instrument is placed near the edge of a table top in such a way that a seated operator can conventiently manipulate the various controls.

The forge is designed for use in working with glass, either pyrex or soft glass, which is the most commonly used material for preparing microtools. For certain special purposes, however, certain metals and other fusible material may be worked. The general operation of the microforge and methods and techniques of manufacture of various types of microtools (2, 75) will all be described for the use of glass, but the general principles are also applicable to other fusible materials.

Preliminary Setting of the Working Parts.

It will be assumed that the instrument is set as previously described with the air hose on the microscope connected to the air line of the base unit and with the electrical cord on the microscope attached to the socket on the housing. After checking that all the individual switches are in the OFF position, one is ready to start operation, and the master electrical supply cord is plugged into the power line.

The starting material for the forming of microtools are fine glass capillaries or solid glass threads of less than 1-mm diameter and 7- to 8-cm length. These are made from glass tubing or glass rods by drawing them in an alcohol or gas flame to fit a metal sleeve of 1-mm diameter and then cutting into pieces of the proper length.

A capillary or thread is selected out of the stock and is inserted into a metal tube or sleeve allowing about 2.5 cm or more to extend beyond the sleeve. The glass capillary or thread is then sealed into the sleeve by applying a small quantity of suitable material, wax or De Khotinsky cement, to the junction of sleeve and glass. The sleeve serves to hold the microtool in the microforge and may subsequently be used for attaching it to the micromanipulator. The sleeve is inserted into the spring clamp which is located on the periphery of the circular disk of the microforge, Fig. 103. The sleeve and the clamp are adjusted manually so that the free end of the glass capillary or thread is within about 6 mm of the platinum-iridium filament and approximately in the position shown in Fig. 103.

The light bulb for illumination of the microscope is lit by turning on the master switch on the right-hand side of the instrument panel, Fig. 102. The intensity of this light is adjusted by flicking the light switch up or down. Light of low intensity is usually desirable when working with a very bright filament, and vice versa.

If the platinum-iridium filament is not in the microscope field, the filament knob located on the right side of the instrument is turned till the filament appears under the microscope and is brought into focus.

When air cooling is required to control the temperature of the rough tool during the forging process, the air pump switch located on the instrument panel, Fig. 102, is turned on and the flow of air to the jets is then adjusted by means of the valve located on the left side of the microscope. In more intricate operations, it is often necessary to apply heat only to a small working area to avoid melting at undesirable spots. The pump switch and the valve are usually kept in the off position till it is required to apply spot cooling to the work.

The capillary or thread held in the instrument clamp is brought into the microscope field by turning the knobs controlling the horizontal and the vertical position of the clamp. The tip of the glass capillary is then focused. Turning the electrode unit in a small horizontal arc by using the index finger and thumb of the left hand will bring the apex of the V-shaped filament into focus together with the glass. The focus of the microscope may have to be readjusted to obtain good images of work and filament.

The switch of the heating element, located on the instrument panel, is then turned on and the filament starts to glow. As the pointer of the rheostat scale on the panel, Fig. 102, is moved towards the upper end of the scale by turning the rheostat knob, the color of the filament changes gradually from very dark red to almost white. For convenience in adjustment, the scale is divided into twelve divisions. In practice the filament is brought to the desired temperature simply by noting the color of the filament under the microscope. The temperature, however, may be set with the use of the scale, before the heating element is switched on.

The glass rod or capillary is then moved as previously described till the tip is close to the hot filament. From this point the forging process has to be selected to produce the desired type of microtool. All subsequent operations are performed in the field of the microscope.

Temperature and Behavior of Glass.

Five thermal phases of the filament may be distinguished and are utilized. According to the amount of heat applied as the rheostat pointer is moved up the scale, the glowing filament appears dark, dark red, red, pink, and finally white. When glass is brought in contact with the dark filament, the glass becomes malleable and can be bent. In case of the dark red filament, glass just adheres to the filament and may be pulled out in short lengths. When the filament is red, glass readily adheres to it and is easily tractable. When the filament is pink, the glass becomes more fluid and thick glass threads may be bent. In the last phase, when the filament is white, the glass is very fluid and forms spherical globules on the filament. This very hot wire may be suitable for remote fusion.

Practically all normal operations are carried out with the pointer between 6 and 11 on the rheostat scale. As a rule, the pointer is not to be moved beyond the 11.5 mark because the filament would then become dazzling white and would quickly melt.

Basic Operations.

There is a number of more or less distinct basic operations which may be used, either singly or in combination, for the manufacture of various types of micromanipulator tools. A variety of microtools made by the microforge are illustrated in Fig. 105. This figure also demonstrates some important applications of the basic operations.

Fracturing or Breaking-off. This is a useful operation permitting sharp sectioning of capillaries and microspheres. It is used in the manufacture of cup-shaped glass containers and various microscalpels, needles, probes, and micropipets of relatively large diameter ranging from about 15 to 150 μm.

Fig. 105 A shows how a capillary is sectioned by pulling apart at the required point. The filament *f*, covered by a tiny mass of glass *v*, is soldered to the capillary as shown in the figure. The heat is turned off. The capillary is then broken off by retracting the filament and thus pulling the adhering piece of capillary apart from the length held by the clamp.

Fig. 105 B, illustrates the sectioning of microspheres. The technique is the same as the one just described for sectioning capillaries. The segment attached to the filament is taken along by the retracting filament after turning off the heat. This is useful in making cup-shaped containers of the type shown in Fig. 105 C. The stem of the cup is bent at an angle for positioning in the operation chamber. Fig. 105 D, shows other forms of cupping glasses, p. 159.

Fig. 105 N shows how the fracturing technique is used for making microscalpels. A glass globule *s* is affixed to the filament. A platinum wire, embedded in a glass capillary as shown in the figure, is made to enter the

surface of the heated glass sphere. The heat is turned off and the microtool
is then pulled away from the glass sphere. Thus a circular cutting edge
breaks off and remains attached to the wire.

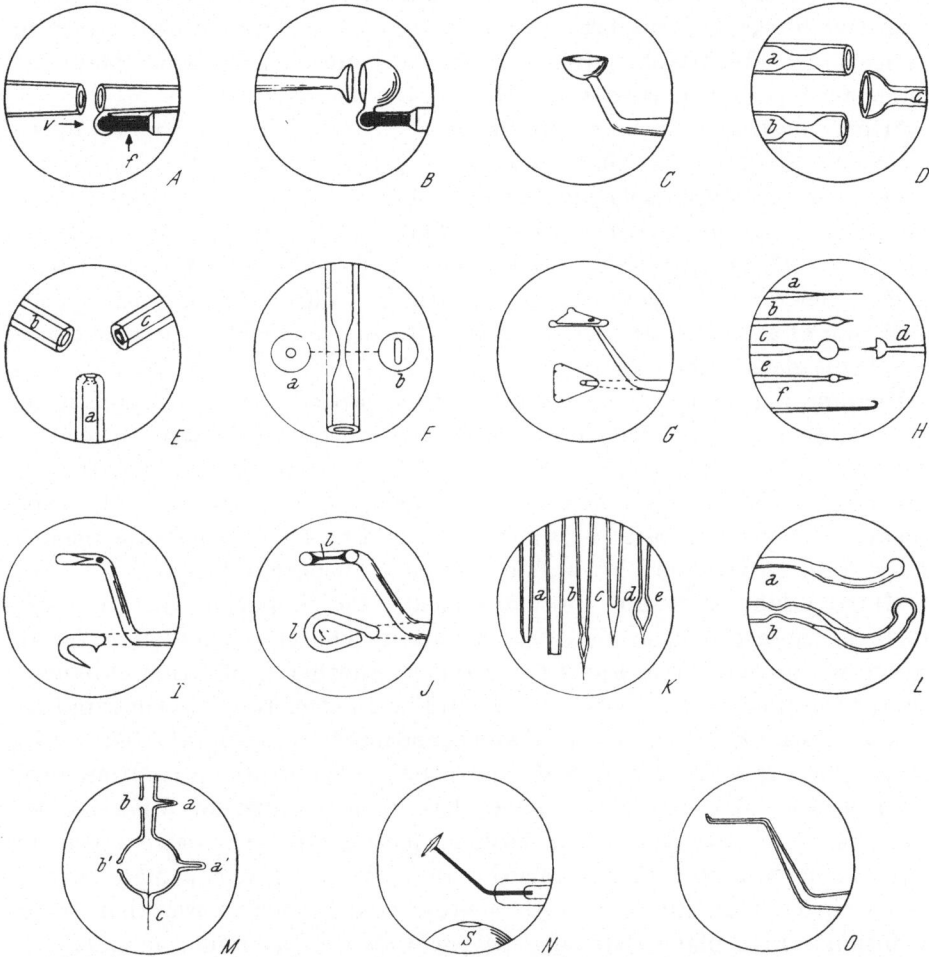

Fig. 105. Types of Microtools Prepared by the DE FONBRUNE Microforge. *A*, sectioning
of capillaries; *B*, sectioning of microsphere; *C*, cupping glass; *D*, three types of cupping
glasses; *E*, orifices of capillary tubes; *F*, local constriction of capillary; *G*, spatula; *H*, microneedles; *I*, retaining hook; *J*, spatula loop; *K*, micropipets; *L*, probe and cannula; *M*, perforated glass sphere; *N*, microscalpel; *O*, microgage. After A. S. ALOE Co. (2).

Remote Fusion. This operation is used in bending threads or fine
capillaries to a specified angle. It may be employed in the manufacture
of curved microtools of about 10-µm diameter as described later in connection with the manufacture of micropipets, Fig. 110. Remote fusion
is also applied in fire polishing and constricting capillaries and in developing
expansions in capillaries. Figs. 105 *Db* and 105 *Ea* show capillaries with

constricted orifices produced by remote fusion. Fig. 105F shows capillaries with a local constriction having either circular or an elongated bore.

Contact Fusion. The method is used in preparing microneedles of various sizes and shapes, straight or bent, and also in the manufacture of special hooks for supporting or restraining a microorganism under investigation. Fig. 105H shows a variety of microneedles with plain and pediculated points, a needle e with an eye, and a hook f.

Microforging. Microforging is used in the manufacture of microelectrodes, fine hooks as that shown in Fig. 105I, the SCHOUTEN loop and other glass loops for isolating bacteria shown in Fig. 105J, and tubular hooks and loops for use in cytology. The manufacture of the SCHOUTEN loop, Fig. 107, is discussed in detail later. Other applications of microforging include the manufacture of spatulas. Fig. 105G shows a spatula consisting of a triangular piece of glass with fire-polished corners and fixed to a bent glass rod.

Microforging in combination with other procedures as contact fusion and remote fusion is often required for the forming of many special instruments. Fig. 105He shows a needle with an eye. Its manufacture requires a combination of microforging, remote fusion, and contact fusion. A spatula loop is illustrated in Fig. 105J. Light passing through the opening of the loop is darkened by a thin film of transparent lacquer.

Micro Glass Blowing provides means for preparing special micropipets as those shown in Fig. 105Kd and e. The two pipets have lateral orifices as those shown in Fig. 105Hb and c. Another application of blowing is in the forming of small cups as shown in Fig. 105Dc. It also permits preparing special canulae with expanded and constricted areas, Fig. 105Lb, making special glass spheres, Fig. 105M, and other instruments blown from glass.

Fig. 105M, shows a glass sphere having a number of different perforations. Perforations a and a' have projecting tips, whereas b and b' are mere openings in the slightly raised wall. At the point c, a metal wire is passed through an aperture with a projecting tip which was then sealed by fusion. Flat planes can also be perforated in a similar manner.

Distant or Contact Stretching. This operation is used in forming ordinary microdropping pipets having tips of an exactly predetermined bore. Such pipets with terminal orifices are shown in Fig. 105K, a, b, and c. Micropipets with fine tips of 1- to 15-μm bore may be prepared. These techniques also permit the lengthening of glass threads, capillaries, and of calibrated capillaries such as the microgage shown in Fig. 105O. The horizontal part of this gage is a capillary of uniform bore.

Preparation of Microtools.

Under this heading, techniques suitable for the manufacture of some specific microtools are described. These are the standard microneedles, the

Schouten loop, and micropipets.   The technique of curving or bending microtools by unequal fusion will also be described to the end of illustrating how the various operations may be applied to produce specified microtools. After the basic operations have been practiced, it should be possible for the operator to fashion specialized micropipets, microneedles, hooks, loops, probes, and other microtools to satisfy his own particular needs.

Preparation of Standard Microneedles; Contact Fusion.   A piece of a Pyrex or soft glass thread, 7 to 8 cm long, is sealed into a 1-mm tube of stainless steel as previously described, allowing about 3.5 to 4 cm of the

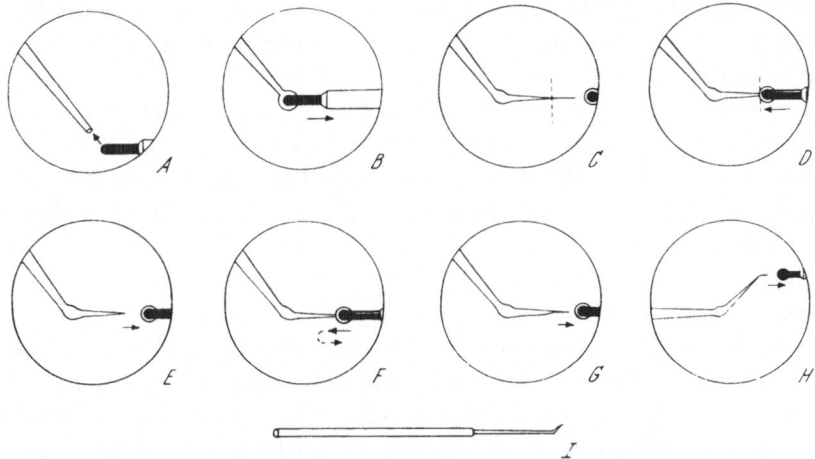

Fig. 106.   Preparation of Microneedles.   *A–E*, drawing-out of shaft; *F–G*, drawing-out of tip; *H* drawing-out of horizontal tip; *I*, finished microneedle with horizontal tip mounted in sleeve.   After A. S. Aloe Co. (2).

thread to project from the sleeve.   The sleeve is then inserted into the microvise, and the position is adjusted so that the thread appears under the microscope in approximately the position indicated in Fig. 106*A*.

The platinum-iridium filament is brought into the field of the microscope.   The air pump is turned on with the air valve fully open.   With the pointer of the rheostat scale set approximately in the middle of the scale, the heating element is turned on.   The position of the rheostat pointer is then adjusted by turning the rheostat knob till the filament takes on a deep red hue.   When the air valve is shut off, the filament should become light red in color.

The above steps and all other necessary preliminary tasks should be performed as already described under Preliminary Setting of the Working Parts.   All subsequent steps are performed in the microscope field.

Practically all types of microneedles are made with the filament deep red to light red in color.   The shaft of the microneedle is drawn out without blowing. The tip is then made with the air valve open so that melting of the glass microtool during this process is confined to a limited area.

Fig. 106 illustrates successive steps, recommended for drawing out micro-needles, from the initial fusion to the final shaping of the microtool. To draw out the shaft of a microneedle, the air valve is first shut off. By means of the knurled knobs, the light red filament is brought so close to the tip of the glass thread, Fig. 106 A, that the tip of the thread melts and the molten glass flows onto the filament in the form of a tiny spherical globule, Fig. 106 B. The filament is allowed to remain in touch with the glass thread for a few seconds, and then the filament is rapidly retracted to form a shaft, Fig. 106 C. If the shaft is longer or shorter than is desired, it can easily be brought to the desired length by touching with the molten bead on the filament as illustrated in Fig. 106 C, D, E.

After the shaft has been drawn out, a fine tip is fashioned for the micro-tool. During this operation, melting of the piece of work should be confined to a limited area. For this purpose, the air valve is fully opened to provide the necessary cooling. The filament, still covered with the globule of molten glass, is just brought into contact with the tip of the shaft, Fig. 106 F. After allowing the filament to remain in this position for a few seconds, it is rapidly retracted to form a fine tip, Fig. 106 G.

For use in moist chambers or in hanging drop preparations, it may be necessary to have microneedles with a horizontal tip, that is, a tip parallel to the shank of the microtool, Fig. 106 H. Such a tip can be formed by turning the circular disk on which the microclamp is mounted so as to bring the shank of the microtool to the horizontal position parallel and slightly below the filament. The desired horizontal point is then formed by repeating operations F and G. Such microneedle complete with shaft and shank mounted in a metal sleeve is shown in Fig. 106 I.

Preparation of the Schouten Loop; Microforging. Glass loops are used, for instance, in developing pure bacterial cultures by isolation of single microorganisms, p. 304. Such loops are best prepared from Pyrex glass. The platinum-iridium filament used in their preparation must be clean and free from all other glass mixtures.

A Pyrex glass thread is mounted in a metal sleeve as described for microneedles. The sleeve is inserted into the microvise and adjusted so that the thread is in the same horizontal plane as the filament and in line with it. The subsequent operations are performed in the field of the micro-scope.

The various successive steps recommended for the forming of a SCHOUTEN loop are illustrated in Fig. 107. The shaft of the microloop is drawn out with the air valve shut. Cooling is needed only when very fine loops are being forged. By means of the knurled knobs, the light red filament is adjusted so that it gets in contact with the tip of the glass thread, Fig. 107 A.

Using the technique already described for drawing out shafts of micro-needles, a shaft for the microloop is obtained, Fig. 107 *B*. With the temperature of the filament slightly reduced, the shaft may be drawn out, if necessary, to the diameter required for the loop, Fig. 107 *C*, *D*. The length of the drawn-out shaft should, at this point, be at least equal to that required for the loop. If the shaft is longer than is desired, the temperature of the filament is lowered to dark red. With the dark red filament, the shaft is reduced to the desired length as illustrated in Fig. 107 *D*, *E*, *F*.

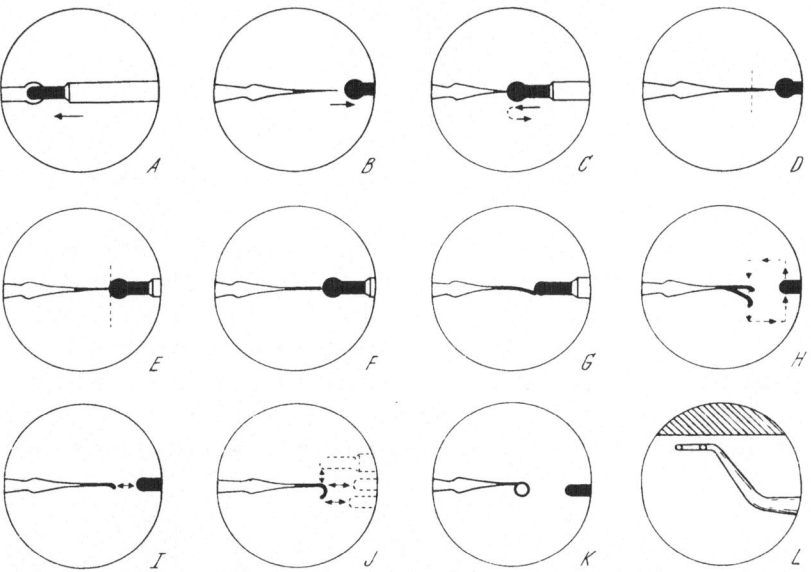

Fig. 107. Preparation of Microloops. *A–D*, drawing-out of the thread; *E–F*, adjustment of shaft to the desired length; *G–K*, forming a loop; *L*, profile of bent microtool with loop positioned in moist chamber under cover slip. After A. S. ALOE Co. (2).

To form the loop, it is necessary to lower the temperature of the filament carefully until it takes on a dark red color and the thread will bend without melting when the filament is brought into contact with it, Fig. 107 *G*, *H*. The filament is then moved in the direction of the arrows as illustrated in Fig. 107 *H*, *I*, *J* until the complete loop is formed, Fig. 107 *K*.

For convenient positioning of the microloop in the moist chamber and in order that the loop may be properly adjusted to the lower surface of a cover slip, the final step is to bend the microtool as shown in Fig. 107 *L*.

Loops different in shape from that illustrated in Fig. 107 *K* and Fig. 105 *J* may be made. As has been mentioned, it is also possible to make finer loops by using essentially the same technique and cooling.

By incomplete closing of the loop, microhooks of varied shapes may be obtained. In forming microhooks, it is necessary to draw out the very fine point of the hook before bending. The point is drawn out as outlined for microneedles. Bending to the desired shape or angle is then performed as described above.

Bending of Microtools by Unequal Fusion; Remote and Contact Fusion. One of the most important operations in the forming of micro-tools is the bending to predetermined angles. Such curving or bending is often necessary for using microtools in moist or oil chambers.

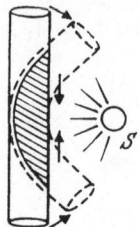

Fig. 108. Bending of a Glass Thread by Unequal Fusion. After A. S. ALOE Co. (2).

When a piece of a glass thread, freely suspended, is exposed to heat from an incandescent body located near by, Fig. 108, the thread will bend symmetrically and take the form shown by the dotted lines. This happens because melting has occurred only in the portion of the thread indicated by the shaded area. The principle is applied to bending microtools with the microforge. Satisfactory bends may be obtained with glass threads of 3- to 4-µm diameter, but the procedure must be adjusted to the diameter of the thread.

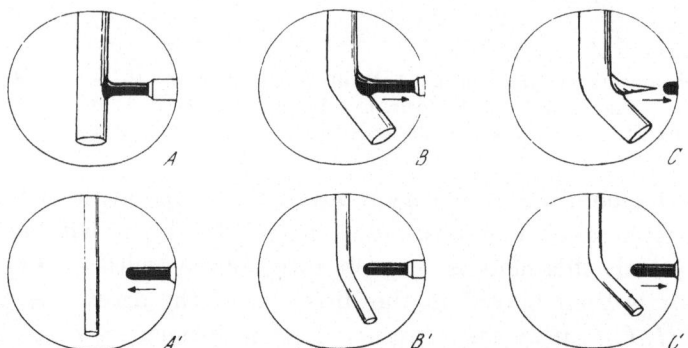

Fig. 109. Bending Glass Threads by Unequal Fusion. A–C, bending thick threads by contact fusion; A'–C', bending finer threads by remote fusion. After A. S. ALOE Co. (2).

Bending Thick Threads by Contact Fusion: 250-µm Diameter or More. A piece of glass thread mounted into a metal sleeve is inserted into the instrument clamp and is brought into the microscope field. With the air valve closed, the red hot filament is brought into contact with the glass thread as shown in Fig. 109 A. The temperature of the filament is gradually raised until the thread starts to bend. The rate of bending is controlled by progressively withdrawing the filament in such a way that the desired angle is obtained, Fig. 109 B, C. A thorn of glass forms at the bend of the thread as the filament is retracted, Fig. 109 C. After the heat has been turned off, this thorn may be broken off with the cold filament.

In this and all other bending operations care should be taken that the filament is perfectly clean and free from glass or metal particles remaining from previous use. Furthermore, it is of particular importance that the filament be accurately centered in relation to the thread.

Bending Medium-Size Threads by Remote Fusion: 50- to 250-µm Diameter. The general technique is similar to that described for bending thick threads except that a light red filament is used. Furthermore, the filament is not brought into actual contact with the thread. As the light red filament approaches the thread, the latter will bend slowly in the direction of the filament. The filament is rapidly retracted as soon as the desired angle is obtained, Fig. 109 A' to C'.

The work should be done in a room with minimum air currents or under a hood.

Bending Fine Threads by Remote Fusion: Less Than 50-µm Diameter. Efficient cooling is essential. The operation is started with a red filament and with the air valve fully open. The fine thread is very cautiously approached with the filament, Fig. 109 A'. As the filament is brought slowly near the thread, the latter will bend in the direction of the filament, Fig. 109 B'. If the thread does not bend, the temperature is gradually raised by turning the rheostat knob until bending takes place. As soon as the proper angle is formed, the filament is rapidly withdrawn, Fig. 109 C'.

It may be preferable to approach the thread with the filament at a temperature lower than that necessary for bending and then to regulate bending by means of the rheostat. The temperature is raised to start bending. The rate of bending is then controlled by gradually decreasing the temperature. Finally, the bending is stopped by shutting off the heat.

It is sometimes preferable to use a filament with a flattened tip as shown in Fig. 104 B and B_1.

Preparation of Micropipets by Stretching Tip: 1- to 15-µm Bore. A technique that can be applied to the forming of micropipets with tips ranging from about 1-to 15-µm diameter will be decsribed. The procedure, however, is not suitable for making pipets with extra fine tips of less than 1 µm.

A set of weights ranging from about 50 mg to 15 g, each attached to a fine wire loop as shown in Fig. 110 C, D, should be available. The bore of the drawn-out tip will depend largely upon the weight used, Fig. 111.

A piece of capillary of suitable diameter is mounted into a sleeve. This mount serves to hold the capillary during forming and, subsequently, in use with the microsyringe and the micromanipulator. The sleeve is mounted so that the capillary appears as illustrated in Fig. 110 A in relation to the filament.

The filament is first heated until it is dark red. It is then brought into contact with the tip of the capillary until the glass melts, and is then moved upward as indicated by the arrows to form a hook, Fig. 110 A, B.

The capillary is turned for 180°. A suitable weight, chosen from the set to give the required bore of the tip to be drawn, Fig. 111, is picked up with forceps and placed on the hook as shown in Fig. 110 C.

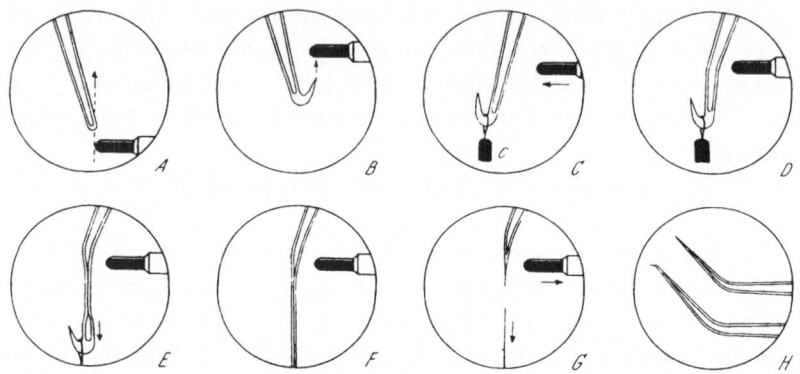

Fig. 110.   Preparation of Micropipets. $A–C$, forming a hook for supporting the weight; $D–G$, drawing-out tip by distant stretching; H, two micropipets with different types of tips After A. S. ALOE CO. (2).

The temperature is raised until the filament is red. The filament is then moved towards the capillary until bending starts, Fig. 110 C, D.

The filament is kept in its position and the temperature is raised so that the capillary is slowly drawn out, Fig. 110 E, F. The diameter of the drawn part gradually decreases until a break occurs near the filament, Fig. 110 G. When the break occurs, the filament is instantly withdrawn.

Depending on the type of work, it is sometimes desirable to bend the micropipet as shown in Fig. 110 H. This can be achieved by applying the principle of unequal fusion.

It has been found out that the tip diameter of the micropipets obtained by stretching is directly proportional to the weight applied. This is illustrated in Fig. 111, in which the diameter of tips is plotted against the weight used.

The quality of glass has an influence on the shape of the tip. Soft glass for example produces a tip having the shape of a truncated cone as shown in Fig. 105 Ka, whereas Pyrex glass will produce a sharply broken tip, Fig. 105 Kb.

c) Storing and Cleaning of Glass Microtools.

A supply of microtools may be kept on hand on suitable stands securely protected in dust-proof, spill-proof containers. Probably the most convenient stands may be prepared from suitable blocks of plastic or hardwood by drilling holes sufficiently wide to receive the shafts, usually about 0.8 mm in diameter, or the stems of the microtools. Such blocks may be covered with beakers of suitable height or with small bell jars of a diameter to fit the blocks. A serviceable wooden block for storing microtools is shown in Fig. 96a. A storage block made of Leucite similar to that described for storing capillary containers, Fig. 78, is very convenient.

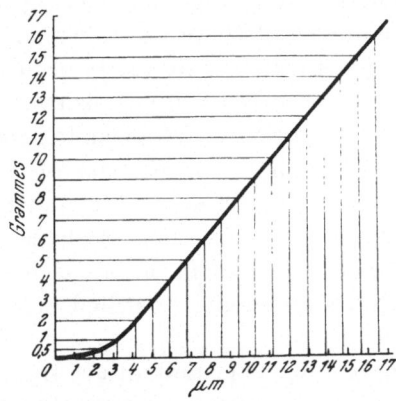

Fig. 111. Diameter of Pipet Tip vs. Weight. After A. S. ALOE Co. (2).

The individual tools being stored should be sufficiently spaced from each other so that they may be easily picked up with forceps without endangering the adjacent pieces. If a variety of microtools is used, it is well to have a number of such blocks, each marked off to accomodate just one type.

The glass tubing from which microtools are to be prepared must be carefully cleaned. It may be sufficient to clean the tubing, after cutting into suitable lengths, inside and outside with soap solution and to follow this by rinsing with tap water and then distilled water. After draining and drying, the tubing is kept inside dust-proof containers ready for use. Alternatively, the pieces of tubing may be more thoroughly cleaned by boiling in soapy water followed by rinsing and immersion for several hours in concentrated chromic-sulfuric acid cleaning solution. After rinsing in running water, the glass is rinsed with ammoniated distilled water. Drying may be done by rinsing with 95% ethanol and then with ether*. The glass may also be dried with a gentle stream of moist air after rinsing with ethanol.

---

* Acetone seems preferable.

Microtools may be used repeatedly. The procedure for cleaning a microtool after use depends upon the contamination to be removed. At times,
distilled water may be sufficient. Micropipets attached to suitable injection
devices in which distilled water is used as a hydraulic fluid may be cleaned,
p. 212, by applying pressure until water from the syringe moves into the
pipet and forms a droplet at the pipet tip. The water droplet is removed
by a strip of filter paper or a small piece of facial tissue. Another droplet
is expelled and removed in the same way. The water meniscus in the
pipet shank is finally returned to its original position by applying suction.

Other liquids may be used for rinsing. They are supplied in the form
of small drops either held on a microscope slide or hanging on a glass rod
or from the tip of a small buret, Fig. 80*f*, serving as a reservoir for the
cleaning fluid. Small volumes of these are taken up into the pipet as described on p. 212 and then expelled onto a strip of filter paper. The process
is repeated using a fresh drop of wash liquid for every rinse.

In working with cells and tissues, cytoplasmic residues adhere to microtools. Successive rinsing by immersing the tips of the tools in strong acid
and strong alkali removes this material. A recommended procedure (4) for
removing protoplasm clinging to microtools involves wiping the tool gently
towards the tip with small pieces of facial tissue moistened consecutively
with distilled water, dilute ammonia, distilled water, xylol, and finally
95% ethanol.

Siliconized micropipets (90, 144) and other siliconized microtools are
cleaned more readily since the water-repellent film, p. 100, 147, reduces
adhesion of aqueous fluids to glass (82).

d) Metallic Micromanipulator Tools.

In addition to the diversity of micro instruments already considered
such as micropipets, microneedles, etc., other micromanipulator tools are
also useful in certain studies. These mainly include different types of
microelectrodes, microthermocouples, and micromagnets. A number of
these instruments will be considered. They are also held and guided in
the field of the microscope by means of micromanipulators and may be
constructed with points fine enough and sufficiently strong to be inserted
into small living cells or into tissue.

Microelectrodes. Several types of metallic and of non-polarizable
microelectrodes have been devised by various investigators (185, 215, 223,
225) for measuring resistances and currents in single living cells and for
stimulating single nerves and muscle fibers.

TAYLOR (223) described a quartz-covered platinum microelectrode
Fig. 112*c*, for the study of electric properties of protoplasm in the interior
of a living cell. The construction of such instruments is based upon the
fact that when a metal wire is inserted into a close fitting capillary of glass

or quartz and the capillary is melted at a certain point over a suitable microflame, a sudden pull will draw the capillary and the metal wire into exceedingly fine points which may be less than 1 μm in diameter. In preparing such instruments certain requirements (221, 237) should be considered in choosing the metal and the glass. If the coefficient of expansion of the metal is less than that of the glass, the latter will shatter on cooling due to its shrinkage onto the metal. The glass should be selected so that its melting point lies between the melting point and the boiling point of the metal. Thus for metals of high melting point such as platinum and

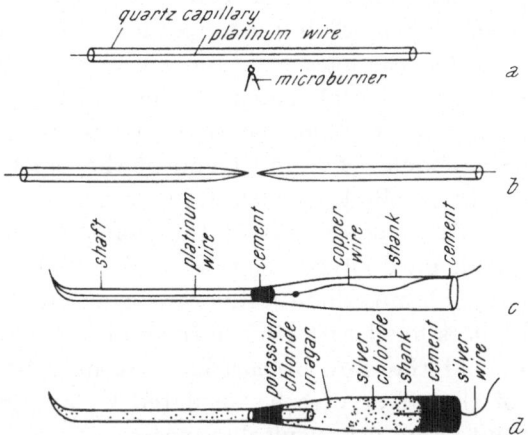

Fig. 112. Construction of Microelectrodes. *a*, platinum wire inside a quartz capillary tubing; *b*, drawn out electrodes; *c*, complete platinum microelectrode with curved fine tip, platinum wire annealed to copper wire and quartz shaft sealed to glass shank; *d*, non-polarizable microelectrode. After TAYLOR (223).

iron, capillaries made of clear fused silica are used. Pyrex glass capillaries are suitable for silver, whereas very soft soda glass can be used for bismuth, solder, and lead.

TAYLOR's platinum electrode, shown in Fig. 112*c*, is constructed by inserting a length of a No. 35 (0.14-mm diam.) pure platinum wire into a tightly fitting quartz capillary tubing, Fig. 112*a*. Using the tiny flame of an oxy-acetylene microburner, the capillary tubing is drawn out into two similar parts, each ending in an exceedingly fine needle point enclosing a platinum core which may be less than 1 μm in diameter. The undrawn end of the platinum wire should extend a few millimeters outside the capillary tubing; using the same microburner, it is fused onto an insulated copper wire of a diameter approximately equal to that of the platinum wire and about 50 cm long. Over the oxy-acetylene microflame, the needle tip is then bent for about 3 mm as shown in Fig. 112*c*. The other end of the electrode is sealed with De Khotinsky cement to a glass shank of suitable diameter to fit into the tool holder of the micromanipulator. With such electrodes, the glass or quartz may be cut back at the tip to expose a very short length of the metal. This operation is performed under the microscope (237) with the aid of a diamond microcutter and a suitable micromanipulator by using a technique similar to that described for the microthermocouples, p. 171. It is possible, however, to construct equally fine electrodes from metallic wires without the glass cover, but they would not be sufficiently strong for use.

A non-polarizable microelectrode, Fig. 112d, used for certain investigations (222) was also described by TAYLOR (223). This electrode is prepared by filling a micropipet with a melted 0.5% solution of potassium chloride in dialysed agar-agar. Filling the pipet may be performed by means of a Luer syringe. After the agar solution has completely hardened, the micropipet is sealed with a dental cement of low melting point into a glass shank of suitable diameter so that the shaft of the pipet extends a few millimeters inside the shank. The inner end of the pipet is then covered with a few drops of the melted agar solution, and the solution is allowed to cool completely. The greater part of the shank is filled with slightly moist, well pulverized, pure silver chloride, and into the salt is inserted the end of a No. 20 (0.8-mm diam.) silver wire about 30-cm long. The open end of the shank is filled with De Khotinsky cement to seal the shank completely and to hold the wire in position. It may be preferable that the exposed part of the silver wire be insulated by applying a coating of De Khotinsky cement or by other suitable means.

TAYLOR used a dry cell of about 1.5 v to provide the electromotive force. The circuit included two resistance boxes with a total of 12 000 ohms and one of 40 ohms, a nitrogen-mercury key, and a specially designed, exceedingly sensitive galvanometer having an internal resistance of 35 417 megohms.

In the two microelectrodes just described, Fig. 112, and in others, the conducting medium which may be either a metallic wire or a solution of potassium chloride in agar is enclosed in a quartz or glass micropipet. SEN (215) devised a different type of stimulating and non-polarizable microelectrodes, Fig. 113, in which the conducting medium is deposited as a continuous coating on the outer surface of a microneedle made of glass or vitreous silica. For the preparation of stimulating electrodes, SEN uses a film of silver. Gold or platinum may also be taken. The metal may be deposited on glass or vitreous silica by spattering in vacuum. A number of microneedles of the desired shape is prepared.

For silvering, the shank ends are warmed in a microflame and embedded vertically in a block of paraffin, or the needles are supported in any other convenient way. They are then thoroughly cleaned by immersion, tips downward, in a beaker of cleaning fluid for a few hours, and this is followed by washing in running water. They are then kept immersed in distilled water until they are transferred to the silvering solution. After silvering, the needles are washed thoroughly and dried in a hot oven.

The silvered needle is then mounted in an electrode holder, Fig. 113, consisting of a glass tubing, 4 mm in diameter and tapering at one end. A length of the glass tubing at the tapered end contains Woods metal with one end of a copper wire lead buried in it. To insert the needle into the holder, the Woods metal is melted by applying gentle heat and the needle is thrust in. On cooling, the needle is firmly held and good contact is secured. Thicker coats of silver or other metal may be obtained by electroplating the already coated microneedle after mounting in the holder.

For the preparation of non-polarizable silver-silver chloride electrodes, SEN first prepares silver-coated microneedles as described above. After mounting the coated needles in the electrode holder, Fig. 113, a thicker coating of silver is deposited electrolytically. A continuous film of silver

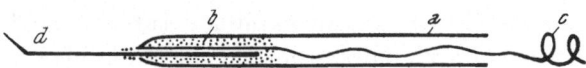

Fig. 113. Microelectrode. *a*, glass tubing; *b*, Woods metal; *c*, copper wire lead; *d*, coated microneedle. After SEN (215).

chloride is then obtained by electrolysis of a 5% potassium chloride bath and using the microelectrode as the anode and a piece of silver as the cathode; 0.5 volt are applied across the terminals, and the current is allowed to flow for only 30 to 40 seconds. The microelectrode is then washed and kept in distilled water for a few hours before use.

Both kinds of SEN's electrodes may be easily prepared in large numbers. The metallic film, however, is apt to rub off. The metallic surface of the shank may be insulated by applying shellac solution with a fine brush. The microtip can be insulated up to within 10 to 5 μm from the point. This is done under the microscope after suitably mounting the micro-electrode, with the microtip pointing upward, on a simple rack-and-pinion micromanipulator. The needle tip is observed with dark-field illumination while the shank is lowered into a small container of shellac solution.

The electric resistance of these electrodes is low. It has been found to be only 0.35 ohms for a pair of electrodes with microtips less than 3 μm in diameter.

Minute bipolar electrodes (237) may be prepared in the way described below for the construction of the microthermocouples. Both cores are made of platinum and the projecting tips of the two wires are left unjoined.

Microthermocouples. The construction of microthermocouples sufficiently small to measure the light absorption of a single plastid and suitable in general for temperature measurements at minute points is described by WHITAKER (237). They may be used for studying the reaction of small particles towards heat and for measuring the temperature at which a specific change does occur.

The instrument is a double electrode in which two different metals are used. The two metals are joined at the tip to make a sensitive point. The microthermocouple is prepared from two similar capillary tubings of the same kind of glass or silica, each containing one of the metals. The two capillaries are first fused together at the region which is to be drawn out to give a rod of nearly elliptical cross section. This rod, enclosing the two metal cores completely separated by glass, is drawn out to a fine microtip as described on p. 169 in connection with TAYLOR's metallic micro-electrode, Fig. 112c. The two capillaries are then joined firmly at several points by applying De Khotinsky cement.

In preparing these thermocouples, due consideration should be given to the selection of the pair of metals to be used. Besides having a suitable emf, the melting points of the two metals should be sufficiently near each other so that they may be drawn out in capillary tubings made of the same kind of glass. Of the suitable pairs, iron and platinum give an emf of about 19 mv per degree. An emf of about 45 mv per degree is obtained by using iron and an alloy of 60% gold and 40% palladium. The pair is drawn out in capillary tubings made of vitreous silica. A more sensitive combination is bismuth against an alloy composed of 95% bismuth and 5% tin. This pair has an emf of about 95 to 100 mv per degree. Such brittle metals may be put in the capillaries in the granulated form, whereupon the points are drawn.

The glass or silica must be removed from the tip of the double electrode to expose a length of the two metallic wires sufficient for joining them. The cutting is done under the microscope with the aid of a diamond micro-cutter if the wires are not too fine and consist of material having some shearing strength: iron, copper, platinum, and gold-palladium alloy. A glass slide, onto which a piece of a razor blade is cemented with its edge in a vertical position, is held by the mechanical stage of the microscope. The double electrode, cemented to a glass holder, is mounted on a suitable micromanipulator, and a diamond splint with a sharp point, cemented to a glass rod, is held also by a micromanipulator. Under a magnification of about 100 diameters, the electrode is manipulated to contact the razor blade at the required point, and the side of the electrode is then scratched by the diamond, with the edge of the blade acting as an anvil. The electrode is then rotated for 180° around its axis, and a scratch is made on the opposite side. The glass or silica is then broken off and removed by carefully pushing with the diamond against the electrode further out toward the tip, while the first scratch is kept on the razor blade.

When very small junctions of 5- to 10-μm total diameter are to be made, it becomes necessary to use very fine wires. It is then difficult to remove

the glass without breaking the wires. It is possible, however, to dissolve
the glass in hydrofluoric acid, and this procedure may be used also when
the wires are made of soft metals that are easily sheared such as bismuth
and bismuth-tin. The electrode is first coated by dipping it into melted
paraffin. The latter is then removed from the tip of the electrode with
xylol and absolute alcohol. The electrode, mounted on a micromanipulator,
is lowered until the tip is immersed to the desired depth in a small drop
of hydrofluoric acid placed on a paraffined slide held by the mechanical
stage of the microscope. The microscope is best protected against the acid
fumes by inserting a glass plate, about 12 cm $\times$ 18 cm, below the objective.
Depending on the size of the tip, a few seconds to 20 minutes may be required to remove the glass. After the acid treatment, the electrode is
washed with distilled water and dipped into a strong alkali to stop the
action of the acid.

Bismuth and bismuth-tin are preferably joined by first bending the fine
wires, being much less brittle than thicker ones, into mechanical contact.
The wires are then melted together under the microscope with the aid of
a small filament mounted on the micromanipulator.

When too fine to be handled mechanically, the projecting ends of the
two wires may best be joined by electroplating. For this purpose, silver
is a satisfactory metal to use since it has a high heat conductivity and is
comparatively inactive. It may be deposited from any standard silver-
plating solution and gives a smooth strong coating.

The two projecting microwires, if not already too close, are first bent
together to come almost in touch with one another. Deposition is carried
out under the microscope. Silver solution is supplied as a drop hanging
from the undersurface of a suitably supported cover glass. A silver wire
mounted on the micromanipulator serves as the positive electrode. The
rate of deposition may be regulated by placing a moistened finger across
an open switch and varying the pressure of contact. The operation is
started by depositing a thin coat of silver on one of the two microwires
by connecting its lead to the negative terminal. Then this first lead is
disconnected and the second microwire is connected to the negative terminal.
Silver will be deposited on the second wire and will continue to build up
until contact is established between the two wires, whereupon silver starts
to deposit also on the first wire. The connecting deposit also grows. When
a sufficient amount of silver has been accumulated, the point is washed
with distilled water and dried thoroughly. To make a neat strong junction,
one further step is necessary. The deposited silver is melted by connecting
the leads of the couple through a spring switch to a single dry cell. One
quick tap of the switch usually melts the deposited silver causing it to
form a solid ball enclosing the two wires. Such a junction will be sufficiently strong to be inserted into cells and tissues.

The resistance of the junctions depends largely upon their size and also
on the steepness and length of taper, quality of joint, and the metals used.
Iron and platinum junctions of about 10-μm diameter have a resistance
of about 10 to 15 ohms, whereas similar junctions of bismuth and bismuth-
tin have much higher resistance.

Temperature lag is negligible because of the minute mass of these junctions. With a sufficiently sensitive galvanometer and under proper conditions, the temperature of a bismuth and bismuth-tin junction of 10-μm
diameter may be measured directly to less than 0.0005° C.

Micromagnets. Micromagnets of the type shown in Fig. 114, consisting of silica microneedles with soft iron cores, were also described by
TAYLOR (223). These magnets have exceedingly fine points just as those
of the platinum electrodes previously described, Fig. 112c, and are suitable
for the study of magnetic properties of living protoplasm inside the cell.

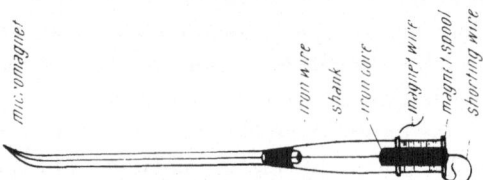

Fig. 114. Micromagnet Consisting of Microneedle with Soft Iron Core. After TAYLOR (223).

They are constructed by inserting a length of a No. 30 (0.25-mm diam.)
soft iron wire into a closely fitting capillary tubing, 30 cm in length, of
vitreous silica. The iron wire should be sufficiently long to extend about
20 cm beyond either end of the quartz capillary. The micromagnets are
then drawn over a tiny oxy-acetylene flame as indicated for the preparation
of the platinum electrodes, Fig. 112a. The resulting exceedingly fine tip
of the iron wire is completely enclosed by the needle point of the quartz
capillary. In the microflame, the tip of the micromagnet is slightly curved
as shown in the figure, and the magnet is then temporarily cemented into
a suitable glass shank so that the free end of the iron wire extends 3 to 4 cm
beyond the base of the shank. The magnet spool shown in the figure is
made of wood and is 1.5 cm in diameter and 2.5 cm in length. It accomodates 1500 ft. (450 m) of No. 36 (0.13-mm diam.) triple insulated copper magnet wire wound on it, and a soft iron core, 4 mm in diameter, one
end of which extends about 3 mm beyond the spool. The free end of the
iron wire is soldered to the projecting end of the soft iron core. The
temporary seal is opened, and the glass shank is slid over the projecting end
of the iron core and permanently cemented on the core with De Khotinsky
wax. The micromagnet is finally cemented firmly into the glass shank
which serves for attachment to the instrument holder of the micromanipulator.

A permanent micromagnet and a needle-shaped electromagnet, both mounted on a micromanipulator, were described by KOCH et al. (138). These magnets are suitable for use in metallurgical investigations for the separation of ferromagnetic from non-magnetic particles. The finely powdered sample is placed on a glass slide and the operation is performed under a stereomicroscope using the technique described on p. 200.

Hot Wires. Electrically heated wires carried and moved by micromanipulators have been devised for different purposes.

A heating element (25), Fig. 76, with a tip made of a copper-nickel alloy resistance wire is described on p. 110. This hot wire was used for performing distillation operations in capillary vessels, p. 219, Figs. 134, 135.

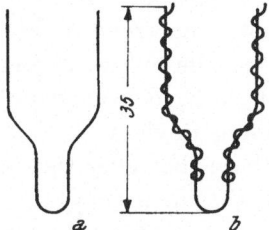

Fig. 115. Hot Wire for Opening Microgram Samples by Fusion with Fluxes, and for Performing Bead Tests. After KOCH et al. (138).

A hot wire constructed from a platinum wire 15 μm in diameter and 5 mm long, connected to a 6.3-v transformer in series with a 22-ohm resistor, was used by ORDWAY (180) for melting small amounts of jeweler's shellac during the operation of mounting under the microscope small crystal specimens in proper orientation for x-ray diffraction studies. Another heating element with a platinum tip was described by TITUS and GRAY (228).

An electrically heated platinum wire mounted on a micromanipulator was described by KOCH et al. (138) for the opening of microgram samples with suitable fluxes under microscopic control. The wire is about 5 cm long and 0.35 mm in diameter, bent as shown in Fig. 115a to allow the formation of a bead in the loop at the end of the wire. The intensity of the current employed is about 3 to 6 amps. so as to bring the loop to the proper temperature. The current intensity is read on an amperemeter. The wire shown in Fig. 115b is used when the melt has a tendency to creep during heating. The melt cools and solidifies as it contacts the thickened part. The wire was also used for the identification of microgram amounts of material by means of the well known bead tests (190). The basic techniques for performing the necessary operations are described on p. 206.

In microbiology, for the separation or cutting out of single cells of hyphae, a suitably mounted platinum wire, about 1 µm in diameter and heated by a weak electric current, has been described (80).

A filament which may be used as a minute source of heat or light (237) may be constructed in the way described for the preparation of the micro-thermocouples, p. 172. First, a minute vitreous silica bipolar electrode with both cores made of platinum is prepared. The tip ends of the wire are joined to form the filament. When the junction of the two wires is white hot, the silica-covered cores are red hot.

7. Auxiliary Equipment.

Microscopical equipment, microtools, and other accessories and supplementary instruments used with micromanipulators have already been dealt with in preceeding sections. In the following pages, certain other auxiliary equipment most commonly used in micromanipulator work, namely various types of micropipet controlling devices and operation chambers, will be considered. A number of these, however, are described in other sections as parts of the assemblies with which they are actually employed. Other miscellaneous equipment, less commonly used or designed for limited purposes, are described only in connection with the apparatus or work to which they apply.

a) Pipet Controlling Devices.

In micromanipulator work, many methods for control of liquid flow in the micropipets have been devised and used. For simple operations which do not require exact adding or withdrawing of minute volumes of liquid, it may be sufficient to use a mouth pipet consisting of a glass mouthpiece and a connecting rubber tubing leading to the micropipet. The mouthpiece is held between the teeth of the operator and the movement of fluid in the micropipet is controlled by blowing or sucking. This may be suitable for the transfer of particles dispersed in a thin layer of a liquid. It has been widely used in single cell isolation work and also for making single grain preparations of microfossils, p. 293. A special, micromanipulator-mounted suction device, connected to a water pump has been described (138) for the collection and transfer of single particles below 0.4 mm in diameter.

For most micromanipulator work, however, in quantitative or even qualitative operations, it is necessary to have more precise and reliable control over the pressure. Several serviceable pipet controlling devices have been described in the literature. Most of these have been used for the purpose of injecting by pressure aqueous or nonaqueous fluids and suspensions into the cytoplasm, vacuoles, or nuclei of living cells or of withdrawing by suction minute amounts of fluids and small particles from the interior of single living cells. They have been also used in single cell isolation to deliver or to collect small droplets in which the bacterial cell is isolated. In other fields such as chemical micrurgy, such devices may be used to control small volumes of solutions and reagents taken up or delivered by the micropipet.

Aside from the fineness of the aperture at the microtip and proper tapering, the functioning of a micropipet depends largely on the accurate control of pressure and upon the surface tension and viscosity of the liquid. It is understood that an advancing piston or plunger merely displaces liquid and that any noticeable pressure in the open pipet must derive from forces originating in the narrow tip. The diameter of the orifice of micropipets varies widely from less than 1 μm to about 100 μm according to the type of work. With micropipets having very small orifices at the microtip, the surface tension and cohesive forces are very high. This sometimes causes difficulties which may be due to the built-up pressure which remains in the system after the formation of a droplet. Certain micropipets (90) are coated with a suitable water-repellent film to eliminate surface tension effects at the microtips, p. 100. This results in accurate control over liquid flow. The menisci of the solution in the pipet invariably follow every movement applied to the pressure control of the microinjection apparatus.

The BARBER mercury pipetting system (10, 12) makes use of a *pusher* of mercury and depends upon the expansion and contraction of a suitable volume of this metal when it is heated or cooled. The mercury is contained in a glass tube, on one end of which the pipet is drawn. The expansion of mercury is controlled by means of warm water. The pipet is held in BARBER's pipet holder, p. 30, Fig. 10, which is clamped to the microscope stage. Though this pipetting method is workable, the pipet is difficult to construct and easily broken. Furthermore, instantaneous control of the driving force of the mercury is difficult to achieve.

TAYLOR (224, 226) described two pipet controlling devices both using mechanical means for varying the pressure on an enclosed volume of mercury. In the first of these devices (224), the mercury, which is contained in a glass capillary tubing, is displaced by a minute plunger consisting of a small steel needle attached to a finely threaded thumb-screw. The needle passes through a close-fitting soft rubber plug into the column of mercury. This system offers a satisfactory control of the pressure in the pipet, but it is difficult to prevent eventual leakage of mercury due to gradual abrasion of the rubber around the metal plunger.

In the second pipetting system described by TAYLOR (226), injection is performed by applying pressure on an enclosed volume of mercury or water by means of two screws pressing against two thick rubber plugs set at the two basal ends of an h-shaped glass shank. The shank is filled with the mercury preferably after having been evacuated. The pipet shaft is sealed with De Khotinsky cement into the free tapering end of the shank. This pipetting system is both easily controlled and accurate. In a modification of this latter device, which was described by McNeil and Gulberg (167), a lever is employed and the screws are removed from the direct line of the micropipet. Thus the transmission of hand movements to the pipet is eliminated, and a better control of pressure is achieved.

Reyniers (192) devised a pipetting system which depends essentially on an apparatus in which air pressure can be built up slowly as required and released instantly. An air column is compressed by the weight of a regulated column of mercury fed through a variable opening. Immediate release of pressure is achieved by means of a specially designed push valve. The device offers satisfactory control of droplet formation or injection.

Chambers (61) used a thin-walled steel cylinder filled with mercury. The driving force required for injection was obtained by compressing the walls of the cylinder. The apparatus was sensitive, but it was rather difficult to secure cylinders with properly resilient walls.

The same author (54, 59) advocated the use of carefully selected Luer-type glass syringes of 1- to 2-ml capacity, filled with distilled water, for controlling delicate movements of a work fluid in the pipet tip. The syringe and pipet are connected by a fine flexible metal capillary tubing about 1 m long and less than 2 mm in outside diameter. A metal adapter joins the syringe to the flexible tubing, and a pipet holder joins the other end of the tubing to the micropipet. The pipet holder with the micropipet attached is carried and moved on a Chambers' micromanipulator, p. 31, 69. The syringe may be fastened to the base of the micromanipulator by means of a special socket or holder. The system is filled with distilled water, which has been boiled prior to use to eliminate air bubbles, to leave a small air gap between the hydraulic water and the injection or work fluid in the pipet tip. The plunger of the syringe may be controlled by hand, but a screw or lever feed may be preferable. The device is easily assembled and gives very satisfactory results. A number of pipet controlling mechanisms commonly used at present, Figs. 11, 28, 82, are based on the general principles used in this device.

The use of pushers of mercury in pipet controlling devices has certain disadvantages. This metal is very easily contaminated, and contaminations alter its cohesiveness and surface tension characteristics. Thus on applying suction to take up a fluid into the micropipet, the mercury column is liable to break. Also, when plunger systems are employed, mercury has the disadvantage of possible leakage around the plunger after a period of use. Nevertheless a plunger system using a pusher of mercury has been used fairly recently by Dorf (81) for titration of microgram samples in coloriscopic capillaries, Fig. 89. This plunger device is described in detail on p. 131.

Oil as a hydraulic fluid in pipetting devices is usually unsatisfactory since it tends to creep and contaminates the injection fluid and the apparatus.

A serviceable instrument for handling volumes of the order of 0.5 nl to several microliters is the drop-retraction unit, recently described by KOPAC (139, 142), which is suitable for comparatively crude extraction of cytoplasm from cells and for qualitative microinjections. The instrument consists of a coarse and a fine volumetric control. The coarse control, used for filling the pipets, consists of a 2-ml Luer syringe having a spring-loaded piston activated by a micrometer head. The fine volumetric control, serving to produce smaller volumes, includes a modified micrometer (205) in which the spindle is replaced by a suitable drill rod which displaces mercury in the steel jacket.

A small capacity microinjection system described by DE FONBRUNE (79) consists of a short tube completely filled with water. Heating the tube electrically causes the water to expand. This particular device suffers from external temperature changes which affect the system seriously. Another small capacity microinjector based on the same principle of the thermal expansion of water was described by WORST (242). It consists essentially of a fine, bent, thick-walled capillary tube which contains water. One end of the capillary is drawn out into a micropipet; the other end, which is closed, is immersed in a test tube containing warm water kept at a fairly constant temperature by means of a special heater. The required change in the volume of the hydraulic water is obtained by heating varying lengths of the closed end of the capillary by the warm water in the test tube. This may be done either by sliding the test tube on a slanting support, or by changing the level of the water in the test tube by means of a syringe operated by a special mechanism. The system is claimed to offer accurate pressure control and possibilities for quantitative injections.

A number of pipet controlling devices are described in this volume together with the particular assemblies with which they were actually used or with which they are supplied commercially. In most of these devices, water serves as the force-transmitting fluid. These systems are used with an air gap in the micropipet between the hydraulic fluid and the work solution in the tip of the pipet and prove quite satisfactory in practice.

An example is illustrated by Fig. 63, showing the device (90) used by EL-BADRY and WILSON for performing chemical operations in capillary cones. The device, p. 98, consists of an Agla micrometer syringe to which the micropipet is attached by means of a simple rubber joint. This micrometer syringe is available commercially (44) and can be easily adapted for the purpose. The micropipet, described on p. 99, has a stem to facilitate its attachment to the glass syringe. The micropipet receives a water repellent coating, p. 100, to eliminate surface tension effects from the pipet tip. This injection apparatus proved very satisfactory in practice. Though

12*

micropipets with tip bore of 20 to 50 μm in diameter were usually employed, pipets with much finer bores, down to a few micrometers in diameter, may be used with this device. The syringe with the micropipet attached is carried directly on the top movement of the micromanipulator.

Several other devices use water as the hydraulic fluid with an air gap between the hydraulic water and the work solution in the pipet tip. The syringe control, however, whether made of metal or of glass, is connected to the pipet holder by means of a long flexible small-bore tubing. Such flexible tubing may either be made of metal or of plastic. The pipet holder with the micropipet attached is carried on the micromanipulator, whereas the syringe is usually fastened to the common base plate or held firmly on a separate base. In all these general features, such devices resemble the injection apparatus originally developed by CHAMBERS (54), a brief description of which was already given.

Of these instruments, the plunger device, Fig. 80, used by BENEDETTI-PICHLER (25) for performing chemical operations in capillary cones is described in detail on p. 112. This device is an adaptation of one originally described by JOHNSON and SHREWSBURY (129). It has also been used by BENEDETTI-PICHLER and co-workers for working in titration cones (154), p. 129, as well as for performing chemical operations with hanging drops (31), p. 134. Commercially available (35, 152) injection devices based on the same general principles are shown in Figs. 27 and 50; one of them is described on p. 52, together with the lever-activated micromanipulator of E. LEITZ (152). Of the same type is the device, Fig. 144, p. 269, used by CADLE (46, 47) for performing chemical micrurgic operations on airborne particulate material collected on object slides, p. 267. Another commercially available microinjector (2) of more or less the same general type, but having a screw-syringe made of stainless steel, Fig. 42, is described on p. 65 together with the pneumatic micromanipulator of DE FONBRUNE. With such devices, micropipets with orifices down to a few micrometers diameter or even less may be satisfactorily used.

In micromanipulator work with comparatively large volumes of solution, simple syringe controls similar to that shown in Fig. 84e, proved very satisfactory. This simple hypodermic syringe, p. 119, is suitable for control of liquid flow in capillary pipets of the types shown in Fig. 84a to 84d, which are commonly used in milligram solution chemistry. Such devices have been employed by CUNNINGHAM and WERNER for the last stages of isolation of the first samples of plutonium and in the investigation of its properties, p. 261. In this work, assemblies of the type shown in Fig. 83, p. 116, were employed. The syringe with the micropipet attached was directly mounted on the micromanipulator and work was carried out under a wide field stereoscopic microscope in microcones, Fig. 85, p. 119, of comparatively large capacity.

Reference may also be made here to the exceedingly sensitive plunger device integrated with the volumetric submicromanipulator, Fig. 13, developed by KOPAC and HARRIS (147) for use in preparing measured volumes of substrate mixtures for microdilatometric studies (144). This highly specialized instrument is described briefly on p. 39. The plunger device has two steel pistons of different diameters and is connected to the pipet holder by means of a long flexible copper tubing. The system is filled with silicone oil for a hydraulic fluid. The micropipets are siliconized to render them water-repellent and to minimize the sticking of cytoplasmic residues to the glass. An electronic circuit permits pre-setting the instrument to the required volume which may be as small as $0.03 \text{ pl} = 3 \times 10^{-11}$ ml when the smaller piston is used.

b) Operation Chambers.

Chemical experimentation with microliter to nanoliter amounts of solution and the investigation of living cells and tissues suspended in suitable aqueous media require that the samples be protected against evaporation or drying out. For this purpose, work is usually carried out inside a suitable chamber, the top and bottom of which are made of glass to permit proper observation of the field of operation under the microscope. The sides of the chamber are lined with wet filter paper or cotton wool to furnish within the cell an atmosphere saturated with water vapor.

The moist chamber is mounted on the stage of the microscope which is provided with an attachable mechanical stage that grasps the chamber and is used for bringing the object of investigation to the required position in the field of the microscope. In certain manipulations the mechanical stage may serve also to bring the working specimen against the microtool.

For introducing the operating tools, the moist chamber is open either at two opposite ends or at one end only, depending on the type of work. Cells open at both ends, Fig. 28, are used when one or more micromanipulator units are mounted on each side of the microscope, Fig. 29. Chambers with a single opening, Fig. 67, are used when a single micromanipulator unit is employed, Fig. 63, or when two or more micromanipulator units are mounted side by side in a convenient position relative to the microscope so that the microtools may extend more or less parallel and close to each other into the moist chamber, Fig. 47, 50.

In the investigation of cells and tissues as well as in certain chemical work performed on droplets under high magnifications, the object to be operated upon is usually mounted in a hanging drop suspended from a thin cover slip serving as roof of the chamber. With this arrangement, the operating tools with their bent-up tips approach the sample from below. Since the microscope objective lies directly above the cover slip with no obstacle between them, the proper observation of the field of operation becomes possible with the highest magnifications and the use of oil immersion lenses. It is only necessary that the cover glass used for the top of the chamber be sufficiently thin to permit the use of high power objectives having a short working distance. With microscope objectives of $20 \times$ or more, the thickness of the cover glass is usually about 0.17 mm and should not exceed 0.2 mm.

For critical illumination, especially with high magnifications, the object of investigation should be close to the focus of the substage condenser. Thus for operating on hanging drop preparations, the height of the chamber that may be satisfactorily used is determined by the working (focal) distance of the condenser and also by the minimum working space which would permit proper manipulation of the operating tools. For this type of work, it is therefore preferable that the base of the chamber be fairly thin to obtain a greater distance for the internal height of the chamber. For best illumination the height of the chamber should be somewhat less than the focal length of the condenser. For ordinary work, however, it may be of advantage to use chambers of greater height, since a condenser functions in some degree also beyond its focal length, which may be sufficient for the purpose.

Various types of operation chambers have been designed for use in different fields of micromanipulator work. Several of these chambers are described in connection with the work to which they apply. These will be mentioned briefly.

For performing chemical operations in capillary cones with microgram to nanogram amounts of material under conditions of low power microscopy, p. 92, two moist chambers, open only at one end, are described. One of these chambers, p. 103, is shown in Figs. 66, 67. The bottom, sides, and back of this chamber are made of transparent colorless Leucite plates, and the cover is made of glass. The cell is simple to construct, spaceous, and light in weight. In Fig. 63, it is shown in position on the stage of the microscope as a part of the complete assembly. The other chamber, p. 104, Fig. 68, has its top, bottom, and back made of glass plates; the sides are made of brass rods, fastened to a brass frame cemented onto the glass plate forming the bottom of the chamber. Both cells are sufficiently large to accomodate a glass carrier for holding capillary vessels and a condenser rod for confirmatory tests. A bright-field condenser which can be focused on a plane about 1 cm above the microscope stage may be used with both chambers to illuminate capillary vessels mounted on the upper surface of the carrier. Both chambers may be used as dry chambers, p. 104.

The metal ring moist chamber of E. Leitz (152), p. 126, has been used for titration of microgram samples in titration cones, p. 125, as well as for performing chemical operations in hanging drops, p. 132. The chamber is shown in position on the microscope stage, Fig. 90, and is schematically shown in Fig. 137. The bottom and top of the chamber are made of very thin round glass plates. This permits the use of high power magnifications for working on hanging drops. The side walls of this circular chamber are formed of a metal ring provided with two openings opposite to one another. An outer metal ring, fitting around the inner one, has two opposite openings of different width so that by rotating the outer ring, the openings in the inner ring can be opened either simultaneously, Fig. 137, or independently. This type of cells can also be used for working under high magnifications with hanging drop preparations of living cells and tissues. A recent form of the chamber (152) is made of chromium-plated metal and can be easily cleaned and sterilized by boiling.

A moist chamber that has been used for titration of microgram samples in coloriscopic capillaries, p. 130, is shown in Fig. 89. As can be seen from the figure, the plunger device to which the titration vessel is directly attached is mounted in the moist chamber. The wide part of the device is cemented into the wall of the chamber by means of De Khotinsky cement, p. 132. The cell is opened on the opposite side for introducing the microburet.

A simple moist chamber, Fig. 28, supplied by E. LEITZ (152), is described on p. 53. The chamber is all made of glass and open on two opposite sides. The inside depth of the chamber is either 3 or 5 mm and its overall height is 7 mm. It is shown in position on the microscope stage, Fig. 27. The cell can also be used as an oil chamber, p. 185, to eliminate the interfering lens action of the hanging drop so as to permit proper phase-contrast and dark-field examination.

Another moist chamber, p. 60, supplied by C. ZEISS, Jena (245), Fig. 38, is particularly suitable for operations requiring uniform and high moisture content. The chamber, also shown in position on the stage of the microscope, Fig. 31, is provided with two small openings, opposite to one another. It may be sealed, if required, by means of two sleeves, Fig. 38, which connect the two openings of the cell with the tool holders.

Several other moist chambers have been described in the literature. Of these, a simple glass chamber of early design was devised by BARBER and described later by CHAMBERS (55). The cell is open at one end and is shown in position on the microscope stage in Fig. 10. The construction of a form of this chamber suitable for cytological purposes, was also described by CHAMBERS (54). The base of the modified chamber is made of a glass slide; the sides and back are made of strips of plate glass. Ordinary glass cement or heated Canada balsam may be used for cementing together the different pieces. A cover slip, 24 mm × 40 mm, forms the top of the chamber. A narrow strip of glass is cemented across the floor of the chamber near its closed end to provide a well which is filled with water. Strips of blotting paper are placed along the sides of the chamber with their ends in the water well. The open end of the chamber may be temporarily closed by a thin, paraffined cardboard trough which is placed over the shafts of the microtools and filled with soft vaseline containing some cotton threads to give it more rigidity. The chamber is thus completely sealed, and a high moisture content can be maintained within the chamber even when a preparation must be kept over night. The vaseline which closes around the shafts of the tools does not interfere with their movements. It may be prevented from spreading on the floor of the chamber by providing a shallow trough for the vaseline to rest upon. It may also be prevented from contaminating the hanging drop preparation by spreading a thin film of molten paraffin around the area to be occupied by the preparation and allowing the paraffin to solidify. Naturally, the above mentioned modern moist chamber of ZEISS which can be completely sealed by using connecting sleeves, Fig. 38, is far more convenient.

BARBER's hermetic chamber, described above, was developed (174) for work on cells in different gases. The open end of the chamber was closed by means of a trough of mercury through which the microtools passed into the chamber. The chamber was provided with an inlet and an outlet for the gas. There was one more inlet for a delivery pipet which served to deposit a hanging drop of the required solution after the chamber was flooded with the gas.

A moist chamber for working at a constant temperature between about 18° C and 48° C was described by RICHTER (194). Temperatures in this range can be obtained and maintained within the moist chamber with a maximum variation of about $\pm$ 0.3° C. The apparatus is constructed of pieces of sheet Plexiglas, having satisfactory thermal, electrical, and optical properties, and is welded together with Plexiglas solvent. The moist chamber proper is open at one side and bounded on the three other sides by a trough filled with tap water. Two brass electrodes extend for a short distance into the water of the trough. Heating is achieved by passing electric current through the tap water in the trough. The generated heat is distributed by circulation of the water within the trough mainly by convection currents and regulated by a gas-mercury thermostat. The chamber is suitable for use with any standard microscope.

A refrigerated micrurgical moist chamber devised by KOPAC (141) permits working at low temperatures which are limited only by the freezing point of the culture medium.

A chamber made of a Leucite body on a glass base is supplied by BRINK-MANN (35). The top of the chamber consists of two regular cover glasses, one placed behind the other, so that material can be transferred from one cover glass to the other.

A special micrurgical moist chamber adapted for working with drops lying on the floor of the chamber on an inverted microscope, Fig. 13, was described by Kopac (142, 144). The floor of the chamber accomodates two square 22-mm cover glasses, placed one behind the other. The chamber is open at both ends. The microtools enter the chamber through the front opening. The top of the chamber can be raised, and the two cover glasses may be inserted on the floor from the rear without disturbing the microtools. With this arrangement, comparatively large lying drops are used with the cells resting on the upper surface of the cover glass. Material can also be transferred from one cover glass to the other.

The use of moist chambers has the disadvantage that the lens effect caused by the suspended drop renders only a small portion at the center of the field of view satisfactory for phase-contrast and dark-field observations since such examinations require good control of the refractive conditions. Consequently oil chambers are employed mainly to eliminate this disturbing effect and to permit satisfactory phase-contrast and dark-field examinations. Furthermore, working with hanging drops under a cover of neutral oil makes cultures adhere to the cover glass and also prevents evaporation of the aqueous drops, see also p. 55. The use of an oil chamber in micrurgy has been developed by de Fonbrune (75). His chamber is quite suitable for routine use and for other problems.

A simple shallow culture chamber as that of Leitz (152), Fig. 28, p. 53, may be readily used as an oil chamber. For this purpose the center of the cover glass which forms the roof of the chamber is first wetted with the culture fluid. A sufficiently large drop of neutral double distilled paraffin oil is then placed over the moistened area so as to cover it completely and to come in contact with the surrounding glass surface. The drop of culture fluid soon spreads and flattens due to the pressure exerted on it by the oil layer. Additional amounts of culture fluid are then injected by a pipet at the moistened area between oil and glass until the entire wetted area is occupied. Sufficient oil is then added to fill the whole empty space of the chamber below the original oil drop. It is possible to renew the culture continuously and supply it with oxygen or nutrients such as glucose by means of capillaries. The technique largely eliminates the lens effect of the hanging drop and has the additional advantage that the culture adheres to the lower surface of the cover glass.

In a simple method for single cell isolation, described by Lederberg (151), a clean cover glass or microscope slide, ruled on one face in 4-mm squares with India ink, may serve as a substitute for an oil chamber. The other face is sterilized in a flame, allowed to cool, and coated with a layer, about 0.5-mm thick, of paraffin oil. White Mineral Oil, USP, may be used. In simple cases not requiring micromanipulatory aids, a capillary pipet of about 100-µm bore, connected to a rubber tubing, and controlled by mouth may serve to deposit a small drop of the properly diluted microbic suspension under the oil layer at the center of each square. The suspension soon spreads and flattens, thus providing satisfactory optical conditions for the examination and the search of each drop by phase-contrast, dark-field, or low aperture microscopy. A mouth-controlled capillary pipet may also be used to recover the selected microbes that may exist singly in the examined drops. To develop large clones from single cells, the slides may be incubated in a simple container of oil, e. g., a staining dish. For prolonged incubation, an oil chamber consisting of a rectangular well built on a glass slide may be used. The well is filled with oil and the cover glass preparation is inverted over it. Single cells and large clones thus developed may be transferred wherever required by using mouth-controlled capillary pipets, preferably drawn out from silica tubing since these, after rinsing, can be sterilized quickly in the flame. More complicated single cell analyses, how-ever, require micromanipulatory aids.

Part II

General Techniques.

A variety of basic micromanipulative techniques will be described, which have been actually used for working with different types of materials and test objects. Certain other techniques will also be described, which in themselves do not involve micromanipulation, but may serve for the preparation of the sample or the object of investigation for subsequent examination by appropriate micromanipulative methods.

These techniques, whether involving or relating to micromanipulation, will be considered within the following general categories:

(1) preparation of samples of non-biological materials,

(2) chemical experimentation with micrograms to nanograms,

(3) working with living cells and tissues.

Other basic micromanipulative techniques, not considered under these three headlines, are dealt with in appropriate places elsewhere in the present volume. Of these, an account of the methods and techniques used for testing microhardness is given earlier in connection with the microhardness testers, p. 82, and illustrated in the part on applications, p. 280. A detailed description has been given already of the various basic operations and techniques involving micromanipulation, which are used for preparing a wide variety of fine microtools under microscopic control using the DE FONBRUNE microforge, p. 152. The essential features of a number of other micromanipulative procedures, also not considered in this part, are illustrated in certain examples given in the part on applications. These demonstrate still further the great versatility of micromanipulative techniques as well as its possibilities.

1. Preparation of Samples of Non-Biological Materials.

Micromanipulative methods are widely used for sampling and investigation of a great variety of material. In many instances, the amount of sample that may be available for investigation and which may be extracted from localized areas in metallurgical products, mineral grains, and other material may be only micrograms to nanograms in mass. Even when comparatively larger samples are available, the constituents of interest may occur only in nanogram amounts. Consequently, exceedingly sensitive methods are often required for sampling and investigation.

The choice of a proper sampling method depends mainly on the nature of the material of interest and of the surrounding medium and also upon the method to be used subsequently for investigation. In most instances, an effort is made to separate the constituents of interest free from extraneous matter and in a form suitable for subsequent examination by appropriate chemical or physical methods which in themselves may or may not involve micromanipulation. Examples of these methods of examination are the capillary cone techniques which may be used with the solution of the sample, p. 208, the chemical micrurgic techniques where tests are performed on individual particles on a glass slide, p. 267, and other suitable physical methods such as microspectroscopy, p. 275, and x-ray methods.

In dealing with massive solids, p. 189, the constituents of interest are usually various microscopic heterogeneities such as tiny inclusions, single grains, very thin individual layers, and small defective or decomposed areas. For the investigation of such constituents, the most reliable sampling method is based upon the separation of these constituents individually, or of an appropriate sample thereof, from the surrounding matrix by mechanical methods. The extracted debris is then collected for examination by an appropriate method as described in connection with loose particles p. 198. Core sampling of paint films which may be composed of several layers of paint may be performed, p. 196, by means of a cut-off hollow hypodermic needle mounted on a micromanipulator and operated as a tiny cork borer. In sampling powders and loose particles, p. 198, the small sample may be spread on a glass surface under the microscope. Particles of interest may be picked out individually with the properly wetted or greased tip of a micromanipulator-mounted glass or silica fiber. Magnetic fractions in the finely ground mixture may be separated, p. 200, Figs. 122a to c, by suitable micromagnets also mounted on a micromanipulator. Collodion may be used for the transfer of loose debris for spectrographic arc analysis, p. 193, Fig. 117. Fine solids suspended in liquids may be centrifuged, dried, spread on a glass surface, and handled as described for powdered materials. For the investigation of various solid particles and liquid droplets suspended in the atmosphere, samples of these materials are usually collected on microscope slides using methods, p. 202, such as impaction, thermal precipitation, and settling. These and other sampling methods not involving micromanipulation are also described, p. 189, since individual particles collected by these methods may subsequently be investigated on the glass slides by chemical micrurgic techniques, p. 267.

The sampling methods, and other operations used in the preparation of samples for subsequent investigation by micromanipulative methods are described under the following main headlines: (a) sampling of microscopic heterogeneities and thin individual layers in massive materials; (b) core sampling of paint films; (c) sampling of powders, loose debris, and insoluble particles suspended in liquid; (d) sampling of airborne particles; and (e) opening of sample.

a) Sampling of Layers and Microscopic Heterogeneities in Massive Materials.

Techniques are considered which may be used in the collecting, mainly by microdrilling, of small heterogeneities, tiny inclusions, and single grains in massive materials such as minerals and rocks, metals and alloys, or slags and refractories. The same general techniques are also applicable to microsampling of very thin individual layers in solders and to other types of investigations.

α) General Methods.

For the investigation of small inclusions and other heterogeneities in massive material, the objects of interest are usually removed from the surface of the properly mounted specimen and subjected to appropriate analysis. The material to be investigated may be separated from the matrix either by mechanical or chemical means.

In the chemical methods of separation, reagents are usually applied which selectively dissolve the matrix material. The inclusions are left behind as loose debris which is separated from the reagent and gathered for analysis. The procedure has the disadvantage that the included material also may suffer some chemical change. Furthermore, there is usually no assurance that the separated residue is not an agglomerate containing some of the matrix material. Alternatively it may be possible to get the inclusion into solution by adding a solvent that will selectively attack the inclusion and not the matrix. Mechanically guided micropipets with inclined tips are useful for applying the reagents to the required localized area on the surface of the specimen and for the removal of the reaction products under microscopic observation.

Electrolysis also may be used for sampling small areas in metallic surfaces. Koch and associates (138) described a small electrolytic vessel in which anodic dissolution of the surface area of interest can be performed under a stereoscopic microscope. The specimen, surrounded with a platinum wire, is mounted in a special holder and its surface is coated with paraffin so that only the area to be electrolyzed is exposed. The holder with the mounted specimen is placed in the electrolytic vessel which is filled with the electrolyte. The operation is observed through the glass cover of the cell. This method also may not exclude the possibility of contamination by the surrounding material.

Isolation of inclusions by mechanical methods has the distinct advantage that the possibility of chemical change is eliminated and that contamination by adjacent material may be kept to a minimum.

If the inclusion of interest is loosely adhering, gentle probing by means of a pointed stylus may be sufficient to separate the inclusion from the matrix. Fine, pointed glass or silica needles may be used as well as suitable metal needles such as steel sewing needles of small size. With small inclusions, the needle is usually mounted on a suitable micromanipulator and the operation is performed under the microscope. With relatively large inclusions, the probe may be manipulated manually, while the hand is firmly supported on a rest.

Surface replicas are frequently used for the removal of loose or weakly attached particles from metallurgical and other materials (67, 182). A solution of plastic or resin is applied to the surface to be investigated. The solution is allowed to dry, and the flexible film is then pulled off the surface of the specimen. The particles embedded in the negative replica obtained can then be studied by appropriate chemical or physical methods. This technique of sample selection is particularly useful since replicas retain information on the original location of the removed particles.

Negative replicas are inverted reproductions of the original surface, and a wide variety of materials is used for making replicas. Collodion, ethylene dichloride, and polyvinyl alcohol in water are some examples. Furthermore several variations are possible in the replica techniques and in the methods of investigation of the removed materials.. It may be possible for instance (67), to remove selectively particles adherent to the surface of the specimen by bonds of various strengths by properly varying the replica material. It may also be possible to use the technique of dissolving the substrate material through the replica by applying an appropriate etching reagent. This leaves any material not attacked by the reagent adhering to the replica where it can be investigated. The replica may be studied microscopically and selected areas thereof may be investigated by suitable microchemical methods such as standard chemical microscopical and micrurgical techniques or by physical means such as x-ray diffraction, electron diffraction, and microspectrography.

Strongly adhering inclusions and other heterogeneities in hard massive material such as minerals and rocks, metals and alloys, slags and refractories may require appreciable force to separate them from the surrounding medium. In comparatively large scale work (64), when the inclusion of interest is not too small, tools used by dentists such as excavators and chisels will serve. Hardened steel sewing needles may also be taken. Writing diamond is valuable for chiseling very hard material. Frequently, as is described below, the sample to be removed from a hard matrix is only micrograms to nanograms in mass. In such instances, the use of microdrills becomes necessary.

β) Samples of Hard Materials by Microdrilling.

Tiny inclusions, micrograms to nanograms in mass, may be only a few tenths of a millimeter to about 10 μm in diameter or less. If they occur in a hard matrix, the use of microdrilling machines is invaluable. Microdrills may also be useful for taking tiny samples from extremely thin individual layers of solders and multiple platings.

When a metal surface is chemically attacked, as for example when a steel tool is attacked by oven gas, different reaction zones are formed. Koch *et al.* (138), used special ball-type microdrills, Fig. 116, mounted on a micromanipulator, for sampling such reaction zones individually under a stereoscopic microscope. The removed samples were then identified separately.

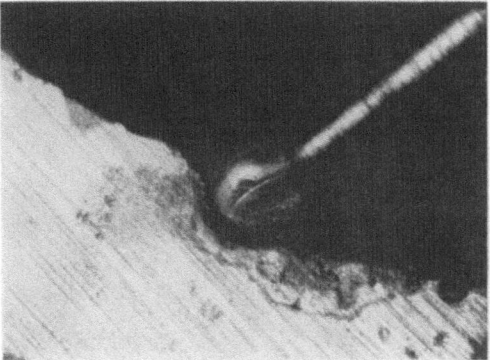

Fig. 116. Removal of a Reaction Zone in an Attacked Metallic Surface. Courtesy of W. Koch.

Small objects are usually mounted and prepared as metallographic specimens, and the samples are taken out from their polished surfaces. The depth and size of the resulting cavities can be controlled. Depending upon the method to be used for analysis, the drilled-out debris may be transferred for examination on the tip of a micromanipulator-mounted fiber, coated with a suitable adhesive, or by means of a suitable stripping film.

In the following, techniques and basic operations involved in the removal of microgram' to nanogram samples by microdrilling are described mainly with reference to the use of the two commercially available microdrilling machines, namely the Najet microdrilling machine, and the Ultrasonic Jack Hammer.

The Use of the Najet Microdrilling Machine. The Najet microdrilling machine Model 7*A* (173) is described on p. 74, and shown in Figs. 51 and 52. As has been already mentioned, the machine operates microdrills of the pivot type ranging in diameter from about 6.4 μm up to about 0.35 mm. The instrument was used by Bryan and associates for extracting microgram (41) to nanogram (197) amounts of samples from metallurgical materials. After drilling out a sufficient quantity of the con-

Fig. 117. Photomicrographs Showing Mounted and Polished Cross Section of a Silver Solder Sheet (consisting of center strip of copper about 125 μm thick between two thinner layers of silver). Drill hole for sampling copper strip made by means of a 75-μm drill operated by a Najet microdrilling machine. *a*, metallic drillings scattered around hole in the copper strip; *b*, drillings embedded in collodion ready for transfer; *c*, drill hole after stripping collodion and chips. Courtesy of F. R. BRYAN.

stituent of interest, the separated material is transferred with collodion to a pure carbon electrode for vaporization in a standard spectrographic arc. The technique permits the use of conventional, commercially available spectrographic equipment for the identification of microconstituents of metallurgical specimens and does not require the specialized apparatus used in earlier work (40, 125, 772, 201, 227).

Figs. 117a, b, c, illustrate the microsampling procedure (41) with the Najet microdrilling machine. The photomicrographs show the mounted and polished cross section of a silver solder sheet. The drilled hole is in the about 125 μm wide center strip of copper between two layers of about 75 μm silver. The hole is made with a microdrill about 75 μm in diameter. After drilling out a sufficient quantity of material, Fig. 117a, the resulting chips are covered with a drop of 10% solution of flexible collodion in butyl acetate. Fig. 117b, shows the metallic drillings covered with the collodion droplet. After a few minutes the dried film with the embedded drillings is ready to be pulled off the surface of the specimen and transferred to the tip of a suitable electrode for analysis. Fig. 117c, shows the drilled hole after removal of the metallic chips with the collodion. By arcing the collodion and chips in the electrode, the simple metallic spectrum of pure copper was obtained. Organic stripping films other than collodion may be suitable, and solvents other than butyl acetate may be used to hasten or to retard the rate of evaporation. Care should be taken, however, that the material does not contribute to the metallic spectra. Alternatively (197) the chips may be picked up with a greased fiber. The electrodes are either spectrographically pure graphite or ultra pure metal (197). The tip of the electrode is moistened with a drop of butyl acetate or a bit of Duco cement to help adhesion. Figs. 151 and 152, p. 276, and Fig. 153, p. 278, illustrate the application of the microsampling technique with microdrills to different specific metallurgical problems.

The constituent to be removed is usually first located under a metallographic microscope before the specimen is transferred to the microdrilling machine. In the drilling machine, the area of interest is first brought into the microscopic field and focused. For precise control of the location of the hole to be drilled, both the area of interest on the surface of the specimen and the tip of the drill should be in sharp focus at time of contact. The specimen should have been mounted so that the bottom of the mount is flat. This is to make the metallographic specimen sufficiently stable and to obtain the necessary friction which will prevent rotation of the specimen under the drill. The surface of the specimen should be at right angles to axis of rotation of the drill. This will prevent sliding and bending of the drill on contacting the specimen. It is advisable to use relatively low spindle speeds so that the resulting drillings accumulate close to the drilled hole. This also has a favourable effect upon the life of the drill.

In sampling inclusions, the drill is usually operated to a depth equal to the drill diameter to avoid contamination from underlying material. The diameter of the hole is usually equal to or slightly larger than the diameter of the drill employed. The surface area to be removed and the proper drill size can be determined by observing the specimen under a known magnification. If the constituent of interest occurs in relatively large accumulations, it may be possible to get a sufficient amount of sample from a single hole in an individual localized area. If the constituent is present only in very minute areas, a sufficient amount of representative sample may have to be obtained from a series of holes of small diameter.

The metallurgical microsampling technique using microdrills was first applied in the field of metallurgical microspectroscopy by BRYAN (39), and later by BRYAN and NEVEU (41) for qualitative analysis of micrograms of drillings with standard commercially available spectrographic equipment. Microdrills ranging from about 25- to 125-µm diameter were used.

A more recent article by RUNGE and BRYAN (197), p. 279, reports the application of the general technique to the removal and identification of nanogram amounts of metallic constituents by using microdrills as small as 6.3 µm in diameter. With a large Littrow quartz spectrograph, arc excitation, and the relatively fast EASTMAN 103-0 emulsion, spectrographic detectability limits of the order of one nanogram were obtained for several metallic elements. Such detectability limits were determined by arcing single metal particles of measured size and estimated weight. For iron, aluminium, copper, chromium, and lead, detectability limits were reached with individual particles of these metallic elements of 7- to 9-µm approximate diameter and 1 to 3 ng in weight depending upon the particular element concerned.

For the removal of a given weight of a particular constituent, the size of the drill is selected accordingly. Of course, the drill size may be limited by the surface area of the constituent. In the instance of iron, it has been calculated. (197) that 12.0 µg, 0.770 µg and 0.095 µg of the metal are removed by microdrills of 127.0-µm, 50.8-µm and 25.4-µm diameter, respectively, if the drill is taken to a depth equal to its diameter. In general, both the microsampling from surface areas 25 µm in diameter or larger and the identification of the drillings are simple to achieve. Both operations are also practicable, though difficult, for surface areas of about 10-µm diameter. On the other hand, the removal of minute quantities of relatively hard metallic constituents with a mass of one to a few nanograms, that is, of the order of the spectrographic detectability limits above indicated, comes too close to the limits of the drilling technique and also closely approaches the limits of sample manipulation. Thus for the removal of an estimated weight of 2 ng of iron, a microdrill of about 6.3-µm diameter would have to be taken to a depth equal to the drill diameter.

Microdrills of lesser diameter are too fragile for sampling metallurgical and similarly hard materials. In addition, the identification of smaller amounts of metallic elements is not practical with conventional spectrographic equipment, though it may be possible to accomplish by using faster spectrographs.

For certain sample compositions, the microspectrochemical procedure used by BRYAN and NEVEU (39, 41) may give semi-quantitative information with the aid of microphotometric measurements, whereas the very sensitive procedure described by RUNGE and BRYAN (197) gives only qualitative results. The general technique, however, has been extended by FELDMAN (98), p. 279, to give quantitative results by means of solution spectrochemical methods and usually using the porous cup technique (97).

For extracting an inclusion (138) a number of tiny holes may be drilled into the matrix close to the inclusion of interest until it gets loose from the matrix. In certain cases, for example when the inclusion is sufficiently thin or weakly attached, one such drill hole may be sufficient to set it free.

The Use of the Ultrasonic Jack Hammer. The other commercially available machine (35), p. 78, Figs. 55 and 56, is the Ultrasonic Jack Hammer. The instrument (132) is suitable for microscopically controlled mechanical isolation of microconstituents from metallographic specimens. It is capable of removing inclusions of 10-μm diameter or larger by means of a pointed stylus oscillating at ultrasonic frequencies. The machine is well suited for removing hard, brittle, metallic inclusions or secondary alloy phases and hard glassy or loose fluffy non-metallic inclusions. On the other hand, direct removal of relatively soft and ductile inclusions is difficult. Such material is not freed or shattered, but is rather cold worked *in situ*.

The extraction assembly is shown in Fig. 55. It consists mainly of the micromanipulator mounted transducer horn assembly, Fig. 57, to which the pointed stylus, Fig. 58, is attached. The micromanipulator permits accurate placing of the stylus point under the microscope. It also enables control of the depth of penetration and the size of the resulting cavity. A second micromanipulator, on the other side of the microscope, carries fibers for collecting and transfer of the resulting inclusion debris. A glass hood of a special design, shown over the metallographically mounted and polished specimen on the stage of the microscope, permits circulation of purified argon gas during operation to prevent air oxidation of the inclusion material, which may be caused by the heat generated at the tip of the vibrating stylus. If the inclusion material does not suffer oxidation, the glass hood may be omitted. In this case, the refracting objective may be replaced by a reflector which offers a larger working space.

The inclusion to be removed is first brought into focus. By operating the proper controls of the micromanipulator carrying the transducer horn assembly, the point of the vibrating stylus is brought over the inclusion

13*

in the field of the microscope and is carefully lowered to bring the stylus
point into contact with the inclusion. To permit location of the stylus
point with precision, the inclusion should be in sharp focus at the time of
contact. The inclusion is shattered at the point of contact, and the resulting
debris is scattered in the vicinity. The size and depth of the resulting
cavity within the inclusion are controlled by manipulating the stylus point
until a quantity of sample, judged to be sufficient for identification by
x-ray diffraction or other appropriate physical or chemical methods has

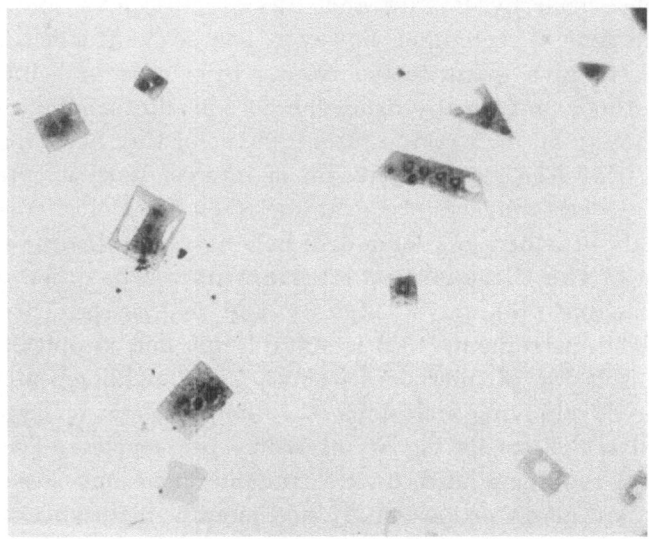

Fig. 118. Uranium Matrix Containing Inclusions of UC. Shown are well defined cavities
within sampled inclusions and scattered inclusion debris. Unetched. Courtesy of G. L. KEHL.

been separated. The resulting loose fragments are collected by means of
glass or silica fibers coated with a thin film of grease or a suitable adhesive.
The pick-up fibers are mounted on a micromanipulator usually placed on
the right side of the microscope, Fig. 55. The debris may also be transferred by means of collodion or other suitable organic stripping films, p. 193.

The technique is suitable for microsampling of a large variety of inclusion material. Evidence of precision of the technique is demonstrated
by the four photomicrographs, Figs. 118 to 121, which show the position
of the resulting cavities within the inclusions from which samples have
been removed. The scattered debris can also be seen.

b) Core Sampling of Paint Films.

Tiny cylindrical core samples of paint films can be taken by means of
a hollow hypodermic needle mounted on a suitable micromanipulator.
The needle is manipulated as a cork borer under low power stereomicro-
scopic observation. For this purpose, GETTENS (106) described a mani-

pulating device operating a cut-off hypodermic needle for taking core samples of valuable pictures and panel paintings (107). This non-destructive sampling method takes amounts of material which are hardly visible to the naked eye. Yet any such sample from a given locality is capable of yielding information concerning the structure and composition of the various base and coating layers, since it shows any stratification that may exist in the paint film.

The hollow needles used by GETTENS are similar to those previously described by LAURIE (149, 150) who manipulated them by hand. The

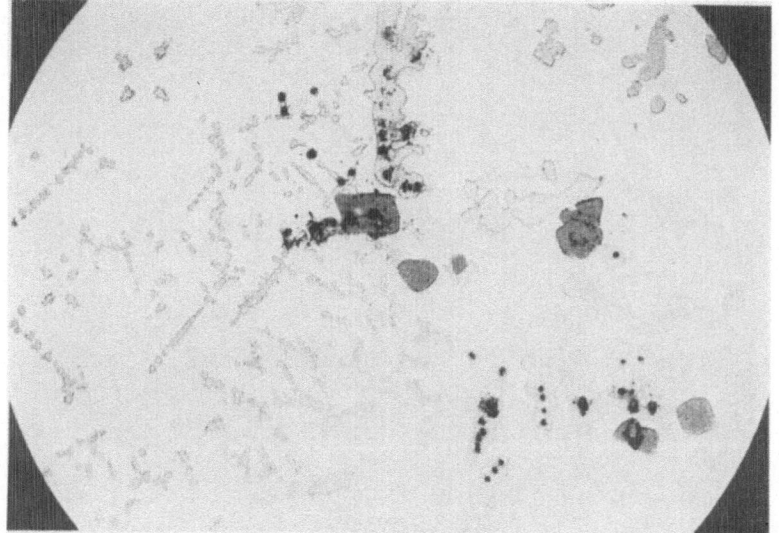

Fig. 119. Inclusions of UO₂ (dark) and U (C, N) (light) in a Uranium Matrix. Sampled inclusions show small areas where material has been removed. Unetched. Courtesy of G. L. KEHL.

needle is about one half inch (12 mm) long, and its point is sharpened on a small hone to form a long, tapered cutting edge. Needles as small as 0.53 mm in outside diameter and 0.29 mm in inside diameter are successfully used. The manipulating device is attached to the nose of a Greenough stereoscopic binocular microscope so that the point of the needle is at the focal point of the microscope. The device consists essentially of two concentric tubes and an inner rod. The hollow needle is mounted on the tapered end of the inner tube which can be freely rotated within the outer tube by means of a knurled knob. A fine wire fitting smoothly into the hollow needle serves to eject the core section cut by the needle and can be thrust forward by means of a screw. The vertical and horizontal positions of the microscope carrying the manipulative device are adjustable within a wide range so that the instrument can be shifted from one part of the large object to another.

For taking a sample, the object is mounted with the paint film in a
vertical plane, and the point of the needle is brought accurately opposite
the locality to be sampled. The needle should be in a horizontal plane,
i. e., perpendicular to the surface of the specimen. In this position the
microscope is inclined to the surface of the object at an angle of nearly 45°,
but this inclination presents no difficulty due to the adequate depth of
focus and the stereoscopic vision. The core sample is obtained by bringing
the needle into contact with the film and rotating it by turning the proper
knob. After the cut is made, the needle is withdrawn by turning the knob

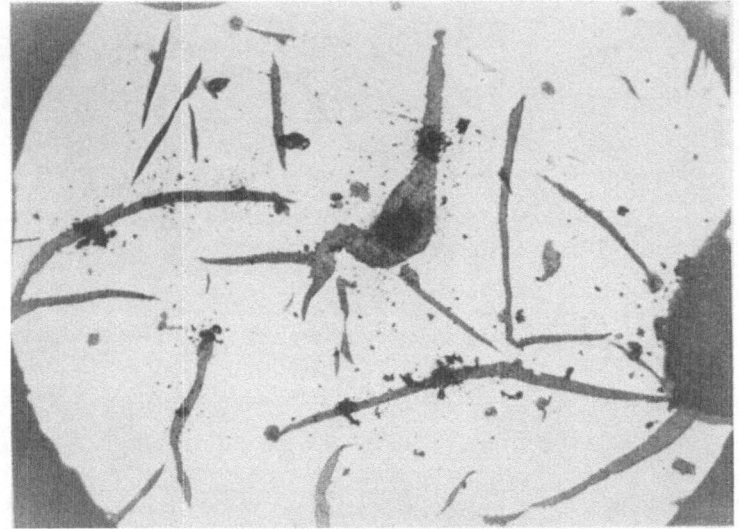

Fig. 120.   Cast Iron Matrix Containing Partially Removed Graphite Flakes.   Unetched.
Courtesy of G. L. KEHL.

in the reverse direction. The instrument is then moved away from the
object, and the core section may then be ejected into a depression ground
into the surface of an object slide which is mounted on a special holder
under the needle. When the object slide is detached from the holder,
a cover glass may then be sealed over the cavity containing the core sample.

Core sampling of old and brittle paint films may present some difficulties
since the film may crumble. This difficulty may be overcome by using
sufficiently sharp needles and slightly moistening the spot from which the
sample is to be taken with a suitable organic solvent (e. g., xylene) a few
minutes before cutting out the core. The tiny hole from which the sample
has been extracted may be rendered invisible by applying with a pointed
brush a small amount of a thin varnish or a solvent.

c) Powders, Loose Debris, and Particles in Liquids.

The choice of suitable methods for the separation and transfer of loose
particles depends mainly on the size and nature of the particles of interest

and also upon the method to be subsequently used for examination. Insoluble particles suspended in liquids may be collected with the centrifuge, dried, spread on a glass slide, and separated as described for loose debris.

Microscopic inclusions mechanically removed from the surface of a specimen or a sample extracted in the form of loose debris from a specific spot in the surface of a test object by microdrilling, p. 191, may be transferred quantitatively for examination. If separation of different components of a finely divided sample is not required, the loose particles may be trans-

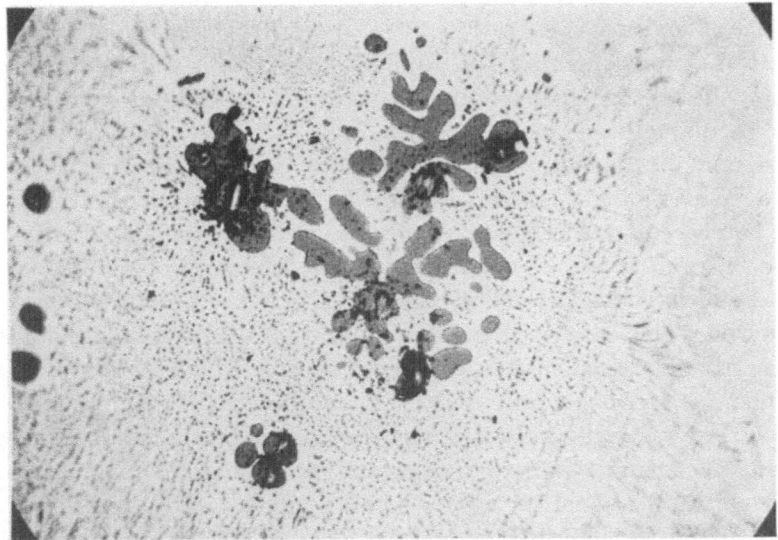

Fig. 121. Copper Oxide Phase in a Matrix of Copper. Shown are small cavities in the primary phase made by the vibrating stylus. Unetched. Courtesy of G. L. KEHL.

ferred with a suitable micropipet or by collecting them on the tip of a micromanipulator mounted needle or fiber. Alternatively the sample may be gathered on a suitable organic stripping film, p. 193.

If certain components of a powdered material have to be investigated separately, particles of interest are sorted out and isolated from the mixture. Particles to be separated may be recognized under the microscope by certain characteristics such as color, crystal form, cleavage, and other physical and optical properties. Individual particles may be picked up by means of suitable microtools mounted on a micromanipulator. Specially designed microforceps have been described (138) for picking up relatively large particles having a minimum diameter of about 100 μm. Fine, pointed metal needles and glass or silica fibers, with tips suitably wetted or coated with a thin film of grease or other suitable adhesive, may be used for the transfer of smaller particles. Such an operation is described on p. 221, and is illustrated by Fig. 139. To facilitate separation of the picked-up particle from the transfer needle, the point of the needle should not be much larger than the diameter of the particle. Very fine fibers made of glass or vitreous silica are easily prepared and are most suitable for the transfer of comparatively small particles.

Suitable micropipets mounted on micromanipulators may also be used for sucking up and transfer of individual particles so distributed on a surface (glass slide under the microscope) that they are sufficiently separated from one another. The micropipet may be connected to a microinjection apparatus or to a rubber tubing with mouth piece. The technique is illustrated in Fig. 163, in connection with its application for picking up and single mounting of microfossils, p. 293. A specially designed suction device mounted on a micromanipulator and connected to a water pump has also been used (138) for lifting and transfer of individual particles below 0.4 mm in diameter.

Difference in magnetic susceptibility of the components of a finely powdered material may serve to separate them. For the investigation of various heterogeneities in steel and other metallurgical materials, Koch et al. (138) described the use of a permanent micromagnet of a suitable power and a needle-shaped microelectromagnet, p. 175, both mounted on a micromanipulator. They are used for the separation of ferromagnetic and non-magnetic fractions in the finely powdered sample which is spread on a glass slide under a stereoscopic microscope. The sample is first removed in a finely divided form by using special microdrills mounted on the micromanipulator. The permanent magnet serves to separate the bulk of the magnetic fraction from the mixture. To this end, a small fragment of a thin cover slip is made to adhere to the lower end of the magnet by means of a small drop of glycerol. The magnet with the cover glass attached is then moved around in a horizontal plane located 1 to 2 mm above the mixture. The magnetic particles of the powder jump to the underside of the coverslip. The magnet is then turned upside down, and the coverslip carrying the magnetic particles is lifted off. The operation is repeated several times until the bulk of the ferromagnetic fraction has been collected. The micro-electromagnet is then used to separate the last traces of the weakly magnetic particles from the non-magnetic residue. The electromagnet operates with ac current having a maximum intensity of 17 milliamp. passing through a coil of 0.05-mm diameter wound round the magnetic needle. The current intensity is read on an amperemeter. To separate an individual particle from the mixture, the coil is supplied with current of maximum intensity, and the particle is approached with the magnet needle. The point of the needle is then moved to a clean area of the glass slide, and the needle is demagnetized by switching off the current. Fig. 122a to c, shows magnetic carbides and non-magnetic spherical oxide in a finely divided mixture before and after separation with the technique described.

d) **Sampling of Airborne Particles.**

Sampling and investigation of solid particles and liquid droplets suspended in air play an important rôle in many fields of study of the earth's

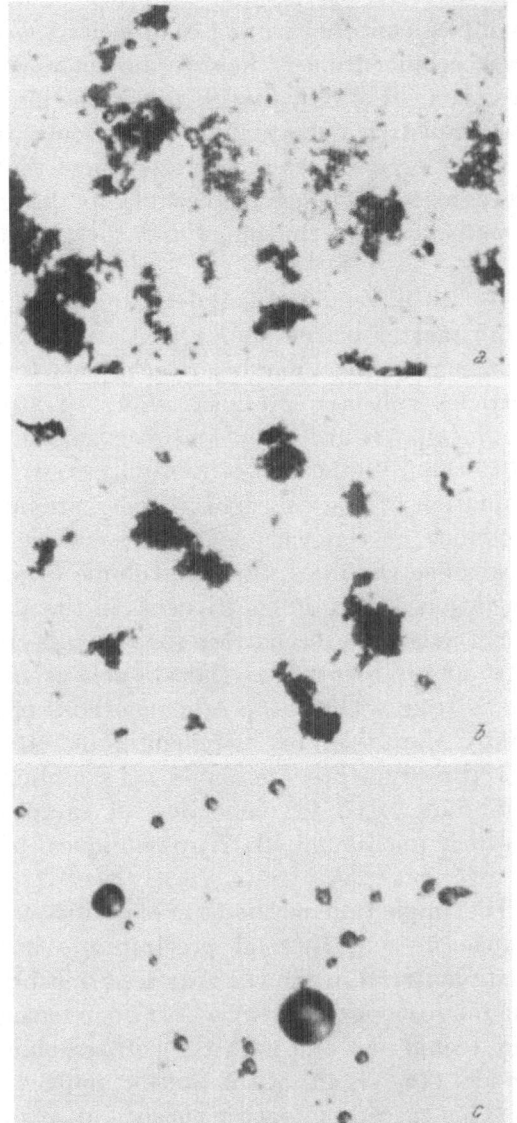

Fig. 122. Separation of Magnetic and Non-Magnetic Fractions in a Powdered Sample.
a, sample before separation; *b*, separated magnetic carbides; *c*, spherical non-magnetic oxide.
Courtesy of W. Koch.

atmosphere. Certain particles are held responsible for the condensation
and precipitation phenomena in the atmosphere. The study of the
distribution of certain components suspended in air may enable establishing

the boundaries and the origin of certain air masses (212). The determination of size (46) and composition of particles collected from the atmosphere have aided greatly in establishing the nature of the material primarily responsible for the decrease of visibility by *smog*. Such material, as for instance in the samples collected from Los Angeles *smog* (196), may consist mainly of hygroscopic droplets having diameters of about the wave length of light. In the field of industrial hygiene, the investigation of dusts released by certain industrial operations is of particular importance. Such dusts, especially the finer particles in the size range that may be inhaled (46), are often recognized as being responsible for health hazards which may have serious effects upon the lungs and the physiological action of the body.

CADLE *et al.* (47, 48) have investigated the nature and composition of airborne particulate matter in the smog-laden atmosphere of Los Angeles County, USA., by using chemical micrurgic techniques for the identification of individual particles collected on glass slides, p. 267. Other optical microscopic techniques (153, 212, 213), and electron microscopy (48) have also been applied to the investigation of these materials, p. 272. Adequate microscopic examination of the collected samples provides information of the relative abundance of the various constituents of the sampled air. If the samples are collected from a known volume of air, information on the actual concentration of certain constituents in the air can be obtained.

A number of methods have been tried for the collection of particulate material from the air on microscope slides, such as impaction, thermal precipitation, and settling. These and other methods of sampling are discussed at length by CADLE in his excellent book (46) on particle size determination. Naturally, these methods of sampling do not involve micromanipulation, but serve for collection of samples for subsequent investigation in which micromanipulative techniques, have actually been used, p. 267.

Impaction: The impaction methods are usually very convenient for general use. Impactors and thermal precipitators have been used for sampling particulate material in the Los Angeles atmosphere. The particles were collected on microscope slides in a form convenient for study with chemical micrurgic techniques and with the petrographic microscope. The use of a plastic model (46, 47, 48) of the SONKIN impactor (218) was found very convenient. This cascade impactor consists of sets of jets and microscope slides arranged in series as illustrated in Fig. 123. The body of the chamber was made of three pieces of machined Lucite. The jets, made of steel, were set in place, and the Lucite pieces were cemented together. The stream of air in which the particles are suspended is forced through the jets and impacts sharply upon the microscope slides, whereupon the particles cling to the slides.

When the aerosol is drawn through a number of jets of progressively decreasing size, the speed of air and impact increases progressively and a rough classification of particles according to size can be obtained. The largest particles are collected on the slide opposite to the outlet of the first and widest jet; the smallest ones, on the last slide opposite to the last and narrowest jet. In the model above described, the sizes of the three jets were 8, 3, and 1.5 mills, i. e., 0.2, 0.075, and 0.04 mm. A motor or hand-operated vacuum pump may be used for drawing the aerosols through the jets. Impactors are highly efficient samplers even for particles in the submicron size range. For particles as small as 0.6 μm, an efficiency of about 100% has been obtained.

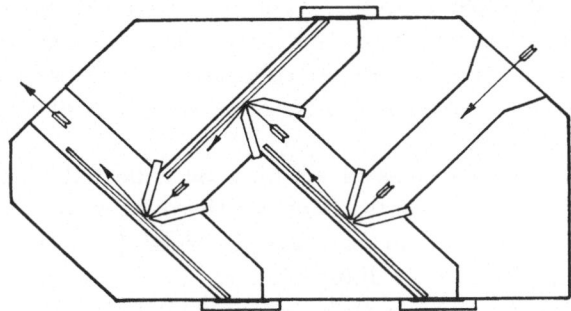

Fig. 123. Diagram of Path of Air Through a Three-Stage Plastic Model of Sonkin's Impactor. After R. D. CADLE (46).

For collection of very fine particles, less than about 2 μm, the glass slides need not be coated with an adhesive since such particles are held firmly on the dry glass surface, once they are deposited. In this instance, an adhesive coating does not increase the sampling yield. For effective adhesion of larger particles, the glass surface should be coated with a thin film of an adhesive material such as glycerine or vaseline. SEELY (212, 213) collects the airborne material by impaction on a gelatine-glycerol coating previously impregnated with a reagent specific for a single ion and thus obtains a modified spot test technique for the detection and identification of particular ions of interest.

CADLE (46) described also a simple single-stage impactor, and another continuous sampling impactor. The latter has a single jet built into the side of an evacuated vessel so that the jet faces a rotating drum. For microscopic work, a cellophane strip, wrapped about the drum, is used for collection of sample. After collection, the strip is removed and cut into suitable lengths which are mounted on glass slides.

Thermal Precipitation: The thermal precipitators have also proved suitable for collection of airborne particles on glass slides. The functioning of such devices depends on the fact that particles suspended in air are repelled when they get close to a hot object. CADLE and associates (46, 48) used a thermal precipitator (51) in which air is drawn through the gaps between an electrically heated tantalum strip or wire and glass slides placed close to the wire in such a manner that repelled particles are driven toward the slides. This instrument is highly efficient for collecting fine particles, even such of the submicron range, but has the disadvantage of being very slow and may be operated at a maximum sampling rate of only about 5 ml per minute. For sampling of very fine particles in the size range of 0.2 to 1.0 μm, however, this instrument was found to be much more efficient than the impactor.

Another thermal precipitator with a heated plate instead of the wire has been reported to be capable of efficient collection of aerosols at a rate of air flow of 1 liter per minute. Thermal precipitators of other designs permit displacement of the microscope slide with respect to the hot wire so that a series of samples can be collected on one slide. Others (99) permit an oscillating movement so that the sample is distributed over the surface of the slide.

Filtration: Solid particles in air may be collected by impaction and retention on various filter mats. Most of these, however, are not suitable for collection of particles for microscopic work (46) since fine particles penetrate into the filtering medium.

For the examination of such embedded particles by microscopic methods, the particles have to be satisfactorily removed from the filtering medium, which is readily accomplished if material forming the mat may be vaporized or dissolved without affecting the particles (35, 36, 167). Filter paper may be cut into small pieces, soaked in a liquid, and beaten with a high-speed stirrer, whereupon particles suspended in the liquid may be separated by decantation from the pieces of filter paper. Such procedure is usually not quantitative, and somewhat soluble particles may lose their original form. Furthermoe, the procedure is tedious and time consuming.

The use of ordinary filter paper and thimbles is suitable only for collecting particulate material occurring in very high concentration as it happens in certain stack gases. The amount of material collected is so large that the small amount of fine particles penetrating the filter material or adhering to it becomes insignificant. After collection, the bulk of the material is simply scraped off and mixed well.

The commercially available (155) *molecular*, membrane, or Millipore filters (46, 49) developed by GOETZ (111, 112, 113) are well suited for collecting particles suspended in air or liquids. Use has been made of this material for examining particles with a variety of microscopic techniques. The filters consist of cellulose acetate and cellulose nitrate and function mainly by a screening action which catches the particles on the surface. They are very efficient for the collection of fine particles of a size range down to less than about 0.2 μm and are particularly useful when the material to be collected is present in low concentration. Membrane filters designed for sampling airborne particles are available in three pore size ranges between approximately 0.1 and 1.6 μm. They are brittle, and special holders are available (44) for supporting them. The refractive index of the filter material is about 1.5, and it is rendered transparent for microscopic study of the collected particles when mounted in a solid medium or treated with a liquid having approximately the same refractive index.

The Millipore filter material was used by CADLE and co-workers (49) as a medium for collecting, mounting and sectioning, also grinding and polishing of airborne particulate material without removing the particles from the filter. The procedure developed for this purpose finally leaves the particles mounted in a thin disk of methyl methacrylate, a material having approximately the same refractive index as that of the filter material. By cementing the disk to a microscope slide, the particles can be readily examined both by means of a petrographic microscope to determine the optical properties of the sectioned particles, and by applying chemical micrurgic techniques (47, 48), p. 267.

The procedure is simple. A piece of filter carrying the particles is placed on the bottom of a hollow metal cylinder with particle side upward. A thick layer of powdered methyl methacrylate is placed on top of the filter which is then heated in a metallographic mounting press to about 150° C, whereupon a pressure of about 6000 psi (400 atm.) is applied. The material is allowed to cool to below 80° C, and the clear plastic is then removed from the press as a cylinder with transparent filter and particles embedded at one end. By means of standard petrographic techniques, this end of the cylinder containing the filter and the collected particles is ground and polished until the particles have received a flat surface. The cylinder is returned to the mounting press, exposed side of particles up, and covered with a thick layer of powdered methyl methacrylate. Heating and pressing are repeated to obtain the particles in a plane bisecting a cylinder of plastic, longer than the first one. Alternatively, the same result may be obtained by cementing a second, particle-free, plastic cylinder upon the polished end of the first cylinder with a solution of methyl methacrylate in chloroform as binder. The original cylinder is sawed through, close to the unpolished side of the particles. The new surface is then ground and polished until thin sections of the particles have been obtained. The largest part of the remaining plastic is then removed by sawing and grinding to leave the particles mounted in a thin disk of plastic which is then cemented to an object slide for microscopic examination.

For mounting particles which would be changed by the heat and pressure used in the procedure just described, CADLE suggests that filters containing such particles be covered by liquid monomers, or polymers, of the selfcuring type.

LODGE (153) utilized the Millipore filter material as a medium for sampling and analyzing airborne particles by the modified spot test method developed by SEELY (212, 213). By filtering a known volume of air, knowledge can be obtained of the concentration of the ion of interest and the particles containing it. Membrane filters may also be used to collect samples of atmospheric particles for examination with the electron microscope (46).

e) Opening of Samples with Fluxes.

The decomposition of samples by fusion with fluxes is often necessary to bring the material into a soluble form suitable for subsequent investigation. To this end, KOCH et al. (138) described an electrically heated platinum wire, p. 175, Fig. 115, mounted on a micromanipulator for the opening of microgram amounts of material under microscopic control. The technique was successfully applied to the investigation of various inclusions and other heterogeneities in steel. After obtaining the required sample in a finely divided form by microdrilling, the opening of the sample is effected by fusion with a suitable flux at the proper temperature in the small loop at the bent end of the wire. Fluxes such as borax, potassium hydrogen sulfate, sodium carbonate, and sodium peroxide are used.

Certain constituents of the sample form characteristically colored compounds with the flux. Colored beads are produced, which serve for the identification of certain constituents of the assay. The technique is a direct adaptation of that of the well known bead tests of mineralogy (190).

To perform a fusion, the dried and finely powdered flux, placed in a tiny platinum dish under the stereoscopic microscope; is first touched with the platinum loop that has been heated to a suitable temperature. A small amount of the flux is taken up by the loop and fused to a clear transparent globule or bead. The powdered sample, contained in another similar dish under the microscope, is then touched with the hot bead so that a small amount of the sample sticks to the bead. By heating the bead at the proper temperature, which depends on both, the sample and the flux, the sample gradually dissolves in the bead. Fusion by means of an electrically heated platinum wire, has the advantage of eliminating exposure to a fuel and its combustion products. Platinum wires are more suitable than silver wires since the latter melt easily and give the bead a dark color. The microscope should be protected against vapors and heat by means of suitable shielding.

With the general working technique just described, some ten fluxes have been tried (138) for bead formation. Clear beads can be readily obtained with borax, boric acid, or potassium hydrogen sulfate. The melts do not creep during heating and scarcely attack the wire. A difficulty is experienced with microcosmic salt since it readily melts to a thin fluid in the tiny dish as soon as it is touched with the hot wire. With alkali carbonates, a brown coloration is obtained at first, but clear beads form on continued heating for some time after decomposition at a temperature just below that necessary for melting.

The melt of sodium peroxide has a tendency to creep on the wire. Creeping is avoided, however, by using the dry salt and a wire of the form shown in Fig. 115b. The melt cools and solidifies in contact with the thickened part of the wire, and a bead is obtained, which is confined within the loop of the wire. Though the peroxide melt attacks platinum crucibles very rapidly, the attack is not so serious on the wire. Therefore, a wire may be used several times for opening samples.

With sodium hydroxide it is difficult to form beads because of the very strong tendency of the melt to creep even when the wire shown in Fig. 115b, is used. Beads can be formed with some difficulty, however, by using the completely dry salt and the lowest temperature possible. It is even more difficult to obtain beads with potassium hydroxide, and its use for opening samples by this technique was found impracticable.

Successful opening of Fe_2O_3, Cr_2O_3, Al_2O_3, TiO_2, and SiO_2 can be achieved in borax beads on the wire. Boric acid is successfully used for opening Fe_2O_3. Sodium carbonate fusion of Fe_2O_3, Al_2O_3, TiO_2, and SiO_2 is also possible. A mixture of equal parts of sodium and potassium carbonate can be successfully used for opening Fe_2O_3 and Cr_2O_3, and sodium peroxide can be used as a flux for Fe_2O_3 and Cr_2O_3. Potassium hydrogen sulfate readily forms clear beads, but it is unsuitable for the opening of any of the oxides just mentioned. This is due to the fact that the melt solidifies on the wire before the dissolution of the sample is complete. If higher temperatures are used so that sulfur trioxide is liberated the melt begins to creep on the wire.

The efficiency of the technique was demonstrated by its application (138) to the analysis of samples of isolated chromium carbide. Microgram amounts of the material were opened on the wire. Iron and chromium were then determined photometrically. The values thus obtained for both elements were comparable to those obtained by opening milligram amounts of the same samples in microcrucibles and determining the two elements by standard microanalytical methods.

Color reactions as well as other phenomena, such as the formation of skeletal or suspended material, are observed in the hot and in the cold bead. The observations may help in the identification of certain constituents of the sample. With borax, sodium and potassium carbonate, and sodium peroxide, characteristic bead colorations can be obtained for the elements iron, chromium, manganese, nickel, cobalt, titanium, vanadium, tungsten, molybdenum, and copper, all of which frequently occur in steel and iron.

2. Chemical Experimentation with Micrograms to Nanograms of Substance.

Detailed description of the apparatus used for performing chemical operations on solutions in capillary cones inside a moist chamber has been given in the first part, p. 92. Two basic microscope-manipulator assemblies for such type of work have been described. One of them, Fig. 63, used by EL-BADRY and WILSON (90), is described in great detail with special reference to qualitative, p. 94, semi-quantitative, p. 119, and gravimetric, p. 119, analysis. The other assembly, Fig. 80, used by BENEDETTI-PICHLER and co-workers has been described briefly, p. 112, mainly with reference to the more fundamental points of difference between the two assemblies.

Various operations that may be performed with either of these two assemblies as well as with the set-up used by THOMPSON and WILSON (229) are more or less the same. The techniques of manipulations involved in the use of all of these assemblies are fairly similar. The following descriptions are, unless otherwise indicated, given with reference to EL-BADRY and WILSON's assembly, Fig. 63.

The assembly, Fig. 83, used by CUNNINGHAM and WERNER (69, 71, 72), suited for work on the submilligram scale with proportionately large volumes of solutions, has been described on p. 116. Since the simple necessary operations are performed in a similar way as with other assemblies, no description of such operations with reference to this assembly is needed. A clear idea of the operations and techniques involved can readily be obtained from the description of the assembly, p. 116, and its use, p. 261.

a) Basic Operations and Qualitative Analysis.

The basic operations of qualitative analysis will serve for general chemical experimentation with solutions and solids.

It is assumed that the apparatus has been prepared and assembled as described on p. 94. The micropipet is properly attached to the glass syringe and in position on the micromanipulator. The meniscus of the water in the shank of the pipet is at approximately one-third of the way from the stem to tip. The ocular micrometer has been calibrated in conjunction with objective and the tube length in use. The moist chamber is on the mechanical stage of the microscope with its open side facing the micromanipulator and micropipet and is fitted with moist filter paper, a condenser rod, reagent containers, capillary cones, measuring capillaries, and a glass rod for thread tests.

Charging Reagent Containers with Solution. For this operation it is convenient to use capillary pipets having a shaft of approximately 0.2-mm bore and about 1 cm in length and a shank about 5 cm long. The pipet is first filled by inserting its tip into the required solution. When sufficient liquid has been taken up, the tip is withdrawn and inserted well into the empty reagent container. Liquid is gradually expelled by blowing into the shank of the pipet until the solution fills a length of the container without bubbles, Fig. 124. Capillary pipets with finer orifices may be used.

Measuring Liquids.   A series of procedures is available.

Fig. 124.   Filling Reagent Container.   After EL-BADRY and WILSON (90).

Withdrawing a Measured Volume from the Container.   By means of the mechanical stage and using transmitted light, the reagent container holding the liquid is brought within the field of view. Its position is then adjusted so as to occupy approximately two-thirds of the field. The focus is adjusted so that a sharp image of the midplane of the capillary is obtained. In this way the walls of the capillary appear as sharp straight lines.

By operating the controls of the manipulator and watching from the side with unaided eye, the micropipet is introduced into the moist cell

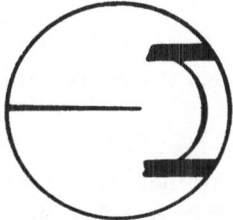

Fig. 125.   Micropipet Ready to Enter Reagent Container.   After BENEDETTI-PICHLER (25).

and its tip is brought close to the opening of the container. By operating the vertical motion of the manipulator while observing through the micro-scope, one brings the tip of the micropipet sharply into focus. This ensures that the tip is now in the same horizontal plane as the axis of the reagent container. By rotating the stage of the microscope, the axis of the container is adjusted parallel to the shaft of the pipet. Finally the shaft of the pipet is brought into line with the axis of the container, Fig. 125, by operating the transverse or N–S (6 → 12) control of the manipulator. The micropipet is now ready to enter the reagent container. With the sagittal, thrust or E–W (3 → 9) control, the tip of the pipet is advanced to lie just within the opening of the reagent container.

The eyepiece is rotated, if necessary, to bring the micrometer scale into the image of the reagent container. The position of the latter is adjusted by means of the mechanical stage so that the meniscus of the liquid in the container coincides with a convenient scale division in the eyepiece micrometer. The pipet is cautiously advanced by means of the thrust control until the tip is immersed slightly further than the length of liquid which it is proposed to remove, Fig. 126.

Suction is applied cautiously by turning the micrometer head. The meniscus recedes in the reagent container. When slightly more than the required amount of liquid has entered the pipet, the micrometer head is adjusted to reach equilibrium and to maintain the meniscus of the liquid in the container perfectly steady at the desired scale division. This state should always be achieved when withdrawing liquid in the pipet. The thrust control is then operated to withdraw the pipet.

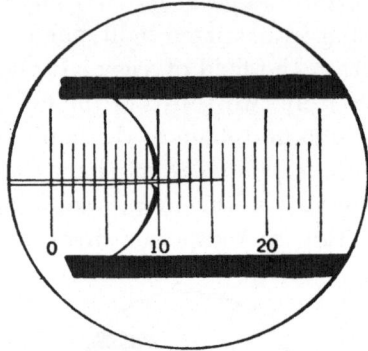

Fig. 126.   Measuring in Reagent Containers.   After BENEDETTI-PICHLER (25).

Accurate Measurement of the Volume. The pipet tip is withdrawn a short distance from the container by operating the thrust control. The tip of the pipet should be visible at a distance of a few scale divisions from the container mouth so that part of the shaft is still within the field of view. With the mechanical stage and direct* observation, the measuring capillary is brought into the position previously occupied by the reagent container. Viewing through the microscope, adjustments are carried out as before to bring the pipet and the measuring capillary into correct alignment. The final adjustment is carried out with the tip of the pipet just inside the mouth of the measuring capillary and the pipet is then advanced by a distance which is assessed to correspond to the amount of liquid to be measured.

When the expelling of the liquid is cautiously started, a droplet forms at the tip of the pipet and grows until it fills the bore of the measuring capillary. Two menisci form, one of which moves toward the opening of the capillary. At this point, the pipet is gradually retracted so that it follows the outer meniscus without being removed from the liquid while liquid continues to be expelled. Pressure is finally adjusted by the micro-

---

*) Not through the microscope.

meter head to produce stable menisci which are the calculated distance apart for the volume required, Fig. 127.

The micropipet is then completely withdrawn from the moist chamber. Excess of solution is expelled and received on a strip of filter paper.

Withdrawing the Measured Sample from the Capillary. In the same way as before, the pipet tip is brought forward and adjusted to come a short distance within the outer meniscus of the liquid, Fig. 127. Very gradual suction is applied by the micrometer head. At the same time, the meniscus is maintained in the same position relative to the pipet tip by advancing the measuring capillary with the mechanical stage. When the liquid has been taken up as completely as possible, pressure on the pipet

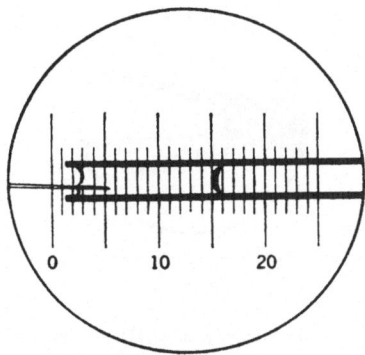

Fig. 127.   Measuring Solution in Measuring Capillary.   After Benedetti-Pichler (25).

is adjusted so as to form a tiny droplet at the pipet tip. This is touched to any drainings still remaining on the wall of the capillary by rotating the stage of the microscope so that the shaft of the pipet takes a slight angle with the wall of the capillary. Application of slight suction, perhaps accompanied by gradual retraction of the capillary, will take up the drainage. Great care should be taken throughout, particularly at this stage, to secure an uninterrupted flow of liquid from the capillary to the pipet so that air bubbles are not trapped. Otherwise these will be expelled during subsequent delivery and may cause spattering of the liquid.

Finally the pipet is withdrawn so that its tip remains visible near the edge of the field of the microscope.

Transfer of the Measured Sample to the Capillary Cone. The selected cone is brought into the field, and the microscope is focused on its axial plane. For this purpose, the point of the taper may be observed. The position of the cone is adjusted with the mechanical stage so that the vessel occupies approximately one-third of the field. The pipet is adjusted and advanced so that its tip comes close to the opening of the cone, and its tip is brought into sharp focus using the vertical motion of the manipulator. The cone is now advanced by means of the mechanical stage so that the pipet touches the wall of the cone lightly near the point of the taper, Fig. 128. Liquid is ejected slowly, and simultaneously the cone is retracted so as to maintain the tip of the pipet just inside the meniscus.

When the meniscus in the pipet arrives at its orifice, the capillary cone contains an accurately measured volume of solution.

Cleaning the Micropipet. Washing with distilled water is carried out by advancing the piston of the syringe until hydraulic water moves into the pipet from the syringe and forms a droplet at the tip of the pipet. This droplet is removed by a strip of thin filter paper. Another droplet is formed in the same way and is removed. The piston is then retracted until the water meniscus in the pipet returns to its original position in the shank.

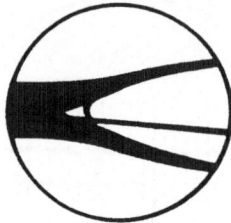

Fig. 128.   Delivery of Solutions to Capillary Cone.   After BENEDETTI-PICHLER (25).

If other liquids have to be used, small volumes of these are taken up into the pipet by slight suction so as to fill the shaft and taper, and the liquid is then expelled on to a strip of filter paper. A fresh drop of wash liquid is used for every rinse. These drops may be supplied on a microscope slide or as a hanging drop on a glass rod. The shaft and the tip of the pipet are so flexible that there is no serious danger of breakage if the tip touches the slide or the glass rod.

BENEDETTI-PICHLER and CEFOLA (29) use a short buret as a reservoir for the cleaning fluid, Fig. 80*f*. The wash liquid is supplied in the form of a drop hanging from the tip of the buret. The tip of the micropipet is introduced into the hanging drop by means of the manipulator. When enough liquid is taken up in the pipet, the pipet is withdrawn and the wash liquid is expelled on to a strip of filter paper. As the manipulator permits a rotating motion in the horizontal plane, the buret with the wash liquid is placed so that it is possible to withdraw the micropipet from the moist chamber and rotate the manipulator to bring the tip of the pipet close to the hanging drop.

Measuring a Reagent. A reagent is brought into the pipet in the same way as described for the withdrawal of a measured volume from a reagent container. This gives an approximate amount of reagent, which is sufficient for most purposes, and it is usually not necessary to carry out an accurate measurement in a measuring capillary.

Addition of Reagent to a Solution in the Cone. The tip of the pipet is set so that it is visible near the edge of the field, and the cone is placed so as to occupy about one-third of the field. With the methods already described for adjustment, the tip of the pipet is brought close to the opening of the cone, and is sharply focused by means of the vertical adjustment of the manipulator. The cone is then advanced so that the tip of the pipet is close to the surface of the liquid already in the cone. By slight pressure, the liquid in the pipet is adjusted to fill the tip completely without forming a droplet. The cone is further advanced to bring the tip into contact with the meniscus, upon which delivery of the reagent commences. By continued slight pressure nearly all the reagent is expelled and the cone is being gradually withdrawn as delivery takes place. Care must be taken that the shaft of the pipet does not touch the wall of the cone since liquid as well as a precipitate may then spread along the wall beyond the level of the meniscus.

When almost all the reagent has been expelled, the cone is retracted slightly to bring the pipet tip out of the liquid. The remainder of the reagent may then either be delivered onto the wall of the tilted cone, close to the meniscus, or it may be discarded.

Stirring the Contents of the Cone. Satisfactory stirring can be obtained by moving the shaft of the pipet, preferably by means of the manipulator, a few times backwards and forwards through the solution. The opening of the pipet is sufficiently sealed by the last traces of a reagent remaining in the tip.

Stirring by expelling a stream of air bubbles from the opening of the immersed tip of the pipet should be avoided since it always results in spattering of the contents over the walls of the cone. Centrifuging may suffice to restore spattered liquid, but much of a precipitate may remain on the walls. Occasionally some of the contents may even be thrown out of the cone by the stream of air bubbles.

With apparatus of the type shown in Fig. 80, stirring can be done by introducing the pipet tip into the contents of the cone and plucking with the forefinger the copper tubing which connects the pipet holder and the plunger device. This imparts a vibrating motion to the tip of the pipet, which gives efficient stirring.

In the course of heating, see below, mixing the contents of the capillary cone can be performed, if required, by touching the capillary containing the cone to the vibrating part of the buzzer used for severe agitation.

Heating the Contents of the Capillary Cone. A little water is allowed to enter the tip of a capillary pipet, about 5 cm in length and about 1 to 2 mm in bore. The tip is then sealed in a microflame. The tube is allowed to cool and, if necessary, the water is collected at the sealed end by shaking or centrifuging.

The capillary cone is removed from the moist chamber. If it has been held on the carrier by vaseline, it is laid on a sheet of filter paper, and the handle is cleaned by rolling on the filter paper. Alternatively (25), vaseline can be removed with a warm cloth while the capillary cone is held by means of cork-tipped forceps. The clean cone is introduced, handle first, into the capillary and allowed to slide down to the sealed end. If necessary, a glass thread which nearly fills the bore of the outer capillary may be used to push the cone down.

The open end of the capillary is then sealed, and the capillary with the cone inside, Fig. 129, is inserted in the appropriate hole of a heating block

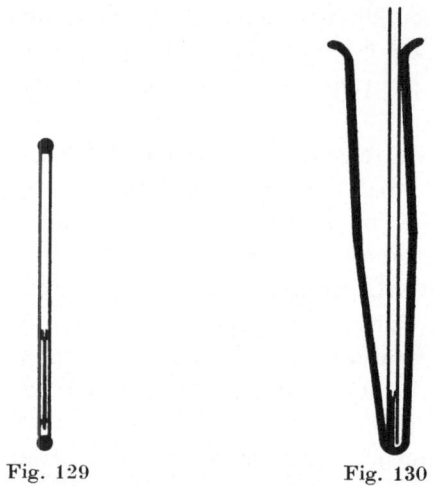

Fig. 129   Fig. 130

Fig. 129. Capillary Cone Inside Heating Capillary.

Fig. 130. Centrifuging Contents of Capillary Cone. After BENEDETTI-PICHLER (25).

at the required temperature. The bore of the heating capillary is of importance. For example, when aqueous solutions are heated for several minutes just below 100° C, wide capillaries appear to encourage partial evaporation from the cone, while narrow capillaries encourage dilution of the solution in the cone.

Because of the possibility of slight evaporation, it may be advisable to centrifuge the contents of the cone by applying a few turns of the hand centrifuge just before inserting into the heating block. This brings a precipitate a short distance below the meniscus, so that subsequent evaporation does not leave dried precipitate sticking to the walls where it might escape detection or estimation.

After cooling, first the upper end and then the lower end of the heating capillary are cut off. Using a glass thread inserted from the lower end, the cone is pushed out through the upper end and is caught with forceps.

Centrifuging the Contents of the Cone. Separation of precipitates from solution is satisfactorily achieved by centrifuging, whereby the precipitates are collected in the taper as shown in Figs. 131 and 136.

The capillary cone is introduced into a capillary approximately 4 cm long and from 1 to 2 mm in bore depending on the size of the cone. The centrifuging capillary, with the cone inside, is held almost horizontally and inserted into an ordinary microcone so that the handle of the capillary cone is pointing towards the tip of the large cone, Fig. 130. The large cone is then placed into the centrifuge.

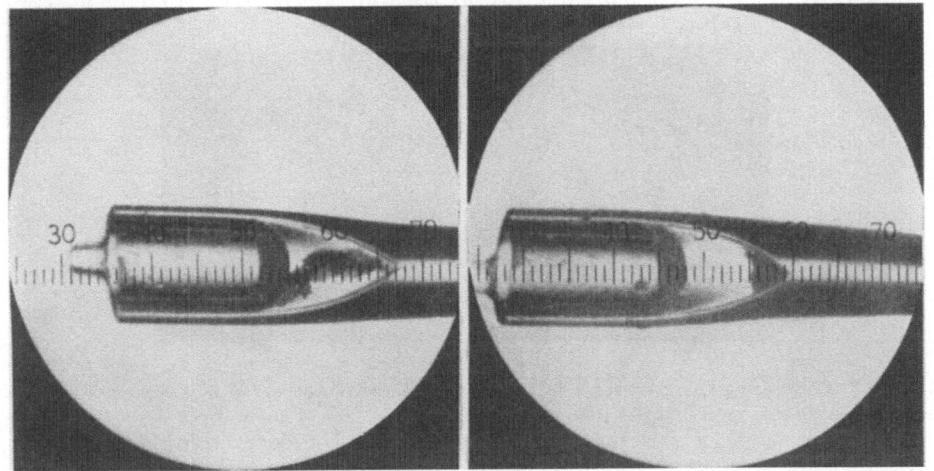

Fig. 131. Photomicrographs Showing Precipitate and Solution in Capillary Cone Before and After Centrifugation (29). Courtesy of BENEDETTI-PICHLER.

If the centrifuge is hand-driven, 20 to 30 strong turns of the crank of the instrument are, in most cases, sufficient. With an electric centrifuge, half a minute at about 2000 revolutions will be quite satisfactory.

After centrifuging, the outer cone is again held horizontally, and the centrifuging capillary, with the capillary cone inside, is withdrawn. The capillary cone may be grasped if the tip of the handle protrudes slightly, or it may be pushed out by means of a thick glass thread.

The centrifuging capillary is useful in two ways. Aside from supporting the cone during centrifugation, it also prevents excessive evaporation of the solution in the cone by limiting the air space around the capillary cone. The same principle is used when liquids in capillary cones are either heated or treated with gases. Of course, capillary cones may be centrifuged in a heating capillary.

Separation of Solids from Liquids. After collecting the solid in the tip of the cone by centrifuging, one adjusts micropipet and cone under the microscope so that the tip of the pipet enters the solution in the cone and gets a short distance below the meniscus. Suction is then cautiously applied, and the tip is kept just below the meniscus until nearly all the liquid has been removed. The last traces of the liquid are removed very cautiously to avoid disturbing the solid or taking any of it up into the pipet. The cone is retracted, the pipet is withdrawn, and the liquid contained in it is either transferred to another cone if it is to be treated further, or is discarded.

Washing Solids. An approximately measured volume of wash liquid is taken into the pipet from a reagent container and then slowly delivered onto the wall of the cone, close to the precipitate, in such a way that the precipitate is covered with wash liquid without being disturbed. The cone

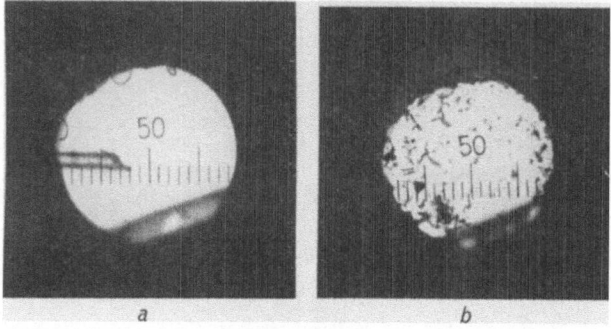

Fig. 132. Performing Confirmatory Tests on Platform of Condenser Rod. Diameter of dark microscopic field about 6 times diameter of plateau. *a*, tip of micropipet above platform; *b*, precipitate of silver dichromate (29). Courtesy of BENEDETTI-PICHLER.

is retracted as the amount of liquid in it increases so that, as usual, the tip is always just below the meniscus. The wash liquid is left for a short time in contact with the precipitate and is then withdrawn by the pipet and rejected. The operation may be repeated as often as desired.

Depositing Solution on the Plateau of a Condenser Rod. An approximately measured volume of test solution is taken up in the micropipet, and the tip is then moved so that it is just visible in the field. The condenser rod is watched with unaided eye and brought close to the tip of the pipet and hence within the field of view. Under the microscope the plateau is sharply focused and moved to the furthermost third of the field away from the pipet. The pipet is raised to the level of the plateau, brought into focus by means of the manipulator, and then raised slightly more so as to bring it just out of focus. It is then advanced to enter nearly one-third of the plateau and lowered so that the opening just comes into contact with the plateau and is therefore in almost sharp focus, Fig. 132*a*.

Liquid is expelled slowly onto the surface of the platform of the condenser rod. The pressure is reduced near the end of the delivery, and when the meniscus almost reaches the tip of the pipet, this is slightly raised, and the pipet is withdrawn.

Addition of Reagent and Observation of Results. The required volume of the reagent is measured off with the pipet. Light is supplied to the condenser rod. The reagent is added slowly to the test solution already on the plateau, whereby care is taken to apply a slight pressure before the tip enters the solution. Liquid thus flows immediately upon touching the pipet tip to the test solution, which prevents deposition of precipitate inside the pipet tip. When the reagent has been delivered, the pipet is withdrawn and cleaned.

For observation of the reaction product, the light coming through the condenser rod should suffice, and other light sources may be turned off. The magnification that may be used is limited by the working distance of the objective and the distance from the plateau to the top of the moist chamber.

Preparing the Plateau for Further Use. After a confirmatory test has been applied, the condenser rod is taken out of the moist cell. The precipitate is removed by dipping the plateau and part of the glass thread into the proper solvent and finally rinsing in the same way with distilled water.

If this treatment fails to clean the plateau completely, a new surface is prepared by cutting the thread a short distance below the original surface. This may be repeated until the thread becomes too short for focusing on the plateau.

Performing a Test on a Thread. A piece of fine cotton thread is attached with gum or starch paste to the sides of a short piece of glass rod so that the thread makes a small loop. Alternatively the thread may be attached to the two ends of a U-shaped glass thread. These two variants are shown in Fig. 77. The thread is then impregnated with the required reagent throughout most of the loop, by supplying the reagent as a drop hanging from the end of a glass rod or from a capillary. The holder is then placed on the carrier so that the thread is horizontal. An approximately measured amount of test solution is taken up in the pipet which is then manipulated so that its tip is almost in contact with the thread. Slight pressure is applied to form a droplet at the opening of the micropipet. The pipet is then advanced to allow the thread to break the surface of the droplet. After a little wait, the whole test solution is discharged slowly onto the thread. Any color produced is observed in reflected light.

Use of Test Paper. The technique (29) can be used for testing the acidity of solutions with litmus paper and also for performing certain spot tests on paper. A narrow strip of the paper, approximately 3 mm wide and 20 mm long, is laid flat upon the carrier and parallel to the capillary vessels so that about one-third of the strip projects beyond the edge of the carrier. The projecting part of the strip is bent downward to permit the tip of the pipet to be brought into contact with the paper. Test solutions and reagents are transferred to the paper by means of the micropipet, and any coloration produced is observed in reflected light of low intensity.

Removal of Gas from a Solution. This is achieved by alternately sucking the solution into the pipet and discharging it. With bromine the color of the solution gradually fades until it finally completely disappears.

Treatment of a Liquid with a Gas. By the following procedure (25, 28), the contents of a capillary cone can be saturated with a gas.

One end of a piece of soft glass tubing of about 4- to 5-mm bore is drawn out to a capillary of approximately 1-mm bore, which is then bent as shown in Fig. 133. A plug of cotton is inserted into the wide tube b to

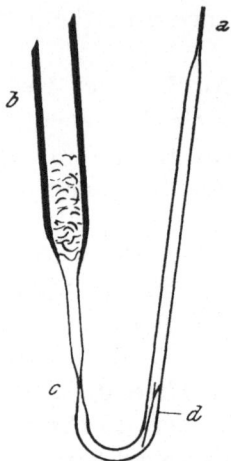

Fig. 133. Treatment with a Gas (25). After Benedetti-Pichler.

serve as filter for the gas. The capillary is cut at a, and it is slightly drawn out at c. If the cone has been held on the carrier by vaseline, the handle is cleaned. The cone is then introduced, handle first, into the capillary at the end a, and is allowed to slide down the capillary to d. The end a of the capillary is then drawn out to a fine tip as shown in the figure.

The tube b is connected to the gas generator, and the gas is passed until the air in the capillary is displaced by the gas. With hydrogen sulfide and other odorous gases this condition is reached when the odour of the gas escaping from tip a becomes distinctly noticeable. Finally the point of the tip a is sealed. If it is required to heat the contents of the capillary cone, the capillary is fused shut at the constriction c, drawn off from the wide part of the tube, and placed for a short time in a water bath at the required temperature. After removal from the water bath, the capillary containing the cone may be allowed to stand for any required time before taking the cone out. By cautiously cutting the capillary open at d and gently stroking with a file, if necessary, the handle of the capillary cone may be made to protrude from the opening of the capillary. It can then be grasped, and the capillary cone is pulled out of the capillary for further operations.

Evaporation in the Capillary Cone. The capillary cone is inserted into a wide capillary, open at both ends, and is transferred to a micro-cone in the same way as for centrifuging, Fig. 130. The microcone is placed in the appropriate hole of the heating block, Fig. 79, at the required tem-perature. For alternatives, the microcone may be placed on a steam bath or simply allowed to lie under a bell jar at room temperature.

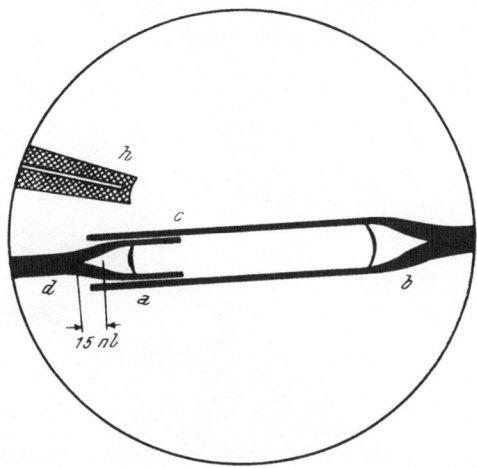

Fig. 134. Distillation from Cone to Distilling Capillary, Microscopic View (25). After BENEDETTI-PICHLER.

Distillation. The capillary cone containing the liquid to be distilled is inserted into a distilling capillary, which is then mounted on the carrier inside a dry chamber as shown in Fig. 75. It is not possible to get the whole length of the capillary part of the distilling capillary into the field of view as shown in Fig. 134, because of the relatively small microscopic field. Thus, the end of the distilling capillary, c, which contains the capillary cone is brought into focus as shown in the photomicrograph, Fig. 135a. The pipet holder, Fig. 80, is removed, and the heating element, Fig. 76, is inserted in the clamp of the manipulator. If an assembly of the type shown in Fig. 63 is used, the heating element can be held by any suitable means in the clamp of the manipulator after removing the micrometer syringe.

The hot point h, Figs. 134, 135a, of the heating element is introduced into the dry chamber and is brought to a position approximately 1 mm to one side of the capillary cone d. It is then focused by means of the mani-pulator to get it in the level of the distilling capillary. The current is turned on, and the voltage is gradually increased. Small droplets start to form slowly on the wall of the distilling capillary, and as the voltage is increased, the meniscus of the liquid in the capillary cone starts to recede towards the point of the cone, which indicates a fairly rapid volatilization. This receding movement of the meniscus should be slow. It can be regulated either by adjusting the position of the hot point or by changing the voltage. Occasionally an air bubble may form at the point of the cone, grow, and push the liquid toward the open end of the capillary cone. This should be prevented by regulating the rate of heating so as to maintain the gas

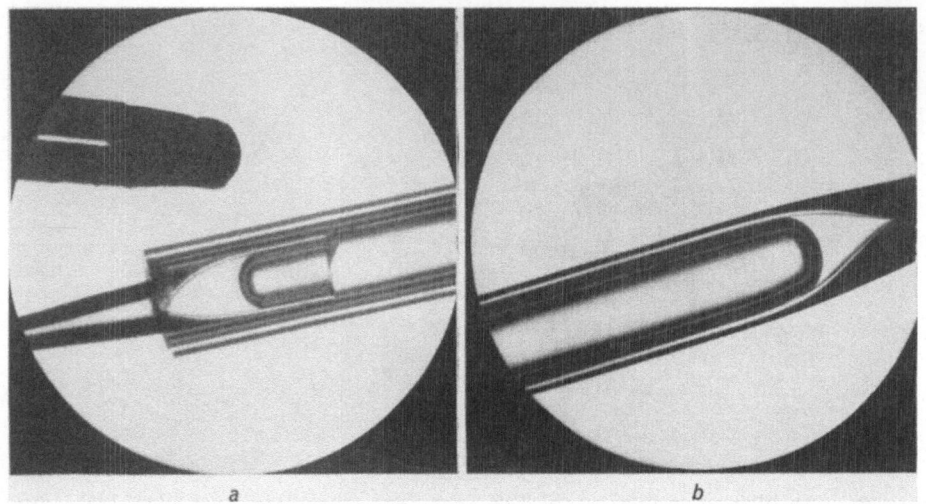

Fig. 135. Photomicrographs Showing Distillation from Capillary Cone to Distilling Capillary. *a*, heating element near end of distilling capillary containing cone with liquid to be distilled; *b*, distillate collected at tip of same distilling capillary (29). Courtesy of Benedetti-Pichler.

bubble at a small size in the point of the cone. When the operation is completed, the hot point is withdrawn from the chamber and the current is disconnected. Inspection of the distilling capillary at the end of distillation will show that the bulk of the distillate collects at the sealed end, Fig. 135*b*.

The capillary cone is withdrawn from the distilling capillary and may be reserved for further operations. The distilling capillary is immediately centrifuged to collect all the distillate in the taper. It is then cut at a point approximately 3 to 4 mm from the taper, and the unwanted portion of the capillary is snapped off with the aid of a cork-tipped forceps. The remaining part of the distilling capillary, which contains the distillate, is placed on the carrier. By means of the micropipet the distillate is then transferred to a regular capillary cone for further operations.

All observations are made with transmitted light.

Transfer of Solids to Capillary Cones. A small amount of the finely-powdered solid, or metal filings, is sprinkled on a glass slide about 2.5 cm square. The slide is then placed level (29) on top of the chamber containing the capillary cone. The micropipet is replaced by another micropipet the tip of which has been sealed and slightly bent. The shank of the pipet is inserted into the holder so that the tip points downwards. The sealed tip of the pipet is coated with a very thin film of grease. The particles of solid are brought into the microscope field, and a selected particle is picked up with the tip of the pipet or needle, Fig. 139a, p. 250. The bent tip of the pipet is then brought into a horizontal position to facilitate introducing the particle into the capillary cone. This is achieved by releasing the clamp of the micromanipulator for a moment and rotating the pipet holder, Fig. 80. The particle is then transferred to the capillary cone, Fig. 139b, and may be stirred into the solution contained in the cone. This procedure is illustrated in Fig. 139, which demonstrates the transfer of a filing of Wood's alloy, about 1 μg in mass, from a microscope slide to a capillary cone (29).

Alternatively (25) the solid particles may be sprinkled on a slide which is then mounted in an inclined position on the top of the chamber by supporting it on a short piece of glass rod. This permits the use of a needle or sealed micropipet with straight tip. The use of inclined slides has the disadvantage that the whole microscopic field cannot be brought into sharp focus at one time.

Instead of using a micropipet with sealed tip, a short length of glass rod of suitable diameter, drawn out to a fine thread, may be used in a similar fashion. Furthermore, it may be preferable to mount the glass slide carrying the solid particles on top of the carrier beside the cone inside the chamber. A thin cover slip of small dimensions, 1 to 1.5 cm square, may be suitable for the purpose. With this arrangement the transfer of particles will be less time consuming.

If it is required to obtain the weight of the solid sample, an empty capillary cone is first counterpoised on a suitable microbalance and then mounted on a clean carrier by means of a rubber band, Fig. 70. A small amount of water or a suitable solvent is transferred to the capillary cone, and the selected particle is introduced as described above. The solvent is evaporated, and the dried sample is weighed, p. 226. The solid particles may also be transferred to a dry cone.

b) Semi-Quantitative Estimation.

As has already been mentioned, the volume of a precipitate can serve as an indication of the approximate amount of the substance present in a solution if a suitable precipitating form is used (25, 92, 93). The estimation of the amount is made by comparison of the volume of the precipitate with that obtained by similar treatment of a known volume of solution of known concentration of the same substance.

Apparatus and techniques are those already described for general experimentation and qualitative analysis. Certain points of importance regarding apparatus suitable for semi-quantitative work have also been mentioned, p. 119.

The volume of a precipitate depends largely upon the particle size, the magnitude of the centrifuging force, and the time of centrifuging. Thus centrifuging must be carried out by following an invariable procedure. In other words, the centrifuge is always made to run at the same speed for the same length of time so that a reasonable reproducibility of results and satisfactory comparison with standard precipitates is possible. With

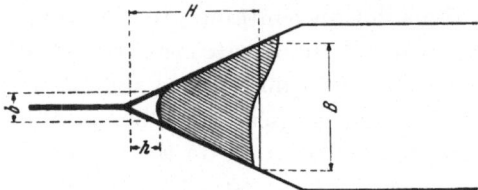

Fig. 136. Measurement of Volume of Precipitate (90). After EL-BADRY and WILSON.

a hand centrifuge it is advisable to count the turns of the crank. Twenty strong turns are usually sufficient. An electrically driven instrument is less liable to variations, and 30 seconds at 2000 revolutions per minute will be satisfactory.

Observation with transmitted light is used for the estimation of the volume of precipitates, and the microscope is focused on the point of the taper of the capillary cone. After centrifuging, the precipitate may take the shape of an irregular truncated cone similar to that indicated by the shaded area in Fig. 136. The approximate volume of the truncated cone is estimated by using the formula (25)

$$V = 0.26 \ (B^2 \ H - b^2 \ h) \quad \mu m^3$$

where the dimensions B, b, H and h, shown in Fig. 136, are measured with the calibrated ocular micrometer and converted to micrometers (μm).

With wide-angle capillary cones (90) and fine precipitates, the tip of the cone becomes completely or nearly completely filled with the precipitate, so that it may be regarded as forming a complete cone. In such a case the formula reduces to

$$V = 0.26 \ B^2 \ H \ \mu m^3$$

V, the volume of the precipitate, may be more conveniently expressed in nanoliters by including the factor 10^{-6} into the above equations.

c) Gravimetric Analysis.

A description of EL-BADRY and WILSON's microgram torsion balance (88) and its construction has been recently given by BENEDETTI-PICHLER (26). Directions for the use of the instrument may be found above on p. 120. The auxilliary apparatus required for performing gravimetric determinations in capillary cones has been described on p. 122.

The basic operations in capillary cones are essentially the same as in general experimentation and qualitative work. Gravimetric determinations, however, require certain precautions in handling the capillary cone and for hindering evaporation of the sample prior to measuring. An account of these will be given and illustrated by a description of a simple gravimetric determination.

α) Handling of the Capillary Cone.

Great care must be taken in working with capillary cones to keep their weight constant within narrow limits. Cones are always handled with forceps kept for this purpose inside a corked tube. Under no condition should a capillary cone be touched by hand. To facilitate handling, the cone is placed on a small watch glass, kept especially for reception and transference of the cone, with the handle of the cone resting on a short length of silica or glass fiber. In this position the handle may be grasped easily with forceps. The watch glass, fiber, drying and centrifuging capillaries as well as the microcone are kept in a covered Petri-dish until needed for actual use.

Placing the Cone on the Balance Pan. To place the cone on the balance pan, the handle is grasped with forceps. The capillary part of the cone is placed on the pan so that the handle extends beyond the edge of the pan.

Placing the Cone on the Carrier. The body of the cone is grasped with the forceps so that the cone is almost in line with the forceps. The handle of the cone is inserted beneath the rubber band of the carrier shown in Fig. 70. The operation may be observed with a hand lens.

Transfer of the Cone to the Drying Capillary. The cone is placed on the watch glass as described above. A drying capillary is grasped by hand, and—with the aid of a hand lens—the open end of the capillary is pushed over most of the cone handle. Upon tilting the capillary gradually to the vertical position, the cone slides inside and drops to the sealed end.

After drying, the cone is easily removed by inverting the drying capillary and receiving the cone, which slides out easily, on the watch glass.

Transfer of the Cone to the Centrifuging Capillary. The handle of the cone is introduced in one end of the centrifuging capillary as in the instance of the drying capillary. The capillary is then cautiously turned to a nearly horizontal position, and its other end is introduced into an ordinary small (about 1 ml) centrifuge cone. When the centrifuge cone is brought to a vertical position, the capillary cone drops so that the handle rests on the bottom of the centrifuge cone, while the capillary cone remains inside the capillary.

After centrifuging, the capillary cone is removed from the centrifuge cone by withdrawing the centrifuging capillary in a horizontal position. The centrifuging capillary is then held nearly vertically over the watch glass and, if necessary, tapped gently, upon which the cone drops, handle first, onto the watch glass.

To avoid all risk of contamination, centrifuging and drying capillaries are used once only and then discarded.

β) Hindering Evaporation of Sample Solution Prior to Measuring.

Special care, more than that usually required in qualitative and semi-quantitative work, should be taken to hinder evaporation of a sample solution, particularly before measuring. As already mentioned, additional filter paper linings are placed around the four sides of the carrier and vertical linings may be added along the three closed sides of the chamber; the whole lining is kept well soaked with water. The assembly of capillary vessels necessary for the work, the counterpoised cone, and the containers charged with the required reagents are placed on the carrier inside the moist cell before indroducing the cone with the sample. The moist chamber is closed, and additional water is added, if necessary, to the filter paper linings. The whole assembly is allowed to stand for some time, whereupon a reagent container is charged with the solution of the sample and placed on the carrier as quickly as possible. Without waiting, an approximate volume of the sample from the container is delivered to the measuring capillary, where the length of the column of liquid is accurately read without delay.

γ) Gravimetric Operations.

For carrying out a gravimetric determination, it will be assumed that the auxiliary apparatus has been prepared and assembled as already described, p. 122. It will further be assumed that the balance (26, 88), is properly set, the dial calibrated, and that the precision of the instrument has been determined.

The procedure for carrying out a simple gravimetric determination may be divided into the five series of manipulations presented below.

Selection and Preparation of the Capillary Cone. By means of a hand lens, a set of capillary cones, all approximately equal in size, is selected. From this set, the largest cone is chosen and placed permanently on the far balance pan as a counterpoise; the body is placed on the pan so that the handle protrudes beyond the edge. The cone to be used in the determination is placed similarly on the near pan. The pan arrest is released, and the beam arrest is partially released to determine whether the operating cone is approximately counterpoised or is distinctly too light. This can be observed through the pan well with the rotating door partially opened. The operating cone should not be heavier than the counterpoise. The pans and beam are arrested.   If the operating side is distinctly lighter, the 2-cm piece of platinum wire (counterpoise) is placed on the pan beside the operating cone. The taring is repeated, possibly with heavier pieces of wire, until the beam assumes an approximately horizontal position or the operating side is a little heavier than the counterpoise or far side. This ensures that the load difference is not more than 200 μg. The images of the two ends of the index fiber are then observed on the ground glass screen, and the released beam is brought approximately to the horizontal position by rotation of the calibrated dial, whereupon the balance is arrested.

The operating cone is then removed to the watch glass and transferred to a drying capillary. It is dried by placing the drying capillary in the narrow hole of the heating block at the required temperature which is that at which the precipitate will subsequently be treated.

Determination of the Balance Reading with Empty Cone. The cone is returned to the watch glass and from there transferred to the balance pan. After acclimatization, the zero point is accurately determined by the triple reading technique. This is the balance reading for the empty cone.

Carrying out the Precipitation. The cone is transferred via the watch glass, onto the carrier inside the moist chamber, which is already equipped with a measuring capillary, containers with the required reagents, and filter paper linings thoroughly soaked with water. The top of the moist cell is closed, and after some time, a reagent container is charged with the sample solution and placed as rapidly as possible on the carrier beside the measuring capillary.

The moist cell is placed on the stage of the microscope. An approximate volume of the sample solution is delivered to the measuring capillary, and the length of the column of liquid is read accurately and as quickly as possible. The measured sample in the capillary is then delivered to the weighed cone, and care is taken that the shaft of the pipet does not come into contact with the walls of the cone, a precaution which is still more important later when adding the reagent and the wash liquids.

Precipitation is carried out by addition of the appropriate reagents in suitable amounts, measured approximately in the reagent containers. The cone is transferred—always via the watch glass—to the centrifuging capillary and, after centrifuging, returned to the carrier.

Subsequent Treatment of the Precipitate. The supernatant liquid is drawn off the precipitate with the micropipet. Care must be taken not to disturb the precipitate. Water or other suitable wash liquid is then added with the pipet, again without disturbing the precipitate. After allowing it to stand in contact with the precipitate for an appropriate time, the wash liquid is removed. After a suitable number of washings, the cone is transferred via the watch glass to a new drying capillary and dried.

Weighing the Precipitate. The capillary cone containing the dried precipitate is transferred from the drying capillary to the watch glass and then the balance pan. The beam is brought to the horizontal and the required setting of the calibrated dial is accurately determined by triple reading. This is the reading of the balance for the cone containing the precipitate. The difference between this reading and the reading of the balance with empty cone corresponds to the weight of the precipitate. The precipitate is again dried and weighed, and this is repeated to constant weight.

The cone is then removed and the dial of the balance is restored approximately to its zero point with empty pans. This is achieved by rotating the dial back until the torsion force is practically zero. If there is any doubt as to the number of revolutions required for this, it is safer to empty the counterpoise pan also and to restore the beam to the horizontal position with unloaded pans. In general, it is good practice to empty the pans and check the balance after a number of determinations.

d) Titration of Microgram Samples.

The apparatus used by LOSCALZO and BENEDETTI-PICHLER (154) as well as that used by DORF (81) for titration of microgram samples are described in detail on pp. 123 to 132.

α) Working in Titration Cones.

It is assumed that the apparatus has been prepared and assembled as already described, p. 125. The moist chamber, p. 126, which is provided with two diametrically opposite openings in the side walls, is properly placed on the mechanical stage of the microscope to admit the titration cone on one side and the microburet on the other, Fig. 137. The chamber is fitted with wet cotton, a carrier, and reagent containers charged with the needed solutions. The microscope should have a built-in rotating stage and an attachable mechanical stage to hold the moist chamber. It must be fitted with a calibrated ocular micrometer.

The buret, Fig. 87 A, p. 126, is properly attached to the pipet holder, Fig. 81 and 87 Ch, which is mounted on the manipulator and connected to the leveling device, Fig. 88.

The titration cone, Fig. 87 B, with shank similarly inserted in a pipet holder mounted on a manipulator or a suitable substitute is properly connected to the plunger device, Fig. 80 d, which has been completely filled with water as already described, p. 114. The meniscus, hw, Fig. 87 B, of the hydraulic water in the shank of the titration vessel is advanced to a point about 2 cm from the titration cone proper.

For the calibration of the buret, the same set-up of apparatus is required, but the place of the titration cone is given to a calibrating capillary sealed into a glass tube, Fig. 137 Ah, which is mounted in the clamp of a manipulator or a suitable stand. The calibrating capillary is introduced into the moist chamber opposite to the micropipet as shown in the figure.

Operation of Buret. The operation of the buret depends on the utilization of surface forces. When the tip of the empty buret is introduced into an aqueous solution, the flow of the solution into the buret starts immediately, and the liquid would fill the whole buret unless the flow is stopped by raising bulb b of the leveling device, Fig. 88, to give the air body separating the two liquids a pressure equal to that produced by the surface tension at the interface ms, Fig. 87 C. In this way, the flow of liquid is stopped when the advancing meniscus reaches a point approximately 1 to 2 mm above the reference mark. The pressure at this time, the equilibrium pressure in centimeters of water column, can be read directly on the scale of the leveling device.

Another meniscus is formed at the tip tp when the buret is withdrawn from the solution. The surface force originating at the new meniscus is considerably stronger than that at the meniscus mc since the bore at the tip is much smaller than that at ms. Thus, to expel liquid from the fine tip of the microburet into air, a pressure considerably higher than the equilibrium pressure is required. This pressure is equal to $2\,\gamma/r - 2\,\gamma/R$, whereas the equilibrium pressure is equal to $2\,\gamma/R$, where γ is the surface tension, and R and r are the two radii of the bore at mc and tp, respectively.

To start a titration, a suitable operating pressure is established by raising bulb b of the leveling device and adjusting its height. Such operating pressure must be higher than the equilibrium pressure and lower than that required to expel solution into air. The operating pressure can thus be regulated as required within a wide range to give the desired rate of outflow. When the tip of the buret is inserted into the solution to be titrated, the outflow of the standard solution starts immediately. It may be stopped at any instant by simply retracting the tip into air.

Satisfactory precision can be obtained with the buret by properly choosing objective magnification and micrometer rulings to get a displacement of the meniscus of the standard solution in the shank of the buret equal to nearly the whole length of the micrometer scale. Furthermore, by projecting the image of the buret on a ruled screen and comparison with the image of a stage micrometer scale, corrections for the distorsion near the edge of the microscope field can be obtained. A column of standard solution up to nearly twice the length of the scale may be utilized and measured if the initial reading is taken with the meniscus nearly a scale length above the reference mark of the buret. The meniscus may then be allowed to drop a like distance below the reference mark for the final reading.

Calibration of Buret. The total capacity of the buret, that is, the capacity of the part of the shank to be calibrated, is approximately 50 nl. Thus the customary procedures frequently used for calibration of burets of larger capacity (25), which depend on measuring and subsequently weighing a column of water delivered by the buret, do not seem practical. Even if a suitable microgram balance (26, 88, 136) were available, appreciable errors would be caused by evaporation of the water delivered during transfer and weighing.

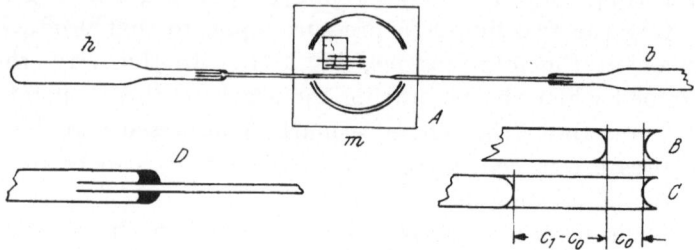

Fig. 137. Calibration of Microburet. b, buret in holder; h, calibrating capillary in glass tube holder; m, moist chamber with carrier and reagent containers; D, use of cement for sealing calibrating capillary into tube. After LOSCALZO and BENEDETTI-PICHLER (154).

To minimize errors due to evaporation, the whole calibration is performed in the moist chamber by collecting the liquid delivered by the buret in a dry calibrating capillary of accurately predetermined bore and calculating its volume from the linear measurement of the column formed. For several reasons, the most suitable liquid seems to be the standard solution which is to be dispensed by the buret. The calibrating capillary, Fig. 137 $A h$, has been described on p. 128.

Fig. 137 A, is a schematic illustration of the set-up used for calibration. It shows the moist chamber m, provided with a carrier and reagent containers, in position on the stage of the microscope with the microburet b and calibrating capillary h introduced on opposite sides of the chamber.

Manipulations involved in bringing a microburet or micropipet and capillary vessels in the plane of the microscope field, in adjusting their position relative to vessels, in taking up and delivery of solutions, in the linear measurement of liquid columns in capillaries, as well as in other operations have all been described in considerable detail in connection with work in capillary cones. For calibration of a microburet, the buret is manipulated in the field of the microscope relative to reagent containers and calibrating capillary by means of the manipulator and the mechnical stage of the microscope.

To start with, the tube holder h should have acquired room temperature. First, a quantity of standard solution is taken from the reagent container into the buret. By means of the leveling device, the operating pressure above the liquid in the buret is adjusted to be slightly higher than the equilibrium pressure. By operating the controls of the manipulator and the mechanical stage of the microscope, the tip of the buret is made to touch the inside wall of the calibrating capillary close to its opening. The liquid will flow from the buret into the calibrating capillary, and two menisci will soon form. One meniscus remains at the opening of the capillary, and the other travels toward holder h. When the distance between the two menisci is approximately equal to the apparent diameter of the bore of the capillary, Fig. 137 B, the buret is withdrawn out of the capillary. The buret is refilled from the reagent container to bring the meniscus of the standard solution in the buret close to the reference mark. The leveling bulb is then raised to re-establish the selected operating pressure in the microburet.

The buret is moved to bring the reference mark on the shank into the field of the microscope. The distance b_0 between the meniscus of solution and the reference mark is measured with the eye-piece micrometer scale. The buret is then retracted for some distance so as to permit the opening of the calibrating capillary to be brought into the center of the microscope field. The drop of liquid in the calibrating capillary is sharply focused, and the distance C_0, Fig. 137 B, between the two menisci of the standard solution in the capillary is measured. The buret is then moved forward and its position is adjusted by means of the manipulator so that its tip appears sharply focused close to the opening of the calibrating capillary. The tip of the buret is then advanced and inserted into the drop of standard solution in the calibrating capillary. The standard solution in the buret starts to flow into the calibrating capillary, and the meniscus inside the capillary starts to move inward towards holder h, while the outer meniscus remains stationary at the opening of the capillary. The flow of liquid from the buret into the capillary is allowed to continue until the distance between the two menisci in the capillary is about three-fourths of the whole length of the ocular micrometer scale. The buret is then withdrawn, and the distance c_1, Fig. 137 C, between the two menisci in the calibrating capillary is measured. The final buret reading b_1, that is, the distance between the new meniscus in the buret and the reference mark, is read as quickly as possible.

The distance $c_1 - c_0$, Fig. 137 C, is the linear displacement of the meniscus in the calibrating capillary corresponding to the volume of standard solution delivered by the microburet. This volume is calculated as the volume of a circular cylinder of a height equal to the distance $c_1 - c_0$ and a diameter d_c which is the diameter of the bore of the calibrating capillary. It is not necessary to make corrections concerning the curved surfaces of the menisci in the calibrating capillary, though the small increase of pressure of the air enclosed in the capillary and the holder will produce a slight change in the curvature of the meniscus at the opening. This slight change, however, can be disregarded.

The volume v, in microliters, of standard solution delivered by the buret per scale division of the eyepiece micrometer is calculated by applying the equation:

$$v = 7.854 \times 10^{-10} d_c^2 \, u^3 \, \frac{c_1 - c_0}{b_1 - b_0},$$

where b_0, b_1, c_0, c_1, and d_c are measured in divisions of the ocular micrometer scale and u is the value in microns or micrometers of one division of this same scale.

In an actual calibration (154) of a buret having a bore of 0.156-mm diameter with three different calibrating capillaries of 0.120-, 0.124-, and 0.133-mm bore and following the calibration procedure described above, the value of one division of the micrometer scale in terms of volume of water delivered by the buret was determined with a relative precision of about 0.016. The results were 67, 65, 65, 67, 64, and 64; on the average (65 ± 1) 10^{-5} µl delivered per scale division.

Relatively large amounts of water remain behind on the walls of the drained buret. For a microburet having a bore diameter of 0.11 mm at the calibrated part, the amount of residual water is approximately 10%, so that the delivered volume is only about 90% of the water contained. It was found experimentally that the percentage of residual water increases gradually as the diameter of the bore is increased and reaches a maximum of about 24% when the bore is made equal to 0.26 mm. The fraction of residual water may be determined by cutting the buret after calibration near the reference mark and measuring the diameter of the bore. The capacity of the used length of tube is then calculated and compared with the volume of liquid actually delivered.

The rate of outflow of liquid from the buret, as indicated by the rate of travel of the meniscus in the calibrated part of the shank during delivery, varies from 0.005 to 0.08 mm per second. Although the amounts of residual water are relatively large, no significant drainage error is introduced, provided the meniscus in the capillary buret travels at the same slow rate in both calibration and use. For comparison, it may be mentioned that the rate of outflow of the horizontal buret of HYBBINETTE (126), 1- to 2-mm bore in the calibrated portion, is about 0.3 mm per second, whereas the corresponding figure for the conventional macroburet, 12-mm bore, is about 5 mm per second.

Titration. The assembly used for titration is the same as that used for calibration, except that the calibrating capillary is replaced by a titration cone, Fig. 87 *B*, which is inserted into a pipet holder connected to a plunger device filled with water. The meniscus *hw* of the hydraulic water is brought to a point approximately 2 cm from the titration cone proper. The working solutions are held ready in reagent containers on the carrier in the moist cell. The tip of the microburet and the opening of the titration cone are made to face one another in the field of the microscope so that they may be readily brought into contact whenever desired.

The solution to be titrated is taken with the buret from the reagent container. The required volume of this solution, which is computed from the displacement of the meniscus in the calibrated portion of the shank, is delivered to the titration cone. The buret is rinsed with water and then with standard solution. It is filled with standard solution, and the position of the meniscus is read. To deliver standard solution into the titration cone, the buret is moved until its tip touches the drop in the cone. To stop the flow of standard solution from the buret tip, the latter is withdrawn from the drop. There are alternatives since contact of the buret tip with the titrated drop can be made or broken by moving the buret, the titration cone, or the drop in the cone.

Efficient mixing of the titrated solution is achieved by making the drop in the titration cone to move back and forth a distance of 0.5 to 1 mm for 3 to 5 times after each addition of the standard. The turbulence in the moving drop produces the required stirring. The motion of the drop is obtained by retracting and advancing the plunger.

When the end-point signal is received, the buret is withdrawn, and the position of the meniscus is recorded. When not in use, the buret may be kept immersed in the standard solution contained in small vials fitted with cork stoppers with holes to introduce the shanks of burets.

β) Working in Coloriscopic Capillaries.

The apparatus, Fig. 89, used by DORF (81) for titration of microgram samples should be made and assembled as described on p. 130. By using the coloriscopic capillary instead of the titration cone, DORF was able to obtain a light path through the small volume of titrated solution which is equivalent to that obtained in customary macrotitrations. By this means the color change at the end point can easily be detected with the customary concentrations of the usual indicators. This does away with the limitations in the choice of indicators when using titration cones and should encourage the development of titrimetry on the microgram scale.

The general manipulations are essentially the same as used for work in titration cones. Use and calibration of the microburet have been described. Therefore, only the use of the titration capillary has to be discussed.

The moist chamber with the titration capillary attached to the plunger device, Fig. 89, is placed on the mechanical stage of the microscope. The narrow vertical capillary is brought in line with the axis of the microscope, and its opening is focused with a low power objective. All observations of the contents of the titration capillary are made with vertical illumination, but transmitted light may be used during manipulation. The mercury column g in the wide portion of the capillary is advanced by turning knurled screw l until the meniscus of the mercury column appears in focus at the opening of the fine vertical capillary. Plunger m is then retracted by turning screw l to give an air space approximately 2 to 3 mm long at the orifice.

The solution to be titrated is taken with the calibrated buret in the usual way from the reagent container. The buret is read. The opening of the titration capillary is again brought into sharp focus. By means of the controls of the manipulator, the tip of the calibrated microburet n is brought over the opening of the titration capillary and very close to it. By means of the pressure control apparatus, pressure in the buret is increased until a droplet forms at the tip of the buret and touches the opening of the titration capillary at o. A measured volume of solution, as determined by the displacement of meniscus in the calibrated buret, is introduced into the titration capillary a. This is done by turning screw l with one hand so that plunger m is retracted and, at the same time, controlling the pressure in the microburet with the other hand. If the pressure is adjusted with the levelling device, Fig. 88, no further adjustment is required during delivery. Flow of liquid from the buret can be stopped at the required instant by breaking contact between the tip of buret and the drop of solution in the titration capillary. This is achieved either by moving the titration capillary by manipulation of the mechanical stage or by moving the buret by means of the manipulator. Indicators may be added either to the solution to be titrated or the standard solution.

The same buret may serve for adding the titrant. After the buret has been properly washed, standard solution is taken by the buret from the reagent container. The opening of the titration capillary and the tip of the buret are again brought into the proper relative position for delivery. Plunger m is advanced by turning screw l until the column of liquid e forms a convex meniscus as shown at o. This is essential for proper delivery of controlled amounts of solution from the microburet. It is also necessary for proper observation of the color of the titrated solution in the capillary. The pressure in the microburet is then adjusted so that delivery of the standard solution from the buret tip into the hemispherical drop at o starts as soon as contact is established. The flow of liquid is stopped whenever desired by retracting the buret. Mixing is achieved by withdrawing and advancing plunger m to move down and up the column of titrated solution in a. Displacement of the liquid for a short distance, four or five times after each addition of standard solution, is sufficient for complete mixing.

e) Working with Hanging Drops.

The assembly of apparatus, Fig. 90, used by BENEDETTI-PICHLER and RACHELE (31) for working with hanging drops has been described on p. 132. It has been used for the determination of the ultimate limits of identification of simple confirmatory tests of qualitative inorganic analysis.

Tests are carried out by treatment of tiny droplets of solution, located at the underside of the thin glass cover of the moist chamber, with very small amounts of reagents. Micropipets with inclined shaftlets, Fig. 91, are used for approaching the field of operation from below. Magnifications up to about 400 diameters are used, and the field of operation is illuminated by means of special condensers with sufficiently long focal length.

The basic techniques used in this type of work will be described. It is assumed that the apparatus, including microscope with a built-in revolving stage and an attachable mechanical stage for holding the moist chamber, micropipet connected to the injection apparatus, micromanipulators, and moist chamber, has been prepared and assembled as described on p. 133.

α) Estimation of Volume of Very Small Drops.

For the determination of the limits of identification it was essential to develop a suitable method for the reliable estimation of the volume of drops delivered by the micropipet. This may be done by using calibrated micropipets, p. 135. In another method, test drop and reagent drop are deposited separately on the suitably treated undersurface of the cover glass of the moist chamber or suspended there in an inert liquid medium. After measuring the volumes, the two drops are allowed to merge.

Calibrated Pipets and Hanging Drops. Calibrated micropipets of the style shown in Fig. 91 b, p. 135, may be used for the estimation of the delivered volume by measuring, with a calibrated eyepiece micrometer, the displacement of the meniscus in the shaftlet of the micropipet and the diameter of the uniform bore of the shaftlet. The volume is that of a cylinder.

$$V = 0.25 \, \pi \, d^2 \, h$$

The diameter of the bore of the shaftlet is measured with a calibrated ocular micrometer while the shaftlet is in the plane of focus and immersed in a drop of cedarwood oil. The use of cedarwood oil, since it has nearly the same refractive index as glass, eliminates the lens action of the glass body of the shaftlet, and thus prevents exaggeration of the image of its bore. The effect can be seen by a comparison of the two photomicrographs of Fig. 138.

The displacement of the menicus in the micropipet can only be measured with the shaftlet located in a horizontal image plane, the focal plane of the objective. Obviously such a position is not suitable for the delivery of drops to the underside of the cover glass of the moist chamber.

These calibrated micropipets are practical, however, for depositing drops on vertical or inclined surfaces, and they have found some use in developing a method for the estimation of the volume of apparently hemispherical drops.

The microdrops were delivered by a calibrated micropipet onto a glass surface coated with a film of cellulose nitrate, p. 236, and their volumes were estimated by measuring the displacement of the meniscus in the

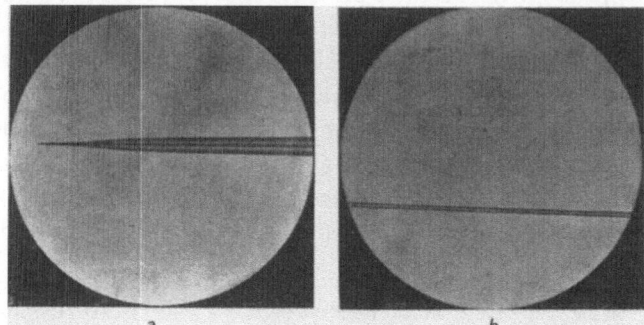

Fig. 138. Calibrated Micropipet. *a*, tip and part of shaftlet showing quick taper at tip, $100\times$; *b*, uniform part of shaftlet immersed in cedarwood oil, outer outlines of glass invisible, only bore filled with 1% glycerol solution is seen, $248\times$. Courtesy of BENEDETTI-PICHLER (31).

shaftlet of the pipet. The volumes of these drops were again calculated from the diameter of the circle of contact of the hanging drop with the receiving surface. When a hemisphere is assumed, the volume becomes,

$$V = c d^3.$$

The values found for the volumes of the drops were compared with those obtained with the calibrated pipet. The two series of data could be made to agree by using for c the empirical value 0.11 instead of the figure 0.262 wich would give the volume of the hemisphere corresponding to the measured diameter of the circle of contact. Though the presence of an appreciable error in both methods of volume determination could be expected, it was possible to prove that the relation between the displacement of meniscus in the shaftlet of the pipet and the diameter of the circle of contact is constant. Thus, either of these measurements may be used to estimate the order of the volume of the hanging drops, and it is possible to estimate from the diameters of the circle of contact the volume of droplets deposited, on the cellulose nitrate coated surface of the cover glass of the moist chamber, with a non-calibrated perpendicular micropipet.

In these experiments, the calibrated micropipets, carried and moved by the left-hand micromanipulator, have been used with their shaftlets in the horizontal focal plane to permit measurement of the length of the column delivered by the pipet. The microdrops were deposited on the vertical surface of a microslide attached to the right-hand micromanipulator. It was found important to deliver liquid with the tip of the pipet perpendicular to the surface of the slide to give the deposited drop a circular area of contact. The microslide, made of a narrow strip of a cover glass— one end coated with a thin film of cellulose nitrate, was attached to the micromanipulator by means of a thin glass thread cemented to the other end of the slide.

Experiments were carried out with 1% phosphoric acid. A large drop of the solution was first delivered to the microslide by means of an ordinary capillary pipet. The microslide was then turned to a vertical position, and the moist chamber was covered. Some of the solution was taken up with the calibrated micropipet and drops were delivered onto the cellulose nitrate film close to the upper edge of the slide. Care was taken not to scratch the cellulose nitrate film with the tip of the micropipet when depositing the drops. The microslide was then rotated to a horizontal position for the measurement of the diameter of the circle of contact. Because of the comparatively short working distance of the objective employed, the micropipet and the microslide had to be operated close to the top of the moist chamber.

It is not recommended to estimate the volume from the dimensions of the profile of the microdrops, seen with the microslide in the vertical position. The side view may not reveal the elliptical shape which may be obtained if the pipet is not properly oriented.

Treatment of Dilute Solutions to Prevent Evaporation. Experience has shown that drops of water, a few microns in diameter, evaporate almost immediately after delivery to the undersurface of the cover glass of the moist chamber. This is due to the high vapor tension of the very small drops. Continuous shrinkage of microdrops of very dilute solutions of electrolytes, of 0.01% and lower, has been observed even in a moist chamber completely closed with water seals and with the inside walls of the cell covered with condensate. The air inside the cell was sufficiently saturated with water vapor to be in equilibrium with the large drops of the condensate, but humidity inside the cell was still insufficient to prevent evaporation of the microdroplets.

Such evaporation was not noticed with equally small microdrops of 1% solution of electrolytes even when using an ordinary metal ring chamber, Fig. 90, p. 126, with one of its working apertures open. Microdrops of such concentration were in equilibrium with the moist air of the chamber. It was subsequently found that the evaporation of microdrops can be prevented by treating water and very dilute solutions of electrolytes with the addition of small amounts of glycerol, approximately 0.05 to 0.2 ml per 10 ml of solution. This treatment prevents also the evaporation of solution from the tip of the micropipet, which must be prevented whenever it leads to an undesirable or misleading concentration increase near the orifice.

Treatment of Glass Surfaces. Suitable treatment of a glass surface which is to receive drops for chemical experiments is very important since the condition of the surface determines the shape and volume of the drops. On clean glass, drops spread out to a thin layer, which is definitely undesirable; the boundary becomes vague, and the dissolved matter is dissipated. Greasing of glass surfaces in any such way as by rubbing with an oily cloth gives to the drop some sort of curved shape which is not reproducible. On a strongly water-repellent paraffin coating, drops of aqueous solutions form almost complete spheres, but are most difficult to deposit since they preferentially adhere to the glass of the micropipet.

Coating a glass surface with a film of cellulose nitrate gives reproducible drops of apparently hemispherical form which is suitable for the estimation of volume, p. 234. Microdrops as small as 2 μm in diameter at the base, 10^{-9} μl $= 10^{-3}$ pl volume, can easily be obtained.

The coating may be prepared from Parlodion, a cellulose nitrate product of DU PONT (85), supplied in strips. A solution, prepared by dissolving one gram of the material in a mixture of 30 ml of ethyl alcohol and 30 ml of ethyl ether, is poured on the clean glass surface. Excess solution is drained off by inclining the surface. The cover glass is then placed on a horizontal support and the solvent is allowed to evaporate. A perfectly transparent and reproducible coating is obtained. Microdrops deposited on a glass surface thus treated, with the tip of the micropipet perpendicular to the receiving surface, assume an apparently hemispherical shape which is reproducible and has a circular area of contact. Concerning the computation of the volume see p. 234.

Estimation of Volume of Spherical Drops. Chemical reactions may be performed by suspending the test drop and the reagent drop close to each other in a thin layer of paraffin oil, and merging the two drops after measuring their volumes. A microdrop suspended in an inert liquid medium assumes a spherical shape, and its volume may be calculated by measuring its diameter with the eyepiece micrometer.

$$V = 0.167\,\pi\,d^3.$$

The relative error of the computed volume should be three times the relative error committed in measuring the diameter. Aggravating factors may be at work, however, as LUNDE (156) found for the estimation of the volumes of spheres of noble metals. The estimation of volumes of small microdrops a few microns in diameter, whether these drops are spherical or hemispherical in shape, is only approximate since the precision of the microscope measurements on which the calculations of these volumes are based is limited by the resolving power of the objective used.

β) Determination of Ultimate Limits of Identification.

The performance of tests directly on the cellulose nitrate coating of the undersurface of the cover glass of the moist chamber proved impractical. The curved liquid-air interface and the large difference between the refractive indices of liquid and air produce reflective and refractive phenomena which make difficult the observation of details in the interior of the drop or of minute quantities of precipitates. Furthermore, after measuring the volumes of the test drop and the reagent drop, the latter has to be taken up by the micropipet and transferred to the test drop. In performing this last operation, it is often difficult to prevent some test solution from entering into the tip of the pipet when it is inserted into the test drop. Trials made to deposit the two drops in close proximity so that they finally touch and merge were not successful.

A suitable technique for performing tests may be based upon suspending the drops in a thin layer of paraffin oil. Better conditions for observation are obtained with this technique because of the smaller difference between the refractive indices of the oil and the work drops.

The cellulose nitrate coating is retained to prevent aqueous test drops which happen to touch the surface of glass from spreading out into a thin layer between glass and oil.

The paraffin oil is brushed on the cellulose nitrate coating to give a film, 0.05- to 0.1-mm thick. The use of a thick layer of paraffin oil is not recommended since the withdrawal of the tip of the micropipet from a large drop of oil would impart violent motion to the floating test drops, which usually does not lead to their merging.

A number of immersion liquids were tried, but paraffin oil proved to be the only one practicable because of its nonvolatility, immiscibility with aqueous solutions, and nonreactivity. Furthermore, the small difference between the refractive indices of the paraffin oil and the test drops permits proper observation of the boundaries of the microdrops. The use of liquids like ethyl oxalate having refractive indices very close to that of the test drops is impractical since the suspended drops would show only very faint boundaries which are difficult to observe.

The addition of glycerol to water and to the more dilute solutions, p. 235, is retained to prevent evaporation of solvent from the tip of the micropipet and to suppress the definite shrinkage that has been observed in the diameter of microdrops while floating in paraffin oil.

Procedure. It is assumed that the apparatus, Fig. 90, has been prepared and assembled as described on p. 133. The microscope is properly adjusted, and the metal ring moist chamber, one of its two apertures open and the cotton linings soaked with water, is in position on the stage of the microscope. Two micropipets of the style shown in Fig. 91a, are properly connected to the injection devices, in position on the two micromanipulators, introduced into the moist chamber, and aligned approximately in the desired position. The tips of the pipets are pointing upwards, and their shaftlets are slightly inclined to one another so that they would form an inverted V if the tips were made to touch one another.

One side of the clean cover glass of the moist cell is coated with a film of cellulose nitrate as described on p. 236. The coated side receives a thin layer of paraffin oil applied with a camel's hair brush to an area of about 1 cm² in the center of the slide. By means of ordinary capillary pipets a drop each of test solution and of reagent are deposited on the cellulose nitrate coating close to the oil film. The cover slip is then quickly placed in position on the moist chamber.

With the use of a low power objective, approximately 5×, the two micropipets are finely adjusted to bring their tips into sharp focus in the center of the field. To this end, an empty area of the cover glass of the moist chamber is brought into the center of the field by means of the mechanical stage. This area is illuminated by the cone of light coming from the sub-stage condenser. The focal plane of the objective is then brought a few millimeters below the cover glass by lowering the microscope tube. By means of direct observation from the side, the tip of one of the pipets is brought into the cone of light coming from the condenser. This is achieved by moving the pipet horizontally by means of the micromanipulator. Observing through the microscope and using the proper controls of the micromanipulator, one brings the tip of the pipet into sharp focus in the center of the field. By means of direct observation, this pipet is then lowered out of focus to a position close to the bottom of the chamber. This is done with only the fine vertical control of the micromanipulator. The tip of the second pipet is then brought into sharp focus in the center of the field by using exactly the same procedure as with the first pipet. The first pipet is then carefully raised by means of the vertical control of the micromanipulator to bring the openings of both pipets, side by side, into the center of the field.

The undersurface of the cover glass is focused, and the drop of the test solution, previously deposited, is brought to the center of the field with the mechanical stage. By means of the micromanipulator, the tip of one of the micropipets is carefully raised to enter the test drop while a slight pressure is applied to the piston of the corresponding syringe. The pressure on the piston is slowly released to allow a small volume of solution to enter the micropipet. The pipet is then lowered. The reagent drop is then brought into the center of the field, and a small quantity of the reagent is taken up by the second pipet.

The chamber is then moved by means of the mechanical stage to bring a part of the oil film into the field of view. The objective is exchanged for one of high power. The tip of the micropipet containing the test solution is carefully raised to enter the oil, and a microdrop of the solution is delivered into the paraffin oil layer by exerting pressure cautiously on the piston of the corresponding syringe. Since the aqueous solution preferentially wets glass, the microdrop adheres to the tip of the pipet, but on withdrawing the point of the tip through the surface of the oil film, the microdrop is stripped off the tip, floats in the oil layer, and assumes a spherical shape. Since the layer of oil is very thin, approximately 0.05 to 0.1 mm in thickness, the suspended microdrop usually remains without motion in the microscopic field. Any slight movement is compensated by means of the mechanical stage to keep the microdrop in the field of vision until the motion stops. Then the diameter of the spherical droplet is measured and recorded.

The tip of the micropipet containing the reagent is carefully raised to enter the oil layer, Fig. 140a, and a droplet of reagent, smaller than the test droplet, is delivered close to the floating droplet of the test solution. The diameter of the reagent droplet is also recorded.

Merging takes place by itself since the smaller droplet is invariably attracted to the larger one, provided that the two droplets were not deposited too far apart. Droplets of like size usually do not merge unless the two drops are brought very close to one another. This may be achieved by stirring the oil with one of the micropipets so as to push the droplets together.

In depositing the reagent droplet, the micropipet must not be brought too close to the suspended test droplet since otherwise the test drop would be attracted to the glass of the pipet.

All available optical means may have to be used for the observation of the outcome of the experiment.

3. Working with Living Cells and Tissues.

Operative work on the unstained and living animal and plant cells and tissues has long been the aim of investigators in various fields of biology. It should be mentioned that microbiologists were the first to conceive and apply micromanipulative techniques. The main objective, as already indicated, had been to provide means for experimental work in the microscopic field on living matter and for the study of the consequences of these

experiments on the living structure rather than the mere observation of
the dead and stained specimen.

H. D. SCHMIDT (202, 203) seems to have been the first to use a microscopic dissector which he described in 1859. Since that time, micromanipulative techniques have attained a high degree of perfection. The development of the varied types of exceedingly delicate microtools and other
auxiliary microdissection instruments went hand in hand with the development of the micromanipulators, and the significant advances in microscopy aided greatly in that direction.

Biological microdissection, for which PETERFI (186) coined the name
micrurgy, has proved of great and growing importance in experimental
cytology, embryology, bacteriology, and cellular physiology, as well as in
other fields of the study of living cells and tissues. The objects of investigation usually are minute structural elements of living matter, localized
tissue and individual muscle fibers, individual cells or parts thereof, individual bacteria and other small organisms. These may be mechanically
probed, dissected, and exposed to various chemical and physical influences
under highly controlled conditions. The development of the stereoscopic
microscopes with sufficiently long working distances then also permitted
the performance of micrurgical operations on intact delicate organs *in situ*
such as specific centers in the brain of small anaesthetized animals. The
meaning of the word micrurgy has later been extended by TITUS and
GRAY (228) to include manipulations under the microscope applied to the
investigation of non-biological material, p. 267.

The choice of a suitable procedure is mainly dependent on the type
of material and the specific problem to be investigated. The basic technique involved in performing operative work on living cells, tissues, and
various microscopic organisms does not vary greatly. It usually consists
of mounting the necessary microtools on one or more micromanipulators
so that the operating tools project over the stage of the microscope into
a moist chamber carried and moved by the mechanical stage of the microscope. The specimen to be operated upon is usually contained in a shallow
drop of a physiologically indifferent fluid, suspended from a thin cover
slip roofing the chamber. The general arrangement, a significant development introduced by BARBER early in this century (8, 10, 12, 55), p. 30,
Fig. 10, is similar to that described for performing chemical operations
on hanging drops, p. 132. The operating microtools, with bent-up tips,
approach the specimen from below, Figs. 47, 91, and 140*a*, so that there
is no obstacle between the cover slip and the microscope objective. This
permits the use of high power objectives including oil immersion lenses,
which is necessary for operative work at cellular and subcellular
levels where the object of investigation may be of the order of a few
microns (μm) in diameter or less. It may be required to dissect a single

living cell, to remove its nucleus, or to inject minute quantities of liquids into it. For such operations the excursions of the points of the operative microtools take place in a space which usually does not exceed a fraction of a millimeter. A wide field of view offered by low power objectives is not needed but for preliminary adjustments. In hanging drop preparations under the high power microscope, the different steps of delicate micro-operations can be properly observed in detail. The various movements of a micromanipulator carrying the microtools added to those of the mechanical stage permit displacing the specimen and the operating tools against each other in any desired directions; they offer the necessary flexibility of motion needed for performing the required micro-operations. In a shallow hanging drop preparation, the specimen of interest tends to retain its position because of the nearness to interfaces harbouring forces that interfere with the mobility. If the hanging drop is shallow enough, it may be possible to dissect a sufficiently soft cell like many marine ova and germ cells, for instance, by pressing it with a suitable microneedle against the undersurface of the cover glass. As a rule, however, if such operations as dissections, injections, or extractions are performed, it is usually more convenient and even necessary to hold the cell or tissue by means of a microneedle and to use another microtool, a needle or a pipet, for performing the actual micro-operation. Certain fluorocarbons are used for hanging drop preparations (145). One advantage of these non-aqueous fluids is that, due to their high density, cells rise in the hanging drop and come to rest against the lower surface of the cover glass, a position which is quite suitable for micrurgic operations. Because of the low surface tension of these fluids, the glass surface should first be coated with a thin film of silicone grease to reduce the spreading of the drops.

Before the technique of working with hanging drops was developed, the specimen to be dissected was usually placed into a drop resting upon a glass surface on the stage of a conventional microscope. In such a set-up, the operating microtools had to be manipulated between the specimen and the microscope objective. This imposed serious limitations upon the initial magnification of the microscope since it necessitated the use of objectives giving relatively low magnification to provide a sufficiently large working distance suitable for the purpose. Furthermore, it did not allow proper observation of the whole field of operation because of the interference of the tool points. This type of arrangements, however, is still used at present in certain operations on comparatively large objects requiring a wide field of view and low to medium magnifications with a stereoscopic microscope. In such investigations use is made also of the long working distance and other advantages offered by the stereoscopic vision.

With the use of the inverted microscope, p. 20, Fig. 9, micromani-

pulative operations may be conveniently performed in lying drops placed
on the thin cover glass forming the floor of the moist chamber on the
microscope stage. Because of the inversion of the microscope, the objective
occupies a position under the cover glass while the fluid drop lies upon it.
The operating microtools, with bent down tips, approach the specimen
from above. As in the instance of working with hanging drop preparations,
this arrangement also permits the use of high magnifications since the
operating tools are not in the way between the specimen and the microscope objective. For the same reason, it also permits proper observation
of the whole field of operation. In this arrangement the substage of the
microscope, which usually includes both the condenser and the illuminator,
lies above the moist chamber. It is possible, of course, to use the various
types of illumination including dark-field. An advantage of the inverted
microscope is that the lying drop may be comparatively large to reduce
the effect of evaporation. In addition, the specimens come to rest on the
upper surface of the supporting cover glass. This makes manipulations
easier (142) since, in contrast to the usual practice when working with
hanging drop preparations, it is not necessary to hold the cells in position
with a microneedle during the performance of such operations as injections
or extractions with micropipets.

It is not intended to outline the numerous micrurgical procedures
applied by various workers in different fields of the investigation of living
cells and tissues. The approaches largely depend upon the material and
the problem to be investigated. A considerable amount of work involving
micrurgy has been published in the literature.

Basic micrurgical operations are adequately described by DE FON-
BRUNE (75) and CHAMBERS and KOPAC (62). A general account of the
development of micrurgical techniques and their applications until 1940
has been given by CHAMBERS (59). A lengthy report concerning a number
of methods for single cell isolation was given by DICKINSON (80). A number
of valuable papers dealing with more recent developments in different
fields of cellular micrurgy, instrumentation, and applications, was published
by KOPAC (140, 144, 145, 146, 147). The same author has also published
an extensive review (142) on cytochemical micrurgy, dealing with different
methods and equipment available for various quantitative cytochemical
investigations on isolated cells or parts thereof. The review also deals with
the preparation and handling of cells for various micrurgical cytochemical
investigations. Included are free cells such as the amoeba, marine in-
vertebrate eggs, and the ascites tumor cells, as well as cells to be separated
from solid tissues.

Among the more prominent advances in cellular micrurgy achieved
during recent years, are the development of the volumetric submicromanipulator, p. 39, and of the television micromanipulator, p. 40. These and

other advances in micrurgical instrumentation and technique enable proper handling and testing of small subcellular fractions.   Such developments are important to the enzyme chemist, for instance, since the living cell is an enzyme-producer in which many enzymatic processes take place. Measured volumes of subcellular components of the order of 1 pl may be removed not only from a single living cell but from parts of a cell, and transferred either to substrate mixtures for testing the enzymatic activity or implanted into another cell.   Direct study of the cytochemical and other properties of nucleoli became also possible.   Nucleoli may be removed micrurgically from the nucleus of the living cell.   These subcellular structures may be transplanted from one cell to another or transferred to the cytoplasm of the same cell.   Micrurgic techniques also enable the isolation of chromosomes and their transplantation to other cells.   Many references involving micrurgy and other micromanipulative techniques, equipment, and applications in various fields are listed in the extensive general bibliography given on p. 315.

For successful work however, the operator, should have sufficient practice in the various basic operations encountered in the setting up of and working with the various equipment.   With certain modifications they are usually applicable to the specific problem under investigation.   Of particular importance is the proper setting up of the micromanipulator, microscope, injection device, and other accessories; adequate illumination of the microscope field; preparation of the specimens for investigation in hanging or lying drops; making proper use of the various motions offered by the micromanipulator; rapid change of microtools and practice of manipulating them in the field of operation.

A few classical examples, due to SCHOUTEN, which are given briefly in the part devoted to applications, p. 304, may reveal some essential features of micrurgy, as well as its potentialities for the study of cells and tissues. The experiments involved, described by WORST (243), are well illustrated in Figs. 173 to 175.   The preparation of the SCHOUTEN loop using the DE FONBRUNE microforge, p. 152, is described on p. 162, Fig. 107; that of standard microneedles, on p. 161, Fig. 106.   A variety of microdissection needles which may be prepared by different methods are also described in appropriate places together with other microtools.

Part III

Applications.

1. Chemical Experimentation and Analysis with Micrograms to Nanograms.

Chemical operations have been successfully performed by numerous workers with amounts of material considerably smaller than 0.1 to 10 μg in mass, while the volumes of the solutions were from 1 nl to 1 μl so that, in general, the customary concentrations were retained. This indicates that chemical reactions take place in the usual way with such minute amounts of material and give the same results as on the large scale. This confirms conclusions derived from theoretical considerations (31) that, when dealing with substances of an assumed average molecular weight of 100, reproduction of qualitative reactions can be expected with amounts of material of the order of 10^{-20} g corresponding to 100 reacting molecules, and of quantitative reactions with amounts of the order of 10^{-16} g corresponding to 1,000,000 reacting molecules. It is assumed that the equilibrium is sufficiently in favor of the expected effect. With these amounts of reacting material, the number of molecules present should be sufficient to statistically ensure the establishment of average conditions to permit the kinetics of reactions to follow the customary observed course for given conditions of temperature and pressure. A chemical reaction should therefore be expected to follow the same course on any small scale down to the above mentioned limits as long as the reactants are given in the same concentrations. This indicates that microchemical work is possible even with quantities considerably smaller than micrograms and nanograms of material.

Whereas no provisions are necessary for ensuring that reactions of classical and microanalytical chemistry take their expected courses as long as not less material is taken than indicated by the above limits, special care must be taken, however, to provide for proper manipulation and observation. A careful appraisal indicates that it may be difficult to practically attain limits of the order of 10^{-20} g with the customary analytical tests even under especially favourable conditions and with the best optical and mechanical aids. This is due to various limitations among which may be mentioned the increase in the relative surface of the solution, the lack of material for the growth of crystals sufficiently large to be observed, the increase of solubility with the decrease of the particle size, the possibility of delayed precipitation, and limitations of microscopic visibility and resolution.   For example, under fairly good conditions (31), the limits of identification that could be attained for barium, p. 257, and iron, p. 258, were 10^{-14} g and 4×10^{-13} g, respectively, which is one million times above the given theoretical limit.

Microgram and nanogram procedures, however, lie well above the practical limits of perception since the quantities of material ordinarily involved are as much as 10^{-9} to 10^{-5} g.

Apparatus and manipulative facilities permit conventional chemical experimentation as well as qualitative and quantitative analytical methods on the microgram scale merely by proportional reduction of the volumes of the solutions so that approximately the same concentrations prevail as on the classical scale.

Assemblies and apparatus have been described for qualitative analysis, p. 92, semi-quantitative estimations, p. 119, and gravimetric analysis, p. 119, performed on samples of micrograms to nanograms in mass. Manipulations involved in performing basic operations have also been given with reference to qualitative, p. 208, semi-quantitative, p. 221, and gravimetric, p. 224, analysis.

Apparatus for work with submilligram quantities, of the order of 1 to 100 μg, in proportionately larger volumes of solutions is described on p. 116. The simplified techniques should also be useful in the investigation of a wide variety of industrial problems.

a) Qualitative Analysis.

Generally speaking, any reliable qualitative scheme of analysis can be successfully adapted for application on the microgram scale using apparatus, p. 92, and the techniques, p. 208, already described for working in capillary cones.   The performance involves certain modifications of the usual practices that are applicable either on the classical or on the milligram scales. Certain techniques are physically impossible on the microgram scale and, in consequence, require complete replacement.   Others, not completely impracticable, are preferably avoided because of the difficulties which may be encountered in applying them to the microgram scale. Such difficulties by no means constitute serious limitations since an alternate solution can be found in practically all such instances.

For example, it is impracticable to bubble a gas through a solution on the microgram scale, and consequently such practice requires complete replacement. By using the device (25, 28) shown in Fig. 133, it is possible to saturate the contents of a capillary cone with a gas, p. 218. In performing confirmatory tests on the plateau of the condenser rod, the use of solid reagents as in slide tests is impracticable since tiny particles of somewhat hygroscopic substances would liquify as soon as they are introduced into the moist chamber; therefore, liquid reagents are usually taken.   Their use is preferable also for the reason that small amounts of liquids can be measured quite readily. Again, with nanogram amounts of material, the use of the crystal form as criterion for the identification (65) is usually not practical; frequently the crystals of the precipitates are so small that their shape can no longer be recognized under the light microscope. A great number of tests that are dependent on the appearance or disappearance of phases or (and) color are already in existence (96, 236), however, and many of these tests are well suited for the microgram and nanogram scales. In selecting confirmatory tests for the microgram scale it is generally preferable to choose selective tests which give intensely colored precipitates to ensure a high degree of sensitivity and specificity.

α) Separation and Identification of Lead, Mercurous Mercury, Silver, and Tungsten.

The ions lead, mercurous mercury, silver, and tungstate comprise group II of a non-H_2S scheme of analysis (87) on the milligram scale, which includes a number of the *less familiar* elements in addition to the *common* elements of the classical scheme. In this scheme, Group I, comprising sodium and potassium, is examined by a separate procedure, and the schematic analysis properly starts with Group II.

This scheme of resolution of the silver group ions was adopted for the analysis of microgram samples (91). The microgram scheme is diagramatically given below. In performing this work, EL-BADRY and WILSON's assembly (90), Fig. 63, was used, but other assemblies of the type shown in Fig. 80 may also be employed. The manipulative techniques required are those already described for qualitative analysis of microgram samples, p. 208.

Reagents. (1) Hydrochloric acid, 5 m. (2) Ammonium acetate, 10% aqueous solution. (3) Potassium cyanide, 5% aqueous solution.

Procedure. Treat 50 nl of the original solution with an approximately equal volume of 5 m HCl in a capillary cone. Seal the cone into a heating capillary, centrifuge lightly, and heat for 5 minutes just below 100° C. Place the sealed capillary in cold water for about 15 minutes. Centrifuge the cone and transfer it to the carrier.

Solution 1	Residue 1
Oxidized Hg_2^{++} and subsequent groups. If the presence of Hg_2^{++} is suspected, confirm on a fresh sample of the original solution (ammonia test).	Wash twice with about 10 to 20 nl of distilled water from a container. Add about 30 nl of 10% ammonium acetate solution, agitate, and centrifuge lightly. Repeat the extraction twice with about 20 nl portions of the ammonium acetate solution. Combine all the extracts in a single receiving cone. Wash the residue with about 20 nl of distilled water, and discard this washing.

Solution 2	Residue 2
Test for Lead.	Add approximately 30 nl of 5% KCN.
(Dithizone test, potassium chromate test.)	Agitate, and heat for a few minutes in a heating capillary at 80 to 90° C. Centrifuge lightly. Repeat the extraction till the precipitate has been completely dissolved. Combine all the extracts in a single receiving cone. Divide the combined extracts into two equal portions.

Portion A	Portion B
Test for Tungsten.	Test for Silver.
(Stannous chloride test.)	(p-Dimethylaminobenzalrhodanine test.)

This microgram procedure for the separation of the silver group ions differs from the milligram procedure (87) in a number of points.

(1) In the milligram procedure, 10 m hydrochloric acid is used as the group precipitant. If this is used on the microgram scale, the hydrochloric acid vapor tends to precipitate the ions of the group when the container charged with the *unknown* solution is placed upon the carrier beside the container of reagent. Such precipitation can occur to an appreciable extent before the sample is measured and transferred to the working cone, and this is clearly undesirable. Because of this, 1:1 hydrochloric acid is recommended as the group precipitant. Even with this modification it is still considered advisable to measure the sample before introducing the reagent container of hydrochloric acid into the moist cell.

(2) After precipitation with hydrochloric acid and heating for 5 minutes at 100° C, as in the milligram procedure, mercurous mercury is completely oxidized to the mercuric state and goes into solution. Oxidation of the ion is enhanced when the original solution contains nitric acid. Consequently it is no longer properly included in Group II of the scheme of resolution. It will pass on to Group IV of the scheme and be detected as mercuric mercury. The presence of mercurous mercury must be confirmed by the ammonia test with a fresh sample of the original solution.

(3) The extraction of lead chloride by hot water, used in the milligram procedure, is not practicable on the microgram scale. Lead chloride reprecipitates in the shaft of the pipet during transference of the water extracts. The water extracts then fail to give satisfactory confirmatory tests. Use of 10% ammonium acetate solution for extraction is satisfactory. Extraction and transference give no trouble, and positive confirmatory tests are always obtained.

(4) Extraction with ammonium acetate in the cold leaves silver and tungsten undissolved.

(5) Unexpectedly, the milligram procedure by which silver chloride is extracted by ammonia and reprecipitated as silver iodide, to separate it from tungsten, proves difficult. In the process of dissolution a considerable loss of silver chloride is found to occur. The reprecipitation as silver iodide from the ammoniacal solution proves to be tedious and time-consuming. An adjustment of the pH, rather difficult to carry out on the microgram scale, is necessary. Finally, confirmatory tests on the very small amounts of silver iodide that are recovered give unreliable results. In contrast, the yellow tungstic acid can be dealt with by a direct reduction of the milligram procedure.

(6) Five per cent potassium cyanide solution can be used as a solvent for silver chloride and tungstic acid regardless whether they occur separately or combined. Both ions can then be identified in the presence of each other in the cyanide solution.

Summarizing, one may conclude that reagents with high vapor tension are likely to cause trouble since they will quickly escape from where they are wanted and, because of the small space inside the chamber, get into capillary cones and adjacent reagent containers and produce undesirable changes. Absorption of the vapors by the wet lining of the chamber may have disturbing consequences. In addition, it may be pointed out that the technique is not designed for work with hot solutions.

Confirmatory Tests.

Dithizone Test for Lead.

Reagents. (1) Diphenylthiocarbazone, 0.05% solution in carbon tetrachloride. The solution is unstable and should be freshly prepared every few days. (2) Potassium cyanide, lead-free, saturated aqueous solution. Ordinary *pure* potassium cyanide often contains traces of lead which give a positive test with dithizone. *Analar* potassium cyanide appears to be free of lead, but it is advisable to carry out a blank test with the potassium cyanide used. (3) Ammonium acetate, 10% aqueous solution.

Procedure. Deliver approximately 20 nl of the test solution with a thoroughly clean micropipet to a capillary cone. Add approximately the same volume of the potassium cyanide solution and agitate the contents of the cone with the shaft of the micropipet. Wash out the micropipet once with distilled water from the syringe, several times with ammonium acetate solution, and again with distilled water. Take care to wash also that portion of the outside of the shaft which will be introduced into the cone.

Take up about 20 nl of dithizone solution into the micropipet. Any change of color from green to red at this stage indicates that the pipet is not completely clean. If the reagent remains green, add it slowly to the contents of the capillary cone.

If lead is present, the green reagent turns red. The reagent and test solution may be agitated lightly by means of the pipet, but if lead is present in any appreciable amount, the change is almost immediate and there is no necessity for such agitation. This test is excellent; its main drawback is its extreme sensitivity. Every precaution must be taken to ensure that pipet, cone, reagents, and wash liquid are completely free of lead.

Potassium Chromate Test for Lead.

Reagent. Potassium chromate, saturated aqueous solution.

Procedure. Treat 20 to 30 nl of the test solution in a capillary cone with an approximately equal volume of the reagent solution. If lead is present, yellow lead chromate separates out. This test may be performed on the plateau of the condenser rod.

Ammonia Test for Mercurous Mercury.

Reagents. (1) Hydrochloric acid, 5 m. (2) Ammonia, 35%, 17 m, Sp. Gr. 0.88.

Procedure. Treat 20 to 30 nl of the sample with an approximately equal volume of hydrochloric acid in a capillary cone. Centrifuge, and replace the cone on the carrier. Remove the supernatant liquid, wash the precipitate with about 30 nl of distilled water from a container, and then add about 30 nl of strong ammonia without agitation. If mercurous mercury is present, the precipitate turns dark grey or black, as seen either by transmitted or reflected light.

p-Dimethylaminobenzalrhodanine Test for Silver.

Reagents. (1) p-Dimethylaminobenzalrhodanine, saturated acetone solution. (2) Nitric acid, 2 m.

Procedure. Impregnate a fine cotton thread fixed on a holder, Fig. 77, with the p-dimethylaminobenzalrhodanine reagent throughout most of its length. Place the holder on a carrier so that the thread is horizontal. With the micropipet add about 20 nl of the test solution to the thread, and follow this with approximately 20 nl of nitric acid. If silver is present, a violet color is developed. This can readily be distinguished from the portion of the thread colored by reagent only, which remains orange. The color is best observed by reflected light.

Even better results may be obtained by applying first the silver solution to the thread, following this with the reagent, and finally adding the nitric acid.

Stannous Chloride Test for Tungsten.

Reagent. Stannous chloride, 25% solution in hydrochloric acid, 10 m.

Procedure. Transfer 20 to 30 nl of the test solution to the plateau of the condenser rod. Allow this to come almost to dryness by partially opening the cover of the moist cell. Then transfer an approximately equal volume of the stannous chloride reagent to the plateau so as to cover the film left by the test drop. If tungsten is present, a blue precipitate is gradually produced. This is best seen by applying illumination through the condenser rod.

β) Separation and Identification of Cerium, Thorium, Lead, Calcium, Strontium, and Barium.

These ions, together with titanium and zirconium, comprise group III of the milligram scale non-H_2S scheme of analysis (87) already referred to, p. 246. THOMPSON and WILSON (229), adopted the scheme of resolution of this group to the microgram scale including suitable confirmatory tests for the six ions cerium, thorium, lead, calcium, strontium, and barium. Titanium and zirconium, a very small fraction of which was found to be precipitated by the group reagent, on the microgram scale, were excluded from consideration in group III since they will be detected in subsequent groups. A few other modifications, based on experimental findings, were introduced into the original milligram procedure to make it suited for application on the microgram scale.

Any of the assemblies shown in Figs. 63 and 80, can be used. The techniques are those already described, p. 208, and applied to the microgram scale separation and identification of group II ions of the same non-H_2S scheme, p. 246.

γ) Separation and Identification of the Common Ions of the Hydrogen Sulfide Group.

BENEDETTI-PICHLER and CEFOLA (25, 29, 30) demonstrated the efficiency of the general technique of working in capillary cones by applying it to the separation, semi-quantitative estimation, and identification of microgram to nanogram amounts of the common elements of the hydrogen sulfide group. The techniques proved well suited for the analysis of complex samples of about 1-μg mass or less requiring lengthy operations followed by sedimetric estimations and confirmatory tests.

This was demonstrated (30) by application of the techniques to the complete analysis of a particle of WOOD's alloy by adopting, with some modifications, the macroanalytical scheme of Swift, in which arsenic is isolated by distillation. Filings of the alloy were spread on a microscope slide placed level on the top of the chamber, and an approximately spherical particle, 66 μm in diameter, was selected and brought to the center of the field. Such a particle would be about 1 μg in mass since the specific gravity of WOOD's alloy is about 7. The selected particle was transferred to the capillary cone by means of a micropipet with sealed, slightly bent tip, Fig. 139, and the technique described on p. 221. The particle in the cone was dissolved in 10 nl of 16 m nitric acid, and subsequent operations were performed with techniques already described in detail, p. 208: passing

H_2S gas into solution in a capillary cone, p. 218; distilling p. 219; confirmatory tests for the ions were performed on the plateau of the condenser rod, Fig. 132, pp. 105, 216.

Limits of identification of a number of confirmatory tests for the ions of the hydrogen sulfide group, performed on the plateau of the condenser rod, were determined (29) with 0.1% solutions of the ions. Most of these tests gave positive results with 1 nl of test solution containing 1 ng or less of the ion concerned.

In the above work, assemblies of the type shown in Fig. 80 were employed, but the assembly shown in Fig. 63 can be used as well.

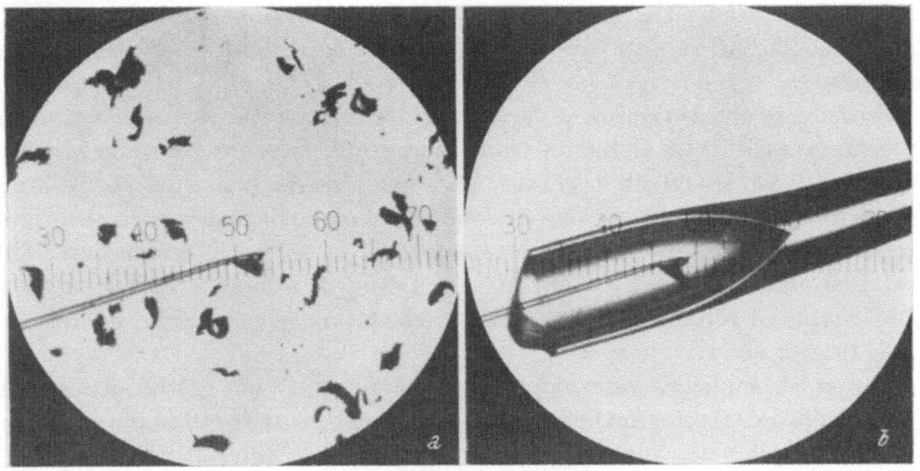

Fig. 139. Photomicrographs Showing Transfer of a Particle of Wood's Alloy of about 1-µg Mass from Slide to Capillary Cone. *a*, selected particle touched with needle point; *b*, particle before deposition inside cone (30). Courtesy of BENEDETTI-PICHLER.

b) Semi-Quantitative Estimations.

In practice, semi-quantitative determinations of microgram amounts of ions by estimation of the volumes of precipitates may be carried out either during the course of qualitative analysis or as independent experiments. The method should be particularly useful where a microgram balance is not available. The method of measuring the volume of precipitates in capillary cones is described on p. 222.

For semi-quantitative estimation it is important to choose a suitable precipitating form that yields a precipitate which is fine and which compacts thoroughly and reproducibly. For example, experience has shown that lead chloride is not suitable because of the long needle-like crystals of the substance which form and fail to pack uniformly in the taper of the cone. In contrast, the fine crystalline lead chromate is a suitable precipitating form (92). It was also found (92) that the apparently gelatinous yellow tungstic acid is quite suitable for semi-quantitative estimation of the tungsten content. Benedetti-Pichler (25, 28) has shown that equal amounts of the elements arsenic, antimony, mercury, silver, lead, bismuth, and copper give sulfide precipitates of approximately equal volumes which are suitable for semi-quantitative estimation, whereas the volume of the sulfides of tin and cadmium, which are frequently precipitated in a gelatinous state, shows wide variations. It thus appears that the appearance of a precipitate is not always a reliable indicator of its suitability for semi-quantitative estimation, which is best determined by trial.

It is only reasonable to expect that voluminous precipitates, i. e., precipitates having a high specific volume, should prove advantageous. This seems confirmed by the good precision obtained with the voluminous lead vanadate. So far, reports are missing concerning the use of organometallic precipitates.

Procedures for obtaining precipitates suitable for the semi-quantitative determination of the elements lead, mercurous mercury, silver, tungsten and vanadium are given together with the average value of the volume of precipitate produced per one microgram of element under the conditions specified and the amount of uncertainty to be expected in estimation. The first four ions, comprise Group II of a non-H_2S scheme of analysis (87). The scheme of separation of the elements of this group on the microgram scale (91) is shown on p. 246.

The stock solutions referred to below contain 10 mg of the element in 1 ml of solution (10 mg/ml). The precipitates are centrifuged before measuring by applying 20 rapid turns to the crank of a hand-driven centrifuge at as constant (reproducible) a rate as possible.

Estimation of Lead as Lead Chromate. The volume of the lead chromate precipitate serves as measure of quantity.

Procedure. Fifty nanoliters of the stock solution (lead chloride in 10% ammonium acetate) are treated, in a capillary cone, with an approximately equal volume of saturated aqueous potassium chromate solution. The cone is centrifuged, and the volume of precipitate is measured.

In a series of ten estimations performed (92), the mean value obtained for the estimated volume of lead chromate was 32 nl of precipitate per microgram of lead. This value was reproduced with a coefficient of variation (relative standard deviation) of about $\pm$ 0.07.

In the qualitative scheme, p. 246, lead is precipitated as the chloride and may then be dissolved in ammonium acetate and estimated semi-quantitatively as the chromate and with the precision quoted.

Estimation of Mercurous Mercury as Mercurous Chloride. The qualitative scheme, p. 246, does not permit collection of Hg_2Cl_2. If the chloride is precipitated without heating, however, it is suitable for the estimation of quantity.

Procedure. Fifty nanoliters of the stock solution of mercurous mercury are treated with an approximately equal volume of 5 m hydrochloric acid in a capillary cone. The fine white precipitate of mercurous chloride is collected in a compact form at the tip of the cone and measured.

The mean value obtained in a series of ten estimations (92) performed according to this procedure was 10.9 nl of precipitate per microgram of mercury. This value was reproduced with a coefficient of variation of about ± 0.12.

Estimation of Silver as Silver Chloride. The silver chloride precipitate, centrifuged without heating, collects as a dense plug quite suitable for the estimation of volume. If the precipitate is heated before centrifuging, it usually gathers in a rather irregular mass, and the reproducibility of volume suffers.

Procedure. Fifty nanoliters of stock solution of silver are treated with an approximately equal volume of 5 m hydrochloric acid. The capillary cone is centrifuged without heating, and the volume of the precipitate is measured.

The mean value obtained in a series of ten experiments (92) according to the procedure was 13.4 nl of precipitate per microgram of silver. This value was reproduced with a coeffiicient of variation of about ± 0.13.

Estimation of Tungsten as Yellow Tungstic Acid. The sedimetric estimation of tungsten as the white amorphous tungstic oxide, obtained with hydrochloric acid at room temperature is not practicable because of the solubility of the precipitate. Satisfactory results may be obtained after converting to the yellow insoluble tungstic acid by heating for 5 minutes just below 100° C, p. 246.

After centrifuging, the yellow precipitate, which is rather gelatinous in appearance, collects in a compact form which readily takes the shape of the cone. The yellow tungstic acid may be purified by dissolution in ammonia, precipitation by hydrochloric or nitric acid, heating, and centrifuging. The volume of the purified acid is very close to that obtained in the first precipitation, but appears to be somewhat better reproducible.

Procedure. Fifty nanoliters of the stock solution of tungsten are treated with an approximately equal volume of 5 m hydrochloric acid in a capillary cone. The mixture is heated for 5 minutes just below 100° C and centrifuged. The volume of the precipitate is estimated.

The mean value obtained for the estimated volume of yellow tungstic acid in a series of ten experiments (92) was 16.8 nl of precipitate per microgram of tungsten. This value was reproduced with a coefficient of variation of about ± 0.11.

After estimation, the precipitate was redissolved by adding approximately 50 nl of 17 m ammonia and agitating slightly in the cold. Excess acid was added, whereafter the precipitate was heated and centrifuged as before.

The mean value of the volume of purified tungstic acid obtained in the series of ten experiments was 16.1 nl with a coefficient of variation of about $\pm$ 0.09.

Estimation of Vanadium as Lead Vanadate. For the semi-quantitative estimation of vanadium on the microgram scale as lead vanadate (93), the mean value obtained for the specific volume of the bulky precipitate in a series of ten experiments was about 115 nl of precipitate per microgram of vanadium (93) with a coefficient of variation of about $\pm$ 0.05. The voluminous orange lead vanadate is precipitated from weak acetic acid solution by adding an excess of neutral lead acetate solution.

The good reproducibility is probably a consequence of the larger specific volume of the lead vanadate precipitate. The agreement might have been still better, but the lead vanadate precipitate has some tendency to stick to the walls of the cone so that the surface is not as clearly defined as with other precipitates.

c) Gravimetric Analysis of Microgram Samples.

The capillary cone technique and the use of a suitable silica microgram balance have led to the development of satisfactory gravimetric microgram procedures (89). The simplified construction of a satisfactory microgram balance (88) should place classical gravimetric determinations on the microgram scale among the readily accessible techniques of the normal analytical laboratory.

EL-BADRY and WILSON (88, 89) applied the capillary cone techniques inside the moist chamber to the gravimetric determination of 10 µg of inorganic ions. The apparatus, p. 92, and manipulations involved, p. 208, have already been dealt with generally, and also with special reference to gravimetric analysis of microgram samples, pp. 119, 224. Due attention should be given to such points of detail as the proper coating of the inside of the micropipet, fire polishing of the end of the cone handle, proper delivery of solutions and reagents, the securing of an atmosphere completely saturated with water vapor, rapid measurement of the sample, and proper handling of capillary cones.

Solutions containing 10.00 µg of determined ion per 200.00 nl are recommended for practicing. With such solutions, series of ten determinations each of the ions lead, silver, and mercurous mercury have been carried out with the use of the procedures outlined below. The mean values found were 10.05 µg lead, 10.07 µg silver, and 9.99 µg mercury.

Experimental study has shown that the washing of the precipitates can be performed without appreciable losses. The drying temperatures used are consistent with the stability and the volatility of the precipitates. Lower temperatures could be used for lead sulfate and silver chloride.

Determination of Lead as Lead Sulfate.

Reagent solution: Sulfuric acid, 4 m.

Wash solution: Distilled water.

Procedure. Approximately counterpoise the selected cone, dry it at 340° C for 10 minutes in a drying capillary, then weigh it accurately and transfer it to the carrier. Supply the carrier with a measuring capillary and with two reagent containers holding the reagent solution and the wash solution. Charge a third reagent container with the lead test solution and transfer it to the carrier. Accurately measure about 200 nl of lead test solution in the measuring capillary and deliver the solution into the cone. Add to the solution in the cone an approximately equal volume of sulfuric acid. After 15 minutes, centrifuge the cone vigorously. If a hand centrifuge is used, 60 hard turns of the crank are required. Remove the supernatant liquid and wash the precipitate twice with distilled water; using enough of it to fill the cone nearly completely. The precipitate should not be disturbed, but each portion of water should be left in contact with it for about 10 minutes.

Transfer the cone containing the precipitate to a new drying capillary. Dry for 10 minutes at 100° C and then for 30 minutes between 330° and 340° C. Weigh the cone with the $PbSO_4$.

Determination of Silver as Silver Chloride.

Reagent solution: Hydrochloric acid, 0.5 m.

Wash solution: Nitric acid, 0.01 m.

Procedure. The procedure is the same as described for the determination of lead.

Determination of Mercurous Mercury as Mercurous Chloride.

Reagent solution: Hydrochloric acid, 0.5 m.

Wash solution: Distilled water.

Procedure. The procedure is similar to that described for lead and silver, but the empty cone is dried for 10 minutes at 120° C. The precipitate is dried for 10 minutes at 100° C and then for 30 minutes at 120° C.

d) Titration of Microgram Samples.

The apparatus, p. 123, and manipulative techniques, p. 226, have been described in detail.

α) Working in Titration Cones.

As has already been indicated, the use of titration cones, Fig. 87, imposes severe limitation on the choice of indicators since the light path through the tiny droplet of titrated solution is too short to permit proper observation of color. In acid-base titrations, the end point cannot be distinctly recognized with indicators like bromocresol green, bromophenol blue, mixtures of methyl red and methylene blue, etc. when used in the customary proper concentrations of 10^{-5} to 10^{-4} mole per liter.

On the other hand, end points obtained by adsorption of the resulting coloration onto the small surface area of microscopic or colloidal particles proved sufficiently sensitive.

Acid-Base Titration. The indicator system iodide-iodate-starch (66) proved successful for the titration of base with acid, due to the high intensity of the blue coloration produced by the adsorption of the tri-iodide on starch. For this purpose *soluble* starch should be used. The use of starch grains in place of soluble starch is not satisfactory since the grains acquire the blue coloration gradually during titration with acid, starting at the edges, so that the end point becomes difficult to recognize.

In a series of titrations (154) of sodium hydroxide with sulfuric acid with the indicator system iodide-iodate-starch, a relative precision of about ± 0.015 was obtained. In this series of experiments one buret was used for each standard, and the amounts of acid and base were given in divisions of the micrometer scale. Consequently, the base-acid ratio was dependent upon the burets used. Besides, not more than one-third of the capacity of the burets was utilized.

If the same microburet is used for measurement of both acid and base, the volume ratio of the two standard solutions as determined on the macro scale with the customary 50-ml stopcock burets can be closely reproduced on the microgram scale. The above titration was repeated in another series of experiments by adding both solutions from the same buret, and a precision of about ± 0.008 was obtained.

Argentometric Titration. Chlorides can be successfully titrated with silver nitrate by means of the adsorption indicator dichlorofluorescein. The color change at the end point can easily be detected on the particles of the precipitate. The color of the particles of silver chloride was observed with dark-field illumination furnished by the Epi-condenser. If dextrin is added, the sharpness of the end point is improved. A dispersing agent, Triton N. E. (195), may be added to prevent the clogging of the buret by the precipitate.

In a series of experiments (154) in which sodium chloride was titrated against silver nitrate by adding both solutions from the same buret, a relative precision of about ± 0.008 was obtained. In this series of experiments the volumes were read as divisions of the micrometer scale. The ratio of volumes of silver nitrate to sodium chloride, determined in the usual way with 50-ml burets, could be closely reproduced on the microgram scale.

β) Working in Coloriscopic Capillaries.

The use of the coloriscopic capillaries, Fig. 89, for titration of microgram samples imposes no limitations on the choice of indicators. Color changes at the end point can easily be observed with the standard concentrations of customary indicators.

In a series of titrations (81) of sodium hydroxide and hydrochloric acid with methyl red indicator, the end points were reproduced with a precision of about $\pm$ 0.012. Approximately 0.3 µl of standard solution was used in each titration. The observation of the color change at the end point may be made more sensitive by inserting suitable color filters into the vertical illuminator and by the use of titration capillaries made of suitably colored glass to produce a background contrast.

e) Working with Hanging Drops: The Ultimate Limits of Simple Tests.

The technique, p. 237, consists essentially of suspending the test drop and the reagent drop close to each other in a thin layer of paraffin oil, measuring the diameter of the spherical drops, and allowing the two drops to merge. It has been used to determine the ultimate limits of identification of two well known qualitative tests of good but not unusual sensitivity. One of these tests, the test for barium by adding sulfate ion, is based on the appearance of a precipitate; the other test, the Prussian blue test for iron, is based on the appearance of a coloration. Under fairly good conditions of observation, limits of detection were reached with amounts as small as 10^{-14} g of barium and 4×10^{-13} g of ferric iron.

It was found that, with the total microscope magnification of $397 \times$, the smallest droplets for convenient work were from 7 to 14 µm in diameter or one-half to one division of the ocular micrometer scale. The use of higher magnifications proved impracticable since the depth of focus became too shallow for convenient manipulation.

Barium Sulfate Test.

Test solutions. Barium chloride, 1%, 0.1%, 0.01% and 0.001% solutions. The two most dilute test solutions were treated with glycerol, p. 235, to prevent evaporation from the supply drops and from the tip of the micropipets and also to retard shrinkage of the microdrops suspended in paraffin oil.

Reagent solution. Sodium sulfate, saturated solution, acidified to pH 1.

With the procedure described on p. 238, a microscope magnification of about 400 diameters, and transmitted light: positive results were always obtained with 1% and 0.1% test solutions. Under the same conditions, positive tests were always obtained with comparatively large drops, 30 to 40 µm in diameter, of the 0.01% and 0.001% test solutions. In all these cases, on merging the two uniformly clear microdrops, a precipitate of barium sulfate, formed after a few moments, could easily be observed. With small microdrops, 8 to 12 µm in diameter, of the same 0.01% and

0.001% solutions, the formation of precipitate was difficult to observe with transmitted light. This is caused by the high curvature of the spherical drops and the difference between the refractive indices of the aqueous solution and of the oil.

Dark-field illumination renders visible particles of barium sulfate which, at the limiting concentration, cannot by seen in the bright field. With dark-field illumination obtained by a ZEISS Epi-condenser W and a total magnification of about 400 diameters, the limit of detection was reached with 10^{-14} g ($= 0.01$ pg $= 10$ E) of barium supplied by a microdrop of 1 pl $= 10^{-12}$ liter of the 0.001% barium solution. Smaller amounts of barium gave tests which were either doubtful or negative.

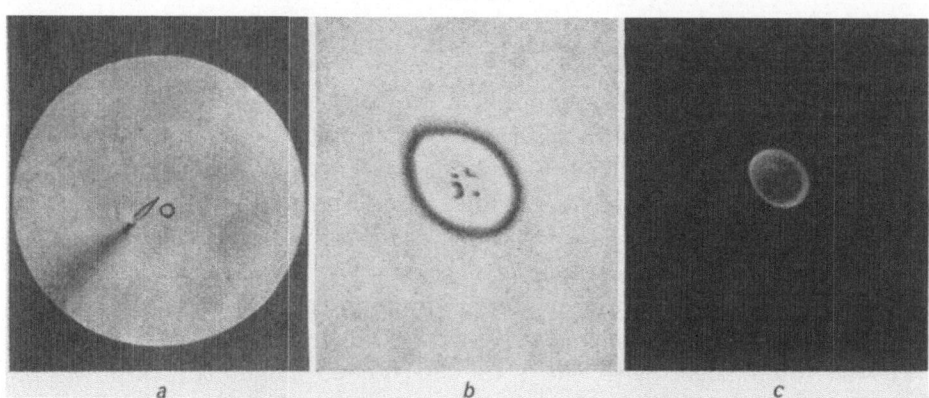

Fig. 140. Performing Barium Sulfate Test on the Underside of the Cover Glass. a, pipet ready to deliver reagent, test drop already deposited, $100\times$; b, the two drops have merged, dark specks of barium sulfate at center, small particles not easily recorded, bright field illumination, $400\times$; c, same drop as in b, reflected dark-field illumination used, apparent shrinkage due to change of illumination and focus. $400\times$. Courtesy of BENEDETTI-PICHLER (31).

Prussian Blue Test.

Test solution. Ferric chloride, 1% and 0.1% solutions.

Reagent solution. Potassium ferrocyanide, 0.0025 m.

With the procedure used for the barium test and with a microscope magnification of about $400\times$ and transmitted light, a dark blue precipitate was obtained with fairly large drops, 30 to 40 μm in diameter, of the 1% ferric chloride solution. The precipitate appeared when the two clear droplets of reagent solution and test solution merged and was only visible with transmitted light.

Small drops of the 1% test solution and comparatively large drops, 15 μm in diameter, of the 0.1% test solution gave blue colorations. The visibility of these colorations was always very poor and easily destroyed on adding excess of the ferrocyanide reagent. The sensitivity was obviously limited by the short light path through the small drops.

The limit of detection for the blue coloration was reached with 4×10^{-13} g (0.4 pg = 400 E) of ferric ion supplied by a droplet 4×10^{-14} liter (0.04 pl) of 1% test solution.

2. Nuclear Research.

Mechanical manipulation under the miscroscope using the general capillary cone techniques for qualitative inorganic analysis of microgram samples, first outlined by BENEDETTI-PICHLER (28) in 1937 and subsequently refined by BENEDETTI-PICHLER and CEFOLA (25, 29, 32), have been successfully applied to the solution of extremely interesting problems in the field of nuclear research during World War II. These techniques proved very valuable in the isolation of the first pure samples of the transuranium elements neptunium (160), plutonium (52, 69, 71, 72, 210), americium (70) and curium (69), with the atomic numbers 93, 94, 95 and 96, and the study of their properties.

Of all the isotopes of these elements. the alpha emitter plutonium isotope of mass 239, having a half-life of about 24,000 years, is of paramount importance, since it is fissionable with slow neutrons. This property makes it comparable in importance with the fissionable uranium isotope U-235.

Early in 1941, G. T. SEABORG and associates had conclusive evidence that plutonium would be produced by two successive beta-transformations upon bombardement of U-238 with neutrons. This indicated the possibility of the large scale production of Pu-239 from natural uranium consisting mainly of U-238 with about 0.7% by weight of the fissionable isotope U-235. If it was possible to cause a chain reaction to occur on a large scale in this mixture, U-238 would absorb the extra neutrons produced in the fission of U-235 to form Pu-239 in substantial amounts. Its isolation could be effected by the comparatively simple chemical separation from uranium and fission products, whereas the isolation of U-235 necessitates laborious and costly physical procedures of isotope separation.

For producing Pu-239 on a manufacturing scale, it was essential at the outset that a chemical procedure be devised for the separation, on a production scale, of the desired isotope Pu-239 from the uranium and the great amounts of highly radioactive fission products that would be produced. Furthermore, such large scale separation procedure, devised for plant operation, should have been studied and tested with the concentrations of plutonium which were anticipated to exist in the manufacturing plants.

The early experiments on the chemistry of plutonium were carried out by means of well known tracer techniques, using carriers, according to customary radiochemical practice. These experiments were carried out with very small amounts of Pu-238 obtained by the deuteron bombardement of U-238 at the cyclotron of the University of California. These preliminary investigations could be performed with milligram techniques because, although plutonium usually existed in less than microgram to nanogram amounts, it was mixed with a great bulk of uranium and fission products. Simple apparatus was employed such as capillary transfer pipets (136, 217) of the types shown in Fig. 84, p. 118, attached to a suitable syringe control, centrifuge cones one milliliter to several milliliters in capacity, and evaporating dishes of varied shapes and sizes fashioned from platinum foil.

A great deal about the chemical properties of plutonium could be learned by these tracer methods of investigation. Early experiments carried out at the University of California by SEABORG and WAHL (211) indicated that the element 94 has at least two oxidation states with chemical properties analogous to U IV and U VI. The same authors were able to show that plutonium in its reduced form only is co-precipitated from aqueous solutions by rare-earth fluorides such as lanthanum fluoride. This important finding formed the basis for the later development of the final procedure for the separation of plutonium from uranium and fission products. Both, the procedure developed for the isolation of microgram amounts of Pu-239 and the production scale separation, developed on the basis of the microgram scale procedure, depend upon the use of the two oxidation states of plutonium.

In summer 1942, the main objectives of the microchemistry group of the Chemistry Division at the Metallurgical Laboratories of the University of Chicago, now the Argonne National Laboratory, were the isolation of microgram amounts of plutonium in the form of pure compounds, the study by direct observation of some of the more important chemical properties of pure plutonium compounds at ordinary chemical concentrations, and the determination of the half-life of the isotope Pu-239.

a) Isolation of Microgram Amounts of Plutonium.

As has already been indicated, one of the primary objects was to establish and demonstrate on a small scale, a reliable chemical procedure for the large scale separation of plutonium from uranium and fission products. It is worthwhile mentioning that, up to this time, no synthetic element had ever been isolated and no weighable amounts of any transuranic element had ever been produced. All deductions as to the chemistry of plutonium had been made indirectly on the basis of tracer experiments in which the concentration of the element was from about 10^{-11} to 10^{-14} M.

The first task was the preparation of weighable amounts of Pu-239. This was done by bombardment of natural uranium with neutrons from cyclotrons with the method described in detail by SEABORG and WAHL (211). Only microgram amounts of the isotope could be produced from large amounts of uranium since the concentration of plutonium in the irradiated uranium, even after bombardment for weeks, did not exceed 0.003 p. p. m. because of the very low neutron flux that could be obtained.

Consequently, the investigation had to be carried out with amounts of plutonium weighing a few micrograms or less. It could not be avoided in the final stages of isolation and in the preparation and study of compounds, to use microgram techniques, particularly if ordinary chemical concentrations of the order of 0.001 to 0.01 m were to be maintained.

Since the very small amounts of plutonium produced were mixed with much larger amounts of uranium and fission products, the great bulk of contaminations had to be separated first, in suitable apparatus of proportionately large size. The scale of work was then progressively reduced to end on the microgram scale with apparatus suitable for the final stages of concentrating and isolating the element.

In these final stages, where the volumes of solutions were reduced to the microliter range, the techniques of qualitative analysis of microgram samples, p. 208, first descibed by BENEDETTI-PICHLER and CEFOLA (25, 28, 29, 32), proved indispensable and were extensively used.

In later work, however, CUNNINGHAM and WERNER's simplified microscope-micromanipulator assembly, Fig. 83, p. 116, was used since it permits working with larger volumes of solutions than could be handled with BENEDETTI-PICHLER's original assembly. Chemical operations are carried out in glass microcones which may be up to 200 µl in capacity, Fig. 85, and measurement and transfer of liquids are performed by means of calibrated capillary transfer pipets of the required capacity attached to a simple syringe control, Fig. 84. Lusteroid microcones made of transparent hydrogen fluoride-resistant plastic may also be used, but these have rather poor optical qualities compared with glass. In addition, the microgram scale techniques developed by KIRK and associates (69, 133, 134) proved of great value.

The fluoride cycle method developed for the separation of plutonium from uranium and fission products and the concentration of the element, was based upon the use of the two oxidation states of plutonium, discovered earlier, and upon co-precipitation with successively smaller amounts of rare earth carriers. The scheme involved the precipitation of plutonium in its reduced fluoride-insoluble state with cerium and lanthanum fluorides as carriers, dissolution of the mixed fluoride precipitate in sulfuric acid and nitric acid, oxidation of plutonium to the fluoride-soluble state by silver peroxide and precipitation of the rare earth fluorides leaving plutonium VI in solution. The plutonium was again reduced to the fluoride-insoluble state, and the cycle was repeated until the desired decontamination was attained. By means of this cycle, also the volume of the solution was successively reduced until the concentration of plutonium became high enough to permit its separation. Assaying for activity was carried out when required to determine recovery in various fractions. Another early discovery, also by SEABORG and WAHL, that neptunium, but not plutonium is oxidized at room temperature by bromate in $1 m H_2SO_4$ to a fluoride-soluble state, formed the basis for the separation of plutonium from neptunium and was also incorporated into the procedure.

On August 18, 1942, about 1 μg of the isotope Pu-239, as plutonium fluoride, was isolated by CUNNINGHAM and WERNER at the Metallurgical Laboratory of the University of Chicago. This small amount had been separated from about 5 kg of uranyl nitrate hexahydrate which had been irradiated with neutrons produced in the cyclotron of the University of California, Berkeley. It should be remembered that this was the first time that an artificially produced element had been isolated. It is also interesting to note that before that time no weighable amount of any synthetic element had been produced (71, 72).

The procedure (72) and a complete flow sheet (71) of the chemical operations used in the second isolation, carried out in September 1942, have also been supplied by CUNNINGHAM and WERNER. Twenty micrograms of Pu-239 were obtained in a high state of purity suitable for gravimetric work. This amount finally separated as a green plutonium hydroxide precipitate which was dissolved in 10 μl of 2 m HNO_3 to give a pale yellow green solution about 0.01 m in plutonium. The starting material was about 90 kg of $UO_3(NO_3)_2 \cdot 6 H_2O$ which had been irradiated with neutrons from bombardment of beryllium with deuterons produced in the Washington University cyclotron at St. Louis.

Until 1943 cyclotron bombardment was the only source of Pu-239. It may be interesting to note that, up to that time, a total of only about 1000 μg = 1 mg of the isotope had been produced (210).

b) Investigation of Properties of Plutonium.

Portions of the 1 mg of plutonium isolated toward the end of 1942 and the microgram amounts of the element obtained earlier were used for the determination of the specific activity and half-life of the isotope, and for the preparation and study of its compounds. Above all, it was possible to test the elaborate separation and isolation procedures which were under consideration for use at the manufacturing plant at Hanford. There, the separation was carried out on a ten billion times larger scale than the microgram pilot experiment. It is surprising to note that the production of the isotope was successful from the start (210).

Many new compounds of plutonium and of other synthetic elements have been prepared and investigated with use of mechanical manipulation under the microscope. Indications and proofs concerning the formation of new compounds were obtained by the appearance of precipitates and colors. If a solid separates, qualitative and quantitative problems could

be solved simply by measuring the decrease of radioactivity in the supernatant liquid after centrifuging the precipitate. X-ray methods, applied
to samples a few micrograms in mass, were useful in the identification of
a number of compounds and for the establishment of their chemical
structure.

Microgram amounts of plutonium metal and several plutonium compounds were prepared by dry chemical reactions involving the solid and
gas phases (69). An attempt was made to prepare metallic plutonium by
reduction in a capillary mercury cathode cell (52). The cell, Fig. 141, was
similar to a short capillary reagent container. For anode, served a second
platinum wire dipping into the solution.

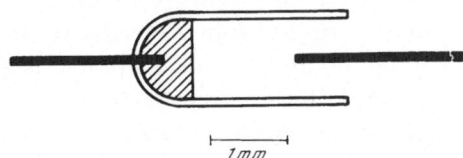

1 mm

Fig. 141. Capillary Mercury Cathode Cell. Two platinum wires, one dipping into the solution
and the other into mercury at the bottom of cell. After CEFOLA (52).

c) Determination of the Specific α-Activity and of the Half-life of Pu-239.

Reliable measurement of the specific alpha activity of Pu-239 by direct
weighing and α-particle counting techniques, and hence the accurate
determination of its half-life, was of primary importance for the use of this
isotope in the atomic bomb and as a fuel in reactors.

The decay constant, defined as the number of disintegrations per unit
time and unit weight, from which the half-life could be calculated, was
accurately determined by weighing microgram amounts of the pure isotope on a suitable microgram balance and measuring the radioactivity of
the weighed sample with suitable α-counters of known counting yields.

The amount of the isotope obtained in the second isolation was of a
high state of purity making it suitable for quantitative studies and for the
precise measurement of the specific α-activity. For this latter purpose,
samples of plutonium compounds, a few micrograms in mass, had to be
weighed with an error not exceeding 0.03 μg.

In the first determinations of the specific activity of the isotope, weighings
were performed on the tiny platinum weighing pan of a simple silica microgram balance of the SALVIONI type (69, 102, 199). A refined balance of
this type was then constructed by CUNNINGHAM and WERNER, and determinations of the specific activity were carried out on samples of 2 to 4 μg.
The construction, calibration, and use of this simple instrument has been
described in detail (71, 72). The tiny platinum pan of the balance weighed
about 150 μg, and the aluminium cradle carrying the pan was approximately

200 μg. In use, for loading and unloading, the cradle and the pan were attached and detached with the aid of a simple micromanipulator.

With only a small amount of the pure Pu-239 available, the first determinations of the specific α-activity were carried out with samples of 2 to 5 μg of the isotope. The determination started with the transfer of a small aliquot of the solution of pure plutonium nitrate in nitric acid onto the weighed platinum pan of the SALVIONI balance. There, the solution was then dried, and the residue was ignited to the oxide, presumably PuO_2. The pan was then placed on the balance to find the weight of the oxide.

The oxide deposit on the pan was not used directly because of the uncertain counting yield due to self absorption and the curvature of the pan. It was dissolved in concentrated sulfuric acid by immersing the pan in 100 μl of the acid and heating to fuming for about two hours. The solution was then diluted to 10.00 ml, and a small aliquot was evaporated to dryness on a platinum dish suitable for the measurement of the α-activity.

More accurate measurements were carried out at a later stage, when larger amounts of plutonium became available. Samples of 30 to 100 μg were weighed on a microgram balance of the torsion-restoration type, p. 120, Fig. 86, which was designed and built by KIRK, CRAIG, GULLBERG, and BOYER (136). Also the alpha counters had been improved. On the basis of these newer measurements, the half-life of Pu-239 was calculated as 24,300 ± 370 years.

The decay constants and half-lives of other long-lived alpha emitting isotopes, Np-237 and Am-241, have been accurately determined. Chemical purity of microgram samples of these isotopes was also achieved by repeated cycles of chemical purifications and microgram techniques as those used for the isolation of pure Pu-239.

d) Proof of the Oxidation Number of Plutonium.

In October 1942, CUNNINGHAM and WERNER (71, 72), by an experiment involving only a few micrograms of plutonium, were able to obtain the first direct proof that the tetrapositive state is a stable oxidation state of plutonium.

The experiment involved the precipitation of plutonium iodate by adding excess iodic acid to about 3 μg of plutonium contained in 1 μl of 4 m HNO_3. The precipitated iodate was washed three times with 20 μl each of distilled water. It was then suspended in a few microliters of distilled water and transferred by the micropipet onto a previously weighed platinum weighing pan of the SALVIONI balance. The weight of the plutonium iodate was determined after drying for four hours at approximately 100° C. The precipitate was then dissolved in a small volume of hydrochloric acid and diluted to 1-ml volume. An accurately measured aliquot of the solution was transferred onto a counting plate, and the amount of plutonium in the aliquot, and hence in the whole plutonium iodate precipitate, was calculated from the measured α-activity. From these data, it was calculated that the molar ratio of IO_3^-/Pu in the plutonium iodate compound was 3.93. This value is sufficiently close to that required for $Pu(IO_3)_4$, to establish that the tetrapositive state is one of the stable oxidation states of plutonium.

The composition of this plutonium iodate compound was checked, at a later date, by determining both plutonium and iodate, plutonium by α-counting and iodate by iodometric drop-scale titration with the technique developed by KIRK and associates (217). It was also established that the oxidation number of the highest oxidation state of plutonium is $+ 6$ and that a lower oxidation state, $+ 3$, exists. In these experiments too, α-counting methods and the microvolumetric technique developed by KIRK were used (133).

e) Solubility Measurements with Plutonium Compounds.

Measurements of approximate solubilities of many compounds of plutonium, particularly those of importance in connection with the choice and development of a suitable chemical procedure for the large scale extraction, decontamination, and purification of plutonium, were started fairly early, soon after the isolation of the first microgram amounts of the pure element.

These first determinations had to be carried out on very small samples. Thus early measurements of solubilities of plutonium compounds, which were carried out by CUNNINGHAM and WERNER (71, 72) in collaboration with M. CEFOLA, utilized only as little as 0.1 µg of plutonium in the solid phase per determination. This was facilitated by the fact that radiometric assay for the still smaller amounts of Pu-239 in the solution in equilibrium with the solid phase could be carried out with certainty. This practice was rendered possible only after the value for the specific α-activity of Pu-239 had been established, which permitted the quantitative determination of extremely small amounts of the isotope by α-particle counting techniques. Thereafter, as little as 0.01 µg of plutonium could be determined with a relative error of less than 2%. Even smaller amounts of approximately 0.1 ng of the isotope, equivalent to only about 10 counts per minute, could be assayed with an error of about 5%.

In carrying out solubility measurements with microgram amounts of material, suitable microscope-micromanipulator assemblies again proved very useful. Early solubility measurements of plutonium compounds were performed in reagent capillaries of vitreous silica, shown in Fig. 142. The capillary part was about 1 cm in length and 1 to 1.5 mm in inside diameter and was provided with a solid handle. These microcones were made from silica capillaries of approximately 1-mm bore, which had been drawn out from larger tubing in an oxyhydrogen flame.

With a syringe-controlled micropipet carried on the micromanipulator, Fig. 83, approximately 0.2 µl of the nitric or sulfuric acid solution containing about 0.1 µg of plutonium was delivered close to the tip of the microcone. The solution was treated with the precipitating reagent and the mixture was stirred with a fine glass thread attached to a magnetic vibrator. The microcone was fitted with a paraffin stopper as shown in Fig. 142 or was placed in a larger closed capillary. After allowing the two

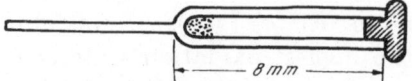

Fig. 142.   Microcone for Solubility Measurements Fitted with Paraffin Stopper.   After CUNNINGHAM and WERNER (71).

phases to stand in contact for about 24 hours, the microcone was centrifuged, and a sample of the supernatant liquid was withdrawn by means of a specially calibrated micropipet having 0.252 ± 0.005 µl capacity. The solution in the pipet was delivered, with rinsing, onto a platinum disk and allowed to evaporate. The residue was dried, ignited, and its α-activity was determined in a counter of known geometry. The concentration of plutonium per unit volume could then be calculated.

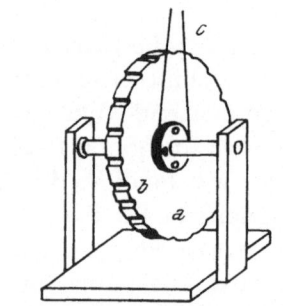

Fig. 143.   Apparatus Used in Microsolubility Measurements. a, Lucite wheel; b, solubility tube; c, drive cord. After CUNNINGHAM and WERNER (69).

The solubilities of various compounds of other synthetic elements also were obtained on extremely small samples, and these measurements also have been greatly aided by the radiometric assaying of solutions. In contrast, solubilities of non-radioactive substances are generally determined with samples several micrograms in mass to permit accurate chemical analysis.

A technique (71) for solubility measurements on very small samples, developed by STEWART and CUNNINGHAM, uses the microsolubility apparatus illustrated in Fig. 143. In essence, sealed glass capillaries b, containing solid and solvent, are held on the periphery of a notched wheel a, made of Leucite, by means of a rubber band fitting tightly around the rim of the wheel. The latter, which may be mounted in an air or water thermostat if desired, is rotated by means of a motor. Continuous mixing of the solid and liquid phases can thus be achieved. After mixing, the solid phase is centrifuged to the bottom of the capillary which is then cut open. Samples of the supernatant liquid are withdrawn for analysis either by the radiometric or some other appropriate method.

3. Identification of Airborne Particles.

Sampling and identification of particles suspended in the air are important for many fields of study of the atmosphere of the earth. Techniques of chemical micrurgy have been extensively applied by CADLE and associates in important studies of air pollution and related problems in the applied chemistry and physics of the atmosphere. These techniques were particularly useful in the determination of the composition of airborne particles collected from smog during investigations of the smog problem in the County of Los Angeles (47, 48). In the investigation of air pollution problems of Los Angeles County, the use of these micromanipulative techniques and other microscopic techniques was not limited to the investigation of substances collected from the atmosphere, but was also applied to the study of gasoline engine exhaust gases (48) to discover what substances were introduced into the atmosphere from this important source.

Particulate materials are not meant to comprise only substances in the solid state. The term is applied to the large variety of particles suspended in the atmosphere, ranging from droplets of dilute solutions to dry crystals, depending upon the salts and other substances involved and upon humidity. The particulate matter collected from the Los Angeles atmosphere, especially that collected during periods of smog, was found to be highly complex and to include a diversity of liquid droplets and both organic and inorganic solid particles. A great proportion of the material consisted of dark brown, gummy, water insoluble organic hydrocarbons. It is desirable that the collected material be examined without delay to avoid, as far as possible, changes which may take place on standing due to such factors as the slow evaporation of certain components and separation of solids.

For the application of micrurgic techniques, the particulate material suspended in the air is first caught on object slides or cover glasses. This is achieved in various ways such as impaction, thermal precipitation, settling, or filtration. These and other methods of sampling aerosols have been described in great detail by CADLE (46) in connection with particle size determination. The principles underlying these methods are briefly considered on p. 202 in the section devoted to sampling. The impaction methods are usually more suitable for general use than the others. A model of the SONKIN impactor (218), made of plastic, is very convenient for collecting the particles on microscope slides (46, 47). It is briefly described on p. 202, Fig. 123.

Particles collected on an object slide in the impactor are covered with a thin layer of mineral oil to prevent evaporation of droplets. The oil is spread over the particles using a glass thread about 1 mm in diameter.

Fig. 144. General View of Set-Up and Work-Bench Arranged for Chemical Micrurgic Work with Particles Collected on Microscope Slide. Assembly and operator in working position. Courtesy of R. D. CADLE.

The slide, supported on a suitable holder or on a dry chamber, oiled surface down, is brought under the microscope. Individual particles are treated with reagent droplets delivered through the hanging oil layer by fine micropipets which are carried and guided with a micromanipulator. The general technique is an adaptation of the methods used by BENEDETTI-PICHLER and RACHELE (31) for working with hanging drops, p. 233, and those described by TITUS and GRAY (228).

Fig. 144, is a general view of the set-up used by CADLE and associates for chemical examination of particles impacted on a microscope slide. The complete assembly is shown in Fig. 145. It comprises a modified TAYLOR micromanipulator (226), p. 32, consisting of two manipulator columns mounted on a common base plate and facing one another. The microscope is placed on the common base plate between the two manipulator columns. The right manipulator column, shown at the left of the figure, carries a microneedle sealed to the holder with de Khotinsky cement. Attached to

Fig. 145. Complete Assembly for Chemical Examination of Particles Collected on Microscope Slide. Modified TAYLOR micromanipulator, microinjection apparatus, and glass slide on support in position on microscope stage. Courtesy of R. D. CADLE.

the pipet holder on the opposite manipulator column is a micropipet with an inclined tip for depositing reagent droplets. Micropipets were made from hard glass tubing by hand-drawing according to the methods described by R. CHAMBERS, p. 142. Pipets with orifices of about 2- to 5-μm bore were found most suitable for working with particles of about 5-μm diameter and larger. Pipets having finer tips produced more uniform droplets, but when the bore at the tip is less than 1 μm in diameter, the micropipet becomes clogged after a short period of use. The pipet holder is connected to a hypodermic syringe by means of a long flexible fine tubing. The syringe, the tubing, and part of the pipet are filled with distilled water. This type of hydraulic system is similar to injection devices already described, p. 112, Figs. 80, 81, 90, and is used to draw small quantities of reagents into the micropipet and to deposit reagent droplets upon the individual particles on the glass slide by gently depressing the plunger of the syringe. Reagents are drawn into the micropipet from drops which are placed, prior to use, on the undersurface of the glass slide close to the oiled area. Also shown in Fig. 145 is a dry chamber supporting the glass slide upon which airborne material had been impacted. The chamber is held and moved by means of an attachable mechanical stage. Arrangement of microtools, slide support, and slide is clearly seen in the close-up figure of the apparatus, Fig. 146.

Fig. 146. Close-Up View of Microneedle (left) and Micropipet Arranged for Examination of Particles. Slide holder and slide with impacted material on underside. Microtools approaching slide from below. Courtesy of R. D. CADLE.

Microscopic examination of the slides may require the use of various types of illumination for proper observation of different components of the sample. Focusing of light on the slide, which is placed on the holder in a position considerably above the microscope stage, is achieved by removing the top lens of the substage condenser to obtain a sufficiently long focal distance. With low power objectives, for example a 16-mm, 0.25 N. A. objective, continuous variation from bright-field to dark-field illumination may be achieved by varying the vertical position of the condenser by means of the rack-and-pinion adjustment. This may help observing certain components of the sample, such as small liquid droplets and certain crystals, which may be difficult to observe by standard methods of illumination. With certain positions of the condenser, such components may stand out brightly and are readily observed.

The two photomicrographs, Fig. 147, and Fig. 148, show positive tests for nitrate (65) and chloride. Though crystals and droplets are larger than usual, the two photomicrographs illustrate the technique very clearly. Tests depending on the formation of precipitates were generally more satisfactory than those depending upon the appearance of color. A number of tests were found suitable for particles down to 5 μm in diameter and probably can be used for submicron-size particles. The two tests just

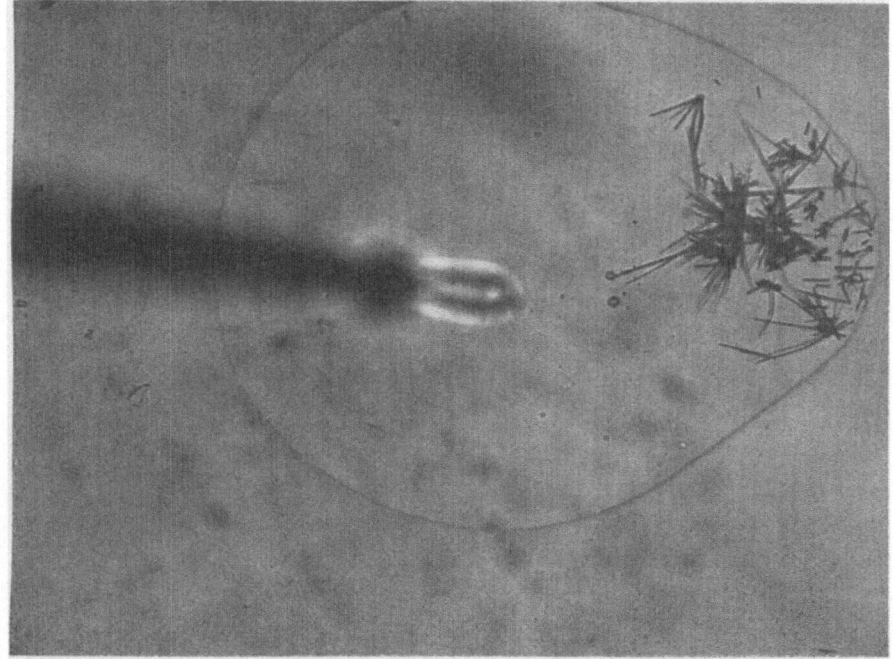

Fig. 147. Precipitate of Nitron Nitrate—after depositing a droplet of nitron acetate solution on a particle containing nitrate ions. Particle was under a layer of oil on microscope slide. Point at which pipet tip entered oil is indicated by halo about the tip. Courtesy of R. D. CADLE.

mentioned are representatives of satisfactory tests. Of other suitable tests may be mentioned the test for sulfate using barium chloride-acetic acid reagent, the test for cyanide with silver nitrate solution, the test for ammonia with NESSLER reagent (128), and for carbonate with concentrated nitric acid and observing evolved bubbles. Though most of these tests are not strictly specific, they usually suffice to indicate the presence of the ion concerned.

In addition to aiding in the performance of chemical tests, the micrurgic technique is helpful in investigation of physical properties which are often useful in identifying the particles or in eliminating certain possibilities. Thus various solvents may be injected into droplets to study whether the solvent and the droplet are miscible. It can be used to deliver liquids of known refractive index onto the particle to determine the relative refractive indices of particle and surrounding medium with the BECKE test (64, 108).

A preliminary examination of the crystalline particles to be tested by the chemical microgram technique may be very informative. This preliminary examination, using a polarizing microscope, may include the determination of such optical properties (64, 108) as isotropism and anisotropism, the extinction angle, the order of interference colors, and whether

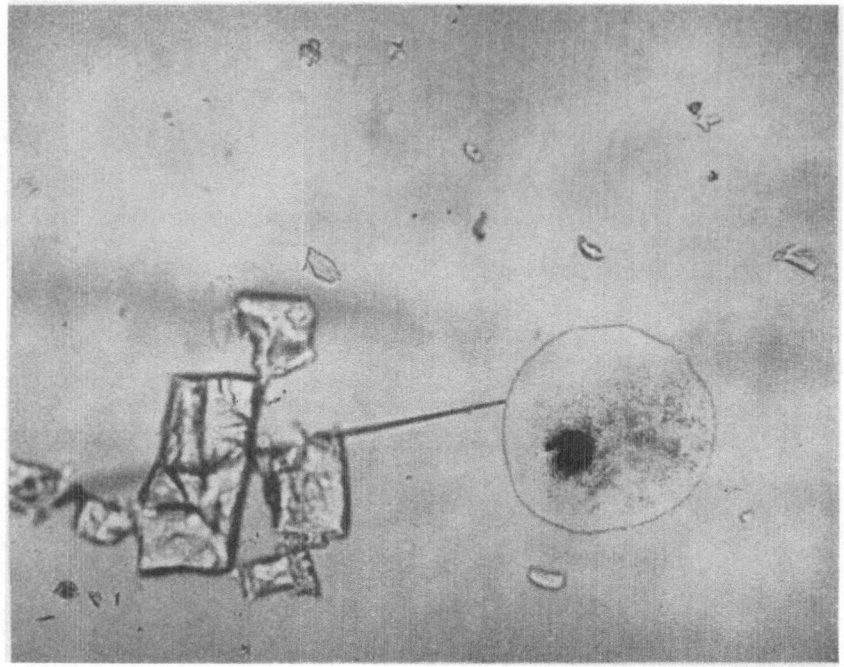

Fig. 148. Precipitate of Silver Chloride—after depositing a droplet of silver nitrate solution on a particle containing chloride ions. Particle was under a layer of oil on microscope slide. Courtesy of R. D. CADLE.

the crystal is uniaxial or biaxial by observing the type of interference figure. The methods of chemical microscopy (65) may be used to determine the over-all composition of the material collected on the microscope slides.

Other methods and techniques, not involving mechanical manipulation, have been used in the study of particles suspended in the air. Thus, the electron microscope proved a valuable tool in examining the physical form of the fine particles collected from Los Angeles smog, which are responsible for some unpleasant properties of such atmosphere (46, 48). It also was able to give some idea of the relative abundance of various types of particles. Due to its great resolving power, the electron microscope is capable of dealing with submicroscopic particles. Its depth of focus is also much greater than that of the optical microscope, and it may be adapted to reveal chemical composition in addition to the physical form of the particles.

CROZIER and SEELY (68, 212, 213) have developed a simplified method of spot-testing (96) suited for the identification of atmospheric particles in the micron and submicron-size range, without involving micromanipulation. This and similar techniques do not deal with individual particles separately, but the entire sample is treated with one reagent, specific with respect to a certain component, and the outcome of the reaction is observed all over the whole sample. Particles are collected, by impaction or any other suitable method, on microscope slides coated with a gelatin-glycerol film containing the reagent. The characteristic reaction product forms spots or halos which are then examined under the microscope, usually with dark-field illumination. This technique was used for the identification of chloride in particulate matter.

The determination of chloride particles suspended in the atmosphere is of special significance since such particles are considered to play an important rôle in the condensation of atmospheric moisture. They occur in appreciable amounts in the atmosphere, especially in air masses coming recently from the ocean, and are among the principal components of sea haze. The procedure permits investigation of a large number of collections made on airplane flights or at a ground station for studying the periodic abundance of certain constituents. During certain flights (212), continuous records were obtained by collecting the particles upon a long moving strip of cellulose acetate coated with gelatin containing the reagent. A microscope built into the apparatus permitted the counting of chloride particles while they were being collected. Thus the boundaries of a chloride-bearing air mass could be defined during flight.

Other modifications of the use of sensitized films were described by CROZIER and SEELY. For the detection of chlorides in rain drops, the droplets were evaporated on a slide and a gelatin film sensitized with the reagent was then applied. Another procedure involved covering the particles on the slide with a plastic film. The reagent, placed later upon the film, diffuses through it and reacts with the particles. LODGE (153) employed the Millipore filter material (155) as a medium for sampling and for testing fine particulate material with the modified spot-test technique developed by CROZIER and SEELY (68).

4. Archeology and Art.
a) Pigments on Chinese Oracle Bones.

In the field of archeology mechanical manipulation under the microscope should receive increased attention in connection with the investigation of valuable objects of art since it permits the use of very small samples for analysis.

In 1937, BENEDETTI-PICHLER (27) demonstrated the value of the technique in isolating tiny particles of pigment in the fossae of the incisions of some Chinese oracle bones used for divination. The specimens, Fig. 149, ascribed to the Shang epoch, which is conventionally dated from 1766 to 1122 B. C., are considered to be of special archeological interest.

Samples of red and black pigments, still retained in the fossae of the incisions, were obtained without damage to the objects by mounting the specimen on the stage of a 40-power binocular microscope. A fine steel needle, mounted on a simple low-power rack-and-pinion micromanipulator

Fig. 149. Chinese Oracle Bones, Princeton University collection, ¾ natural size. Sample of red pigment taken from specimen P 39 from the first horizontal bar below the label, and of black pigment from the large symbol in the right-hand corner of P 13. Courtesy of BENEDETTI-PICHLER.

permitting motion in three directions at right angles to each other, was used to loosen very small amounts of pigment for analysis. After breaking loose the required amount of pigment, the steel needle was replaced by a glass thread, the free end of which was covered with a thin film of glycerol. The loosened particles were then removed from the specimen by touching them with the point of the glass thread. The particles were transferred to a tiny drop of water carried on a glass slide which was placed on the stage of the microscope after removing the specimen. On bringing the point of the glass thread into contact with the water droplet, the particles of pigments became detached from the thread and thus collected into the water droplet.

This made it possible to examine and test the pigments with simple techniques of chemical microscopy. The red pigment was finally confirmed as cinnabar, identifying sulfur as $CaSO_4 \cdot 2\,H_2O$ crystals on the glass slide and mercury by distillation in a capillary according to EMICH (94), Fig. 150. The black pigment was identified as some carbonaceous matter presumably from the blood of animals used in divination. Both pigments appeared mixed with some colorless to yellowish anisotropic particles of quartz or silica minerals, probably loess in which the specimens had been buried.

So little material was taken for analysis that even microscopic inspection of the objects would fail to prove that any amount of pigment had been removed.

b) Core Sections of Paint Films.

Core samples of paint films, which may consist of several superimposed layers of paint, can be accurately extracted under a stereoscopic microscope by means of a hollow cut-off hypodermic needle attached to a suitable

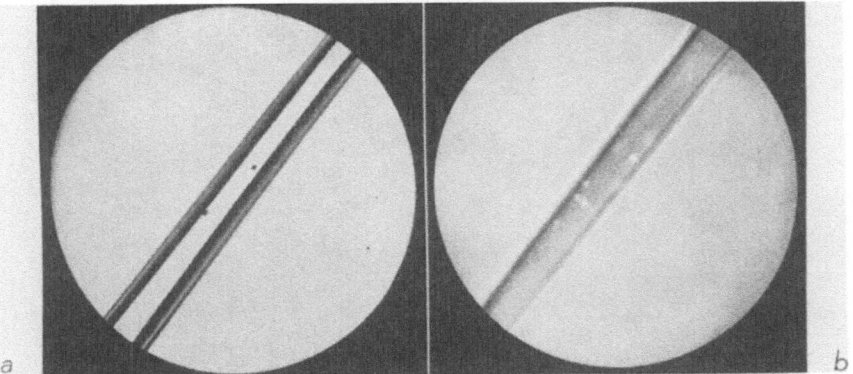

Fig. 150. Distillate of Mercury Inside Capillary. 30×. *a*, transmitted light; *b*, reflected light and black background. Courtesy of BENEDETTI-PICHLER.

manipulating device. This non-destructive sampling technique, developed by GETTENS (106), is already described on p. 196. Core sections thus obtained retain the actual succession of the various strata in their true relation. The various superimposed layers may then be examined individually by appropriate methods. The technique was successfully applied by GETTENS (107) to the sampling of a valuable painting of an old Italian master with the object of establishing the identity of the artist and of the various materials used in the different base and coating layers. Such studies yield important information concerning the techniques employed, and they may also aid greatly in solving problems that arise in the examination and restoration of paintings.

5. Metallurgical Microspectroscopy.

For the identification of microconstituents of metallurgical materials, the sample to be analyzed is often a very small part of the area of the specimen. Several microspectrographic methods have been devised for the analysis of metallic surface areas of very small dimensions (17, 18, 19,

40, 125, 171, 201, 227). One of these methods, the spark technique first described by SCHEIBE and MARTIN (201) and involving the use of a micro-spark source which discharged through a fine quartz capillary onto the area to be analyzed, was later modified by others (40, 171). It has been reported capable of identifying metallic surface areas which may be as small as 20 to 50 μm in diameter.

A major difficulty encountered in the use of all the earlier methods of microsample analysis referred to above is the necessity of constructing

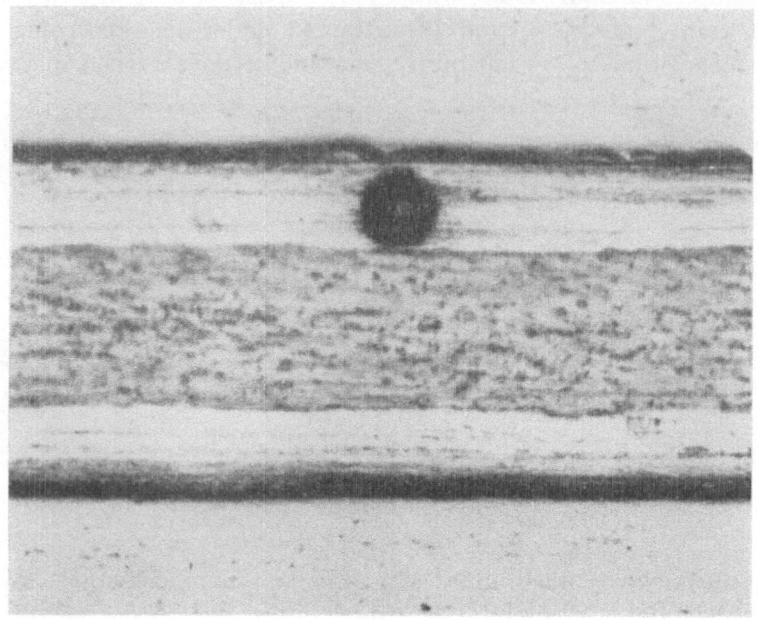

Fig. 151. Sampling Hole Made in a 75-μm Thick Plating with a 50-μm Diameter Microdrill. Drill hole made into the mounted and polished cross section of the plating. Courtesy of F. R. BRYAN.

specialized apparatus. This difficulty was solved by utilizing suitable microdrilling machines, p. 72, which offered a mechanical method for the sampling of extremely small metallic constituents under microscopic control. This permitted the subsequent analysis of such constituents by means of commercially available standard spectrographic equipment. BRYAN and co-workers developed a procedure consisting of drilling out a microsample of the constituent of interest, picking up the drillings with collodion, and burning them in an arc. The specimen is usually mounted and polished as for metallographic investigation, and the sample is taken from the polished surface. The operation is observed under a stereoscopic binocular microscope. This procedure proved both simple and practical and has been successfully applied to the isolation and identification of micro-gram (39, 41) to nanogram (197) amounts of metallic constituents by using conventional spectrographic equipment. The drilling instrument that has been used in these investigations is a NAJET microdrilling machine Model 7A, supplied by the NATIONAL JET COMPANY (173). This machine, Figs. 51 and 52, described on p. 74, is capable of drilling tiny holes by means of microdrills, Figs. 53 and 54, which may be as small as 6.3 μm in diameter. The microsampling procedure is described on p. 193 and is illustrated in Fig. 117a to c, showing a hole drilled in a thin center strip of copper between two layers of silver by means of a microdrill 75 μm in

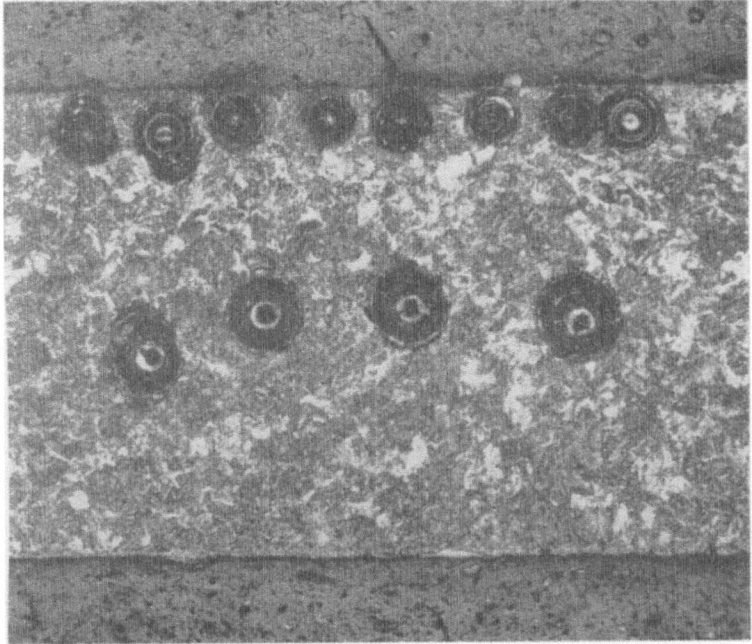

Fig. 152. Mounted and Polished Cross Section of Boron Steel. Samples taken from two series of drill holes to determine relative concentration of boron at the surface and in the center of the specimen. Courtesy of F. R. BRYAN.

diameter. The specimen is a mounted and polished cross section of a silver solder sheet.

This microsampling technique may be applied to a diversity of metallurgical investigations. Thus, microsamples may be removed for identification from thin individual layers of solders and multiple platings by drilling into their mounted and polished cross sections. This is illustrated in Fig. 117, which shows a drill hole made for sampling a layer of copper between two layers of silver in a silver solder sheet. The techniqe is also illustrated in Fig. 151, which shows a sampling hole made by means of a 50-μm diameter microdrill in a 75-μm plating. In this particular investigation (41), carried out to determine the purity of the plating, the elements copper, zinc, and cadmium were identified spectrographically in the sample,

in addition to the major constituent, silver. The alloy was later found to
contain only 15% of each of the elements copper and zinc. It follows therefore that approximately 0.1 µg of each of these two elements could be
identified in the microsample removed from the drill hole.

Evidence of segregation may be obtained by taking microsamples from
a series of closely spaced drill holes and identifying segregates. Diffusion
characteristics in metals may be determined by the drilling out and analysis
of a series of samples. Fig. 152, illustrates how a specimen of boron steel

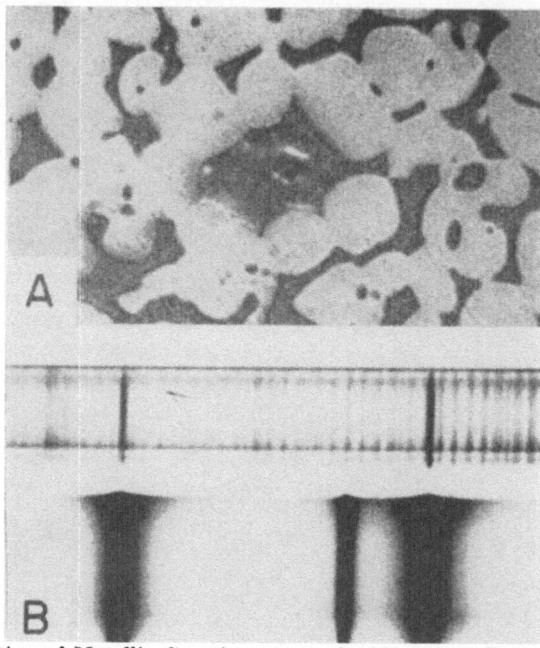

Fig. 153. Identification of Metallic Constituents in the Nanogram Range Using Microdrills.
A, two-phase non-ferrous alloy system showing minor phase sampled by using a 6.4-µm drill.
B, arc spectrogram of microsample (upper), and comparison spectrogram of pure lead
(lower). About 1 ng of the element could be detected. 500×. Courtesy of F. R. Bryan.

containing 0.01% boron was sampled to compare the concentration of
boron at the surface with its concentration in the center of the specimen.
In this experiment (41), a series of samples were taken near the edge of
the cross section by means of a 75-µm diameter drill. The accuracy of the
drilling machine enabled sampling to be confined to within approximately
125 µm of the edge. Drill holes in the center of the specimen were made
by a microdrill about 125 µm in diameter.

Metallic and non-metallic inclusions in metals, alloys, slags, minerals,
rocks, and other material may be separated from the matrix, p. 189, and
identified. The microsampling of a variety of inclusions in different types
of matrix material (132), is illustrated in Figs. 118 to 121. Removal of

samples was performed by means of the ultrasonic Jack Hammer, Figs. 55 and 56, p. 78.

The microsampling of the minor phase in a two-phase non-ferrous alloy system by means of a microdrill under a magnification of 500 diameters, is illustrated in Fig. 153 A. In this particular instance, the distribution of the unknown minor phase necessitated the use of an exceedingly fine drill to obtain a pure sample of this constituent. For this purpose, RUNGE and BRYAN (197) used a 6.4-μm diameter drill operated by the NAJET microdrilling machine, p. 74, and applied the spectrographic procedure, p. 194, developed to permit identification of samples in the nanogram range. The spectrographic equipment used is also briefly indicated on p. 194. Fig. 153 B shows the spectrogram resulting from excitation of the microsample in a graphite electrode. The arc spectrogram contains the two sensitive lead lines at 3639.58 Å and 3683.47 Å, confirmed by the reference spectrum of pure lead shown at the bottom of the figure. It is estimated that approximately one nanogram of lead could thus be detected.

As has already been mentioned, the general procedure, p. 193, of drilling out a microsample, picking up the drillings with collodion and burning them in an arc, was first used by BRYAN (39) and later by BRYAN and NEVEU (41) for the identification of microgram samples. Semiquantitative information, however, may be obtained in certain cases (41) by making microphotometer measurements on the spectral lines of the microsample and corresponding readings on the spectra of reference standards. In the identification method employed, the full slit aperture of the spectrograph was uniformly illuminated giving line definition adequate for microphotometry. On the other hand, the more sensitive procedure, p. 194, developed by RUNGE and BRYAN (197) for the identification of nanogram samples, gives only qualitative information. In this procedure, the full slit height is not illuminated and microphotometry is not involved. The resulting spectrum consists of a series of short broad lines similar to those shown in Fig. 153 B.

FELDMAN (98) uses the microspark technique for examining concentration gradients in the surfaces of metals. The procedure is essentially similar to that of SCHEIBE and MARTIN (201), p. 276. The method can give quantitative information on composition only when metallic standards for the area of interest are available. Since this is seldom the case, FELDMAN, using the NAJET microdrilling machine, extended the technique of BRYAN (39), which has been already described, to give quantitative results. In essence, the quantitative procedure consists of drilling out the specimens by means of 125-μm or 250-μm diameter microdrills. The drillings are then transferred for dissolution into a detachable quartz tube forming the lower section of a quartz reflux dissolver. The reflux dissolver makes it possible to retain boron if present. After the sample has been dissolved, the detachable lower section of the dissolver containing the solution of the sample is removed. An internal standard is added, and the solution is diluted to a volume of 1.0 to 2.0 ml, using calibration marks scratched in the walls of the tube. The constituents are then quantitatively determined by solution spectrochemical methods, usually by the porous cup technique (97).

6. Microhardness Testing.

In the extensive field of testing materials, microhardness testing finds important and wide applications mainly in the investigation of small discrete grains, inclusions, and other fine textures as well as in the examination of thin layers. Microhardness testers, p. 82, are therefore widely used in the examination of raw materials as well as for testing semi-finished and finished products. Thus the application of microhardness testing methods is well known in the fine tools and instruments industries, glass works, cement industry, slag smelting, and in several other industries. The methods are also widely applied in metallography, mineralogy, and allied fields.

The construction and general operation of two commercially available microhardness testers of recent design, namely the Miniload tester (38, 152), p. 85, and the HANEMANN tester (245), p. 89, have been described. The methods most commonly used at present for microhardness testing have also been considered, p. 83. These are the VICKERS, the KNOOP, and the scratch hardness testing. A general account of the various applications of microhardness testing was also given, p. 82.

In the field of mineralogy, hardness is among the more important physical properties commonly used in the identification of opaque metallic minerals. Since this property is fairly constant for the same mineral species, it is suitable for diagnostic use, provided it can be reliably measured. One of the early endeavours to obtain precise and standardized hardness values was made by MURDOCK (170) who, in 1916, used a scratch-sclerimeter equipped with a steel needle for producing standard scratches on the polished surface of mineral specimens under the microscope. This was done in an attempt to establish a more extensive and reliable determinative table for the identification of opaque minerals than those based upon the Mohs hardness scale proposed at a much earlier date.

In 1925, TALMAGE (220), using the instrument previously employed by MURDOCK, but substituting a diamond point for the steel needle, added the accurate calibration of the intensity of the scratch and succeeded in developing a practical and reasonably reproducible method of quantitative hardness measurement based upon the principle of a measured weight on a standard point. Essentially, the instrument used was a rather simple device having a graduated beam carrying a sliding weight. The diamond was attached at the free end of the beam so that the standardized scratching point, directed downwards, could be lowered onto the polished surface of the properly mounted specimen which was held and moved by the mechanical stage of the microscope. Using the mechanical stage, the scratch was produced by moving the specimen under the point of the diamond with the sliding weight set at the proper position on the graduated arm. The entire operation was performed while the surface of the test specimen was viewed in the field of a low power microscope. By this means TALMAGE hardness could be determined by producing a standard scratch, and the test load applied was a measure of hardness. The eyepiece disk was provided with standard-limit scratches with which the scratches produced on the surface of the test specimen could be compared. Using this instrument, TALMAGE successfully established an extensive table in which opaque minerals are grouped into seven reasonably well defined classes starting with the softest or low hardness minerals and ending with the hardest species. These groups are designated by the letters A to G and by a type mineral for each group for comparison and for calibration cf instruments. These are: A, argentite; B, galena; C, chalcopyrite; D, tetrahedrite; E, niccolite; F, magnetite; G, ilmenite. The table largely replaced earlier tables such as those based on the older MOHS scale, all of which were largely dependent upon personal judgment rather than upon accurate quantitative measurement; minerals had been classified into a few poorly defined groups each including a comparatively large number of minerals which made them still less suited for determinative work.

In addition to the classification of minerals by hardness measurement, it is possible to compare the relative hardness of adjacent mineral grains by producing a scratch in the polished surface across the boundary of two minerals and comparing the scratch intensity on the two sides of the boundary.

One of the disadvantages of the TALMAGE method, however, is that, due to the design of the scratching mechanism, it requires a considerably longer working distance than could be offered by high and medium power objectives. The method, being limited to low power magnifications, is not suitable for testing hardness of small mineral grains and inclusions that could only be examined under higher magnifications.

With the highly refined microhardness testers available at present and the improved techniques of preparing polished specimens, microhardness testing methods attained a high degree of precision and are applicable to investigation of minute grains, inclusions, and other small structural elements in a large variety of materials under various microscope magnifications. Of the three methods of hardness testing commonly used at present, namely the VICKERS, the KNOOP, and the scratch hardness methods, p. 83, the VICKERS method is the most widely used since it embraces a wider hardness range and is considered more accurate than all other methods of testing hardness. The Miniload microhardness tester, p. 85, is equipped with exchangeable diamonds for the VICKERS, p. 83, the KNOOP, p. 83, and the scratch hardness testing, p. 84.

Using the Miniload microhardness tester, NAHKLA (172) determined the VICKERS hardness of fifty different species of opaque minerals mainly to show the applicability of the VICKERS microhardness method to the study of polished surfaces of opaque minerals, minute inclusions, and small discrete grains. VICKERS indentations were made on the surface of the specimens which had been polished by standard methods applying both

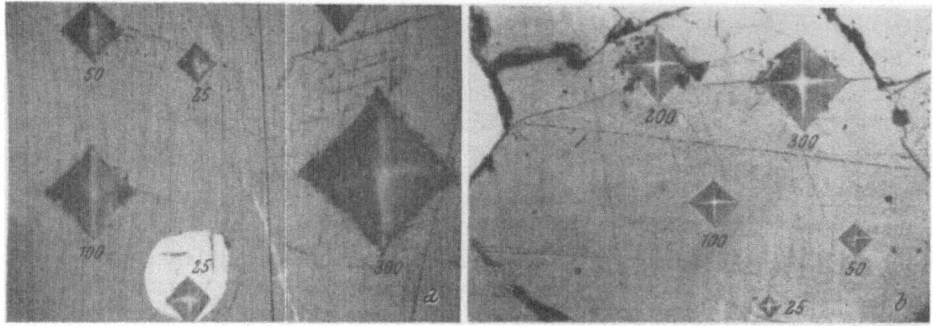

Fig. 154. VICKERS Indentations Made in the Polished Surface of Ore Minerals. Different test loads with the Miniload microhardness tester. *a*, bornite and chalcocite; *b*, pyrrhotite. 400×. Courtesy of F. M. NAKHLA.

hand and mechanical polishing. Different test loads ranging from 25 g to 300 g were used, and the tests were performed following the procedure given on p. 87. Fig. 154 illustrates VICKERS indentations made on polished mineral surfaces under various loads. The indentation diagonals were measured as described on p. 88, and the VICKERS hardness values in Kg/mm² were either obtained directly from VICKERS conversion tables or calculated from the equation $HV = 1854.4 \, p/d^2$, p. 83. The fifty mineral species were arranged according to the VICKERS hardness values thus obtained, and the sequence was correlated with the order of succession of the same minerals in the TALMAGE determinative table (220). This table, as already mentioned, comprises seven classes of minerals graded according to TALMAGE scratch hardness values, starting with the softest and ending with the hardest species. A few discrepancies are observed in the sequence of the minerals arranged according to VICKERS and to TALMAGE hardness values. These may be attributed to the difference in the methods and probably also to the higher accuracy of the VICKERS instrument. Whereas indentation hardness is related to the tensile strength of a material, scratch hardness is more dependent upon wear resistance. Inspite of the considerable variability of VICKERS hardness values obtained for several minerals under different test loads, further study may lead to a better understanding of the test load range and other conditions best suited for testing various minerals. It was noticed for instance that, for most common opaque minerals examined, the hardness values obtained were reasonably stable in the test load range of 200 g to 300 g. More extensive work is needed, however, and sufficient data should be obtained on a large number of minerals so as to develop a new determinative table equivalent to that of TALMAGE, but based on VICKERS hardness.

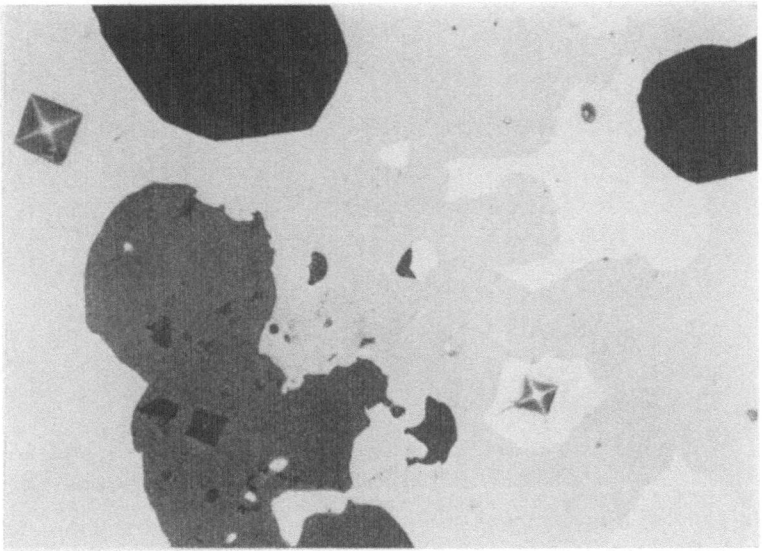

Fig. 155. VICKERS Indentations Made on Lead Glance (medium grey), Copper Pyrites (light grey), and Zinc Blende (dark grey) with the HANEMANN Microhardness Tester. A test load of 30 g was employed for all components. Shown also are some quartz grains (black). 400×. Courtesy of C. ZEISS, Jena (245).

Examination of the TALMAGE table (220) indicates that the majority of the opaque minerals are soft and that hardness of the minerals may be dependent to a certain extent upon the nature of their metallic elements (172). It was noticed, for instance, that minerals of metallic elements belonging to the same group in the periodic table of MENDELEEFF are generally of the same order of hardness. From these and other studies (216) made on polished surfaces of opaque minerals, significant relationships between hardness and texture were also observed. It has been noticed for instance that, in most cases, harder minerals seem to have crystallized earlier than softer minerals and that softer minerals commonly replace harder minerals, whereas minerals of the same grade of hardness frequently replace each other. Among other tendencies commonly observed are that ex-solution textures, graphic intergrowths, and graphic replacement textures are generally exhibited by minerals of the same order of hardness.

Existence of a relationship between hardness and insolubility of minerals
as well as between hardness and abundance of constituent metallic elements
has also been suggested (109).

In addition to determining the hardness of opaque mineral species,
microhardness testers are also used for testing hardness of various rock-
forming minerals. Fig. 155 is another illustration showing VICKERS indentations made for testing opaque mineral species in a lead-copper-zinc
ore, whereas Fig. 156 shows VICKERS indentations made on a number of

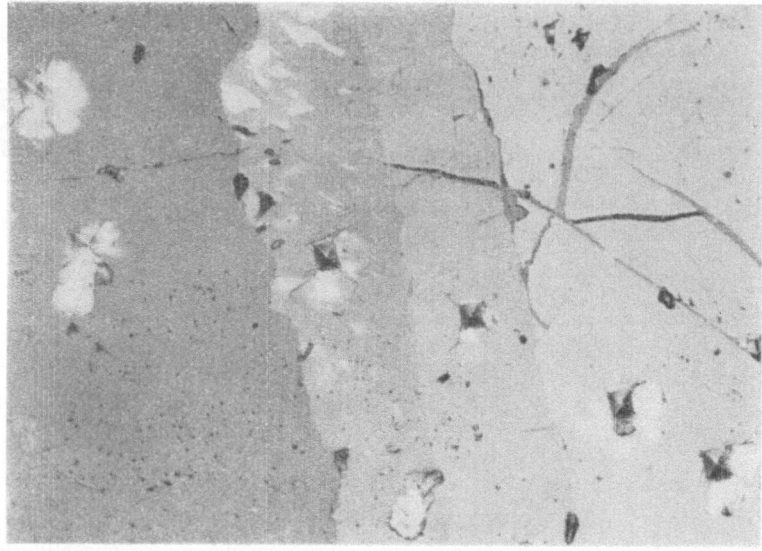

Fig. 156. VICKERS Indentations Made for Testing Rock Forming Mineral Species in a Norite
from Bamle, Norway, with the HANEMANN Microhardness Tester. Components in the sequence
of increasing greyness are olivine, hypersthene, actinolite, and plagioclase. A test load of
30 g was used for all components. 400×. Courtesy of C. ZEISS, Jena (245).

rock-forming minerals in the polished surface of a norite rock composed
of olivine, hypersthene, actinolite, and plagioclase feldspar. In the tests
illustrated by the two photomicrographs, the HANEMANN microhardness
tester Model D 32, p. 89, and a constant test load of 30 g were used for
all components.

The microhardness of single crystals can also be be satisfactorily tested.
Fig. 157, shows VICKERS indentations made on the surface of a single crystal
of nickel with varying test loads and the HANEMANN instrument.

In industry, testing of materials from which various objects are manufactured as well as testing of finished workpieces and semi-finished products is of paramount importance. The testing of microhardness finds wide
applications in several fields of engineering and technology, particularly
for testing texture, foil material, thin surface layers, and tiny objects.

Figs. 158 and 159, illustrate applications to the examination of steel and other metals and alloys.   Fig. 158, shows VICKERS microhardness indentations made in the polished surface of a sample of steel for the purpose of comparing the hardness of the dark cubic martensite with that of the lighter austenite matrix.   For the purpose of comparison, the inden-

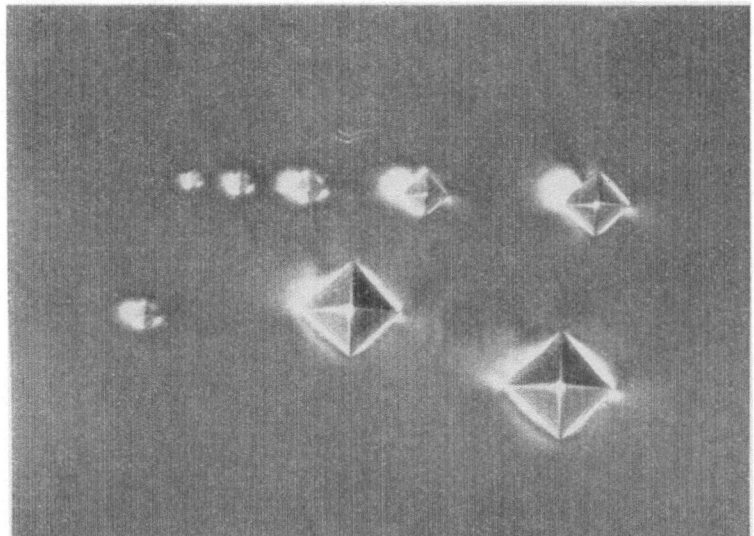

Fig. 157. VICKERS Indentations Made on a Single Crystal of Nickel. Different test loads with the HANEMANN Microhardness Tester. 400×. Courtesy of C. ZEISS, Jena (245).

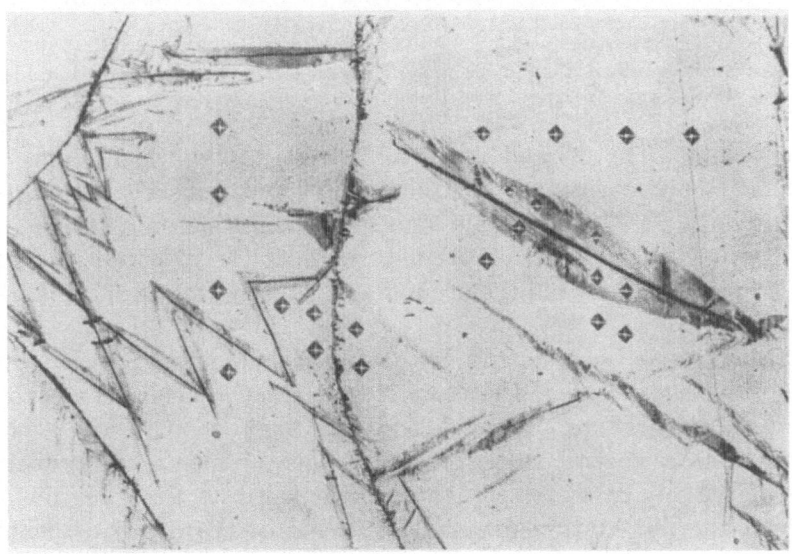

Fig. 158. Polished Surface of Steel Showing VICKERS Indentations. Comparison of hardness of the dark cubic martensite and the austenitic matrix with the HANEMANN Microhardness Tester Model D 32. 400×. Courtesy of C. ZEISS, Jena (245).

tations are made with a constant test load. Fig. 159, shows VICKERS indentations made for testing components of a ternary cast-aluminium alloy. In both examples, the HANEMANN microhardness tester was used.

Since microhardness indentations require only small test loads and may be confined to very small areas of the test specimens because of the great accuracy of locating the spot to be tested, the modern microhardness testers are particularly suited to the non-destructive testing of delicate objects and for the testing of very small parts. Thus the test can be made on watch wheel spindles and even on proper cutting edges of tools and on

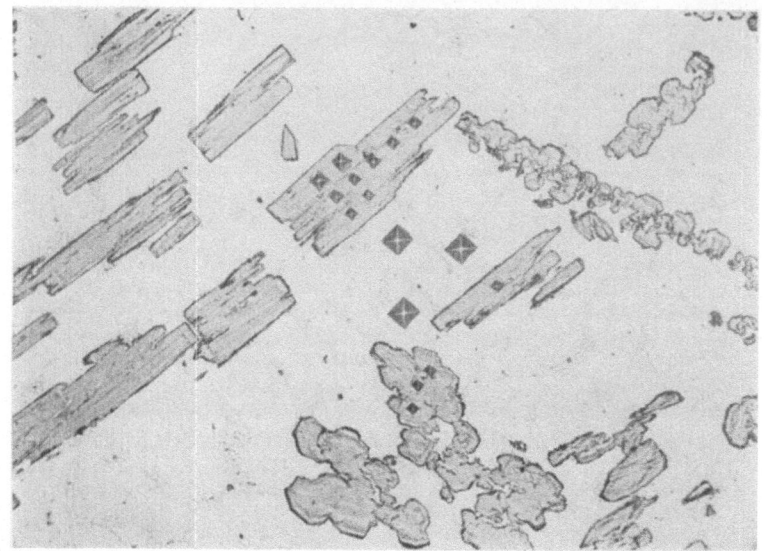

Fig. 159. VICKERS Indentations Made on Components of a Ternary Cast-Aluminium Alloy with the HANEMANN Microhardness Tester. 320×. Courtesy of C. ZEISS, Jena (245).

needle points. This is well illustrated in Fig. 160a which shows several VICKERS indentations made on the ground and polished tip of a record player stylus. Fig. 160b shows a VICKERS indentation on a link of a zipper. Both specimens had been embedded in Plexiglass to permit indentations to be properly made while the ground and polished test surface is perfectly horizontal.

Extremely small objects and very thin sheets must be properly mounted in a suitable material. If the cross section of such specimens is to be examined, it may be ground and polished together with the embedding medium. As a general rule, for proper measurements of hardness, thin layers and sheets to be tested should be at least 1.5 times as thick as the length of the indentation diagonal or 10 times the depth of penetration of the VICKERS indenter. If a cross section is to be examined, the minimum thickness of the layer (or the diameter of the wire) to be tested, should be twice the length of the indentation diagonal.

The Miniload microhardness tester, p. 85, Fig. 59, is suitable for testing small specimens as well as large objects. Suitable mounting devices, designed for use with the instrument, are available for holding test specimens of various sizes and shapes which are not too fine to be clamped. Small specimens such as sheet metals, wires, and bearing points, which are to be examined in cross section, may be mounted between the jaws of a small specimen holding vise which can be tilted about two axes and locked to secure the required position. The complete fixture is clamped to the compound stage b of the tester, Fig. 59. Many tools or work pieces are too large to be held in the small mounting device described or to be placed

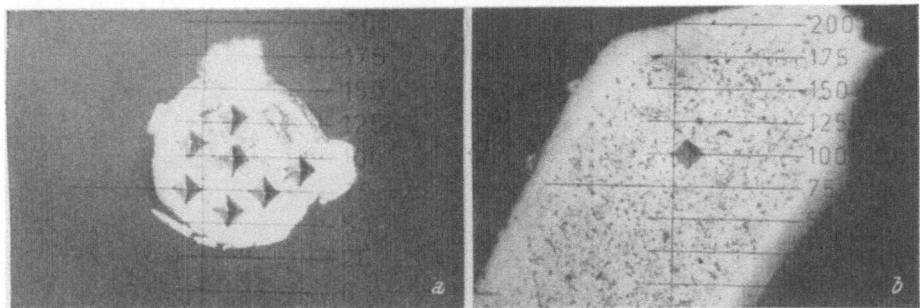

Fig. 160. VICKERS Indentations Produced with the Miniload Microhardness Tester. The divisions are those of the vertical scale of the ocular micrometer. a, indentations on the tip of a phonograph needle; b, indentation on a link of a brass-zipper. Specimens mounted in Plexiglass. Courtesy of E. LEITZ, Wetzlar (152).

on the compound stage of the tester. For the inspection of these large objects, such as hobs, cutters, twist drills, conical and cylindrical reamers, or similar parts, the microhardness tester is swung around the upright column d for 180°, Fig. 59, and the specimen is mounted on a suitable fixture. Fig. 161 shows a lead screw securely resting on two V-blocks and the microhardness tester swiveled into working position. Other adjustable fixtures to suit different kinds of parts are also available. A special model of the instrument can be placed directly and securely on test parts too large to be held by the devices just referred to. This model is suitable for testing both plane and cylindrical parts such as large castings, machine beds, rolls, and steel structures.

7. Mounting Small Crystals for X-Ray Investigations.

Operative work for various purposes on individual particles and single crystal specimens of very small dimensions must frequently involve mechanical manipulation under microscopic control. Thus, metal filings and mineral particles may be oriented for the purpose of taking electron micrographs. Micromanipulative techniques enabling proper and close mounting of very small single crystal specimens, with dimensions of about 100 μm or much less, in a predetermined orientation have also gained much importance with the development of micro x-ray diffraction cameras. Such small crystals may also be ground, cut, and shaped with the aid of mechanical devices under microscopic observation. For x-ray studies, it may be required, for instance, to remove excessive amounts of glass or crystalline material from the crystal to be investigated, or to obtain cylindrical crystal

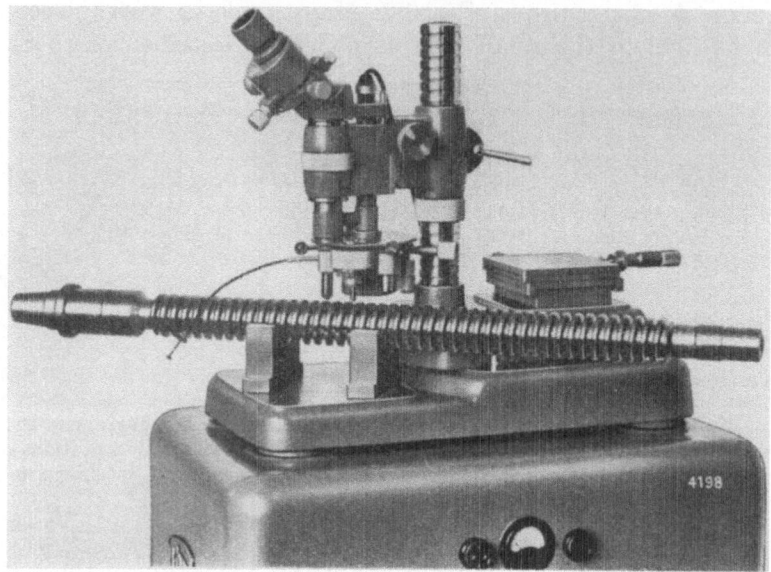

Fig. 161. The Miniload Microhardness Tester Swung Around the Supporting Column for Examining a Large Lead Screw. The screw is mounted on two V-blocks. Courtesy of E. Leitz, Wetzlar (152).

specimens to simplify absorption corrections. By means of suitable micromanipulations (157, 180, 184, 234) such delicate operations may be performed on the tiny specimens, which may be irreplaceable, without fear of losing or destroying them through clumsy handling.

WAIT (234) described a micromanipulative technique for orienting and mounting single crystals which may be as small as 100 cubic microns (0.1 pl) in volume and 0.5 ng in mass for x-ray diffraction investigations. Since in many cases, in the study of inorganic systems, it is not possible to prepare crystals larger than about 30 μm in the longest dimension, the technique permits the more advantageous single-crystal x-ray diffraction studies on systems which previously could be investigated only by the use of the powder diffraction methods. The procedure consists essentially of selecting a suitable crystal on a dry glass slide under the microscope, manipulating it on the slide into proper orientation in relation to a suitable mounting fiber carried on a micromanipulator, and attaching it, usually by means of cedar-wood oil, onto the end of the mounting fiber. The movements offered by the micromanipulator added to those of the revolving mechanical stage of the microscope are adequate for these operations. By operating the vertical movement of the micromanipulator, the fiber with the adhering crystal is slightly raised, and when the oil hardens, which usually takes place in about half an hour, the tool head with the fiber is removed and mounted in the x-ray goniometer by standard methods (43) so that the required axis of the crystal is perpendicular to the x-ray beam.

For the manipulative operations above mentioned, WAIT uses a petrological microscope fitted with a built-in revolving stage and an attachable mechanical stage which provides motion in two mutually perpendicular directions in the horizontal plane. It is also equipped with 25-, 16-, and 6-mm objectives and a $6\times$ eyepiece. Polarizing microscopes are often very helpful since they offer better facilities for choosing well developed crystal specimens in which the crystallographic axes can be easily identified and which are free of twinning and impurities. The microscope is used in conjunction with a micromanipulator which provides motion in the three space coordinates and which is equipped with a suitable tool holder. While many micromanipulators of the more elaborate types may be successfully used, a simple three dimensional rack-and-pinion instrument is preferable if it is mounted on a heavy base and suitable for use at magnifications up to about 200 diameters. The mounting fibers are made of a material with a low scattering power for x-rays. Pyrex glass fibers of about 0.5-μm diameter are satisfactory. These are prepared from thin Pyrex glass rods, about 2 mm in diameter, with the aid of a DE FONBRUNE microforge, p. 152. A short length of the original rod, to which the fiber is attached, is retained to serve as a tool head. Between the fibers and the original rod, a thinner rod, drawn out by hand, is bent to facilitate insertion of the fiber under higher power objectives. The tool head is attached with soft wax to a piece of a glass tubing, about 15 cm long and 2 mm in inside diameter, which is then mounted in the tool holder of the micromanipulator.

Another method involving micromanipulation was described by WAIT (234) for maintaining minute crystals in a controlled atmosphere in equilibrium with the mother liquor. The necessary operations can be performed in a moist cell as that described by BENEDETTI-PICHLER, p. 104. A thin walled Pyrex capillary of about 1-mm inside diameter and 15 mm long, sealed at one end, is placed into the moist chamber. By means of a micropipet mounted on a micromanipulator, a droplet of the mother liquor is introduced into the capillary near its closed end. The crystal is mounted on the free end of a short glass fiber, a few millimeters long, drawn out at one end of a thin glass rod about 1 mm in diameter; a short length of the original rod is retained. The fiber, with the crystal attached, is then inserted into the capillary containing the droplet of the mother liquor of the crystal, so that a part of the rod, a few millimeters in length, fits into the open end of the capillary. The diameter of the rod should be slightly less than that of the capillary to make a close sliding fit. The rod is then sealed to the open end of the capillary. This may be done by applying to the joint a solution of wax in a volatile solvent. A durable glass-to-glass ring seal between the capillary and the rod can be produced by means of the microforge after making a temporary seal by applying a blob of wax.

The device used by MACKAY (157) in mounting small single crystals for x-ray structural analysis in a predetermined orientation is briefly described on p. 38. The device is actually a simple microscope attachment by means of which crystal specimens down to about 15-μm length can be mounted onto glass fibers 15 μm in diameter. The operations may be performed under microscopic magnification of 75 diameters. By means of a fine needle fitted to the shorter arm of the lever, a crystal is selected, adjusted to the desired orientation, and then brought to the intersection of the crosshairs by moving the slide. The microscope objective is then racked up, and the needle is replaced by a bent glass capillary or wire carrying a fine glass fiber of a suitable diameter, the free end of which has been freshly dipped in a suitable adhesive material. Jewelers' shellac may be used for this purpose (157, 180), but oxidizable oils such as linseed oil and cedar-wood oil are preferable since they dry more slowly (158).

The free end of the mounting fiber is first brought into focus at the crosshairs, and the objective is then lowered until the end of the fiber comes into contact with the crystal which adheres to it. The objective is again racked up, and the form and orientation of the crystal with respect to the fiber are checked by examining the crystal from all sides. The fiber, with the crystal attached, is then transferred to the x-ray goniometer. This simple technique is to replace that involving the use of a manually operated dissecting pin, the handle of which is supported and steadied by pressing it down, near its end, upon a finger of the other hand.

Other micromanipulative procedures were developed by ORDWAY (180) for mounting and grinding of small single crystals of refractory compounds to be used for x-ray diffraction investigations. The low power micromanipulator (180, 181) used by ORDWAY to facilitate the necessary operations is briefly described on p. 37. For removing excessive amounts of glass or crystalline material from a desired crystal or for grinding cylindrical crystal specimens to simplify absorption corrections, ORDWAY recommends the use of a small crystal grinder mounted on the micromanipulator. The grinder is made by fitting a simple adapter on the shaft of a miniature Alnico field motor, 27.5 v, to permit the use of standard abrasive tips with 1/8-in. shanks. The specimen is mounted in the usual way so that the portion to be removed is exposed, and the unwanted excess is then carefully removed. A very efficient, smooth, grinding surface is obtained by coating a 1/8-in., 3-mm, polished steel shaft with an abrasive mixture made by stirring a little of 5-μm diamond powder with a few drops of thinned lacquer. This and other devices (184) for mechanical grinding and shaping of crystal specimens replace the slow and often inefficient technique of applying a solvent on a camel's hair brush operated by hand to remove unwanted material. In many instances no proper solvent can be found, and mechanical grinding becomes indispensable.

For cutting and shaping fragile crystals, PEPINSKY (184) strongly recommends the use of the commercially available S. S. WHITE industrial Airbrasive unit (175, 238). This cutting device directs a high velocity stream of gas-propelled abrasive particles against the work surface through a small diameter nozzle, whereby accurate and fast cuts can be made in hard and brittle materials. By means of this unit, fragile crystals can be cut and shaped into cylinders with accurately circular cross sections, down to a fraction of a millimeter in diameter and 1.5 to 2 cm long. Crystal specimens of these dimensions are required for neutron diffraction studies. For shaping a crystal, the specimen is mounted and properly oriented, either optically or by x-ray, on a standard goniometer head which is then placed on the shaft of a small lathe, while the Airbrasive tool is mounted on the tool holder. For best results the lathe should turn very slowly.

8. The Study of Fibers.

Microdissection techniques involving the use of micromanipulator mounted microneedles have been applied to the study of the microstructures of simple ramie nitrate fibers (188), and paper pulp fibers (214).

Direct microscopic observation, examination in dark-field and polarized light, as well as other methods such as maceration and staining indicate that individual fibers of natural cellulose are made of finer fibers or fibrils. The evidences obtained by these methods, however, are not quite conclusive. For instance, natural cellulose, such as wood fibers and cotton fibers, shows striations in bright-field and dark-field illumination, indicating that the individual fibers consist of parallel oriented fibrils. There is a probability, however, that such observation may be due to surface irregularities or diffraction phenomena.

With the aid of micromanipulative techniques, SEIFRIZ and HOCK (214) verified the fibrillar structure of wood fibers and thus proved that the striated structure, seen in natural cellulose under the microscope, is a true property of the fibers and not due to surface convulsions or diffraction phenomena. Two samples of paper pulp were investigated. The first was a dry sample of unbeaten material, and the second was of a wet highly beaten material. Under the microscope, individual fibers of both samples, as all other beaten or unbeaten paper pulp fibers, exhibited many surface striations. This appearance was distinctly more vivid in the dark-field than in the bright-field, and likewise more pronounced in the beaten pulp than in the unbeaten sample. A number of experiments were performed by means of fine microneedles mounted on a suitable micromanipulator. It was shown, for instance, that by applying a lateral pressure to a fiber from the beaten pulp so as to flatten it, the striae could be separated as individual strands.

Further evidence was obtained by inserting two microneedles into the middle of a fiber. By moving the two needles apart, the fiber could be split longitudinally into two. The inner surface of the fiber thus separated appeared perfectly smooth indicating that the fiber is composed of linear strands. Even single fibrils of an estimated diameter of about 1.4 μm could be separated and lifted off. This could be achieved by holding the fiber in position by means of one needle and inserting another needle close to the edge of the fiber. By moving the second needle away from the first one, an individual fibril or a number of fibrils could be lifted off.

More careful dissection revealed further that each of the primary fibrils above indicated is in itself a bundle consisting of much finer secondary fibrils having a roughly estimated diameter of the order of 0.1 to 0.3 μm. These secondary fibrils, the diameter of which is approaching the limit of microscopic visibility, are difficult to be handled individually for further dissection but may, in turn, be composed of finer fibrils or of shorter units of microscopic or submicroscopic dimensions. This view, as to linear and more elementary units of which the secondary fibrils may be built, is supported by evidence derived from other considerations.

9. The Structure of Rubber Latex.

The techniques of micromanipulation have been extended to the field of colloidal chemistry and should prove of increasing importance in the study of colloids, particularly in the investigation of surface films.

HAUSER (120) reported the use of microdissection techniques in studying the structure of the latex particles of rubber. Microcinematograms were taken to show the behavior of particles and the various stages of the operation. When *Hevea* latex, for instance, is examined with dark-field illumination under high magnification, numerous globular particles, varying from about 0.5 to 3 μm in diameter, are seen to exhibit Brownian movement. The smaller particles have a rounded shape, whereas the larger ones are mostly pear-shaped. In vulcanized latex, the particles exhibit a more rapid motion, are mostly circular in outline, and are of a more uniform size than in the unvulcanized latex.

For the microdissection of latex particles of rubber, HAUSER used a fine-pointed microneedle, mounted on a manipulating device, to penetrate the individual latex particles under high power magnification. It was demonstrated that when one of these minute sac-like particles is indented with the point of the needle, it behaves like a toy balloon. The outer elastic skin of the latex globule resists the penetration of the needle point. When the needle finally penetrates the globule, a viscous liquid, which forms the interior of the globule, is seen on the needle point. This experiment helped greatly to understand the nature and structure of the latex globule of rubber and to interpret results obtained in x-ray as well as in other investigations.

10. Micropaleontology: Single-Mounting of Microfossils.

Specially fashioned micropipets, mechanically guided in the field of the microscope, have been used by ANDERSON (5) for making single-grain preparations of various types of microfossils. These are first dispersed in a suitable liquid on an object slide, whereafter a specimen is selected under the microscope, taken up into the pipet, and transferred onto another slide. There, it is then covered with a suitable mounting medium and a cover slip. The technique permits rapid sorting and mounting of various microfossils such as pollen, spores, diatoms, and hystrichospherids. Single mounting of as many as 30 to 40 specimens per hour is possible (6). Semi-permanent mounts in glycerol jelly or permanent mounts in Canada balsam may be made. The technique is more rapid and versatile than the widely used fishing operations such as those described by KLAUS (137) and MADLER (159), or those described by FAEGRI and IVERSON (95) for mounting pollen grains.

The assembly used by ANDERSON is shown in Fig. 162. It comprises an ordinary biological microscope with a revolving nosepiece provided with low and high power objectives. The large microscope stage accomodates a mounting bracket and a LEITZ (152) attachable mechanical stage PIRUX firmly bolted in a vertical position to the microscope stage as shown in the figure. This compound mechanical stage model is provided with two rack-and-pinion cross movements, each permitting a displacement of 3 cm, and an object slide holder. The upper arm of the slide holder is removed, and the micropipet is attached to the lower slide-holding arm by means of a rubber tubing fitting tightly round the arm and the wide stem of the micropipet. This mounting is sufficiently flexible to permit necessary adjustment of the position of the pipet.

The micropipet is constructed from a length of glass tubing 4 to 5 mm in diameter. The tube is bent, and its end is drawn out in the flame into a fine capillary which should be inclined downward toward the tip as shown in Fig. 163 *a*. The angle of inclination is important for proper observation and manipulation of the specimen. If the shaft of the pipet is too much inclined at the tip, it would be difficult to observe the specimen. If the inclination is too small, the surface tension effects between glass and fluid would be too strong for proper control of the specimen. The diameter of the orifice of a tip suitable for most microfossil work is about 100 μm. On a lap using a fine abrasive, the point of the tip is ground, as shown in Fig. 163 *b* to the shape of a wedge with an upper and a lower ground surface.

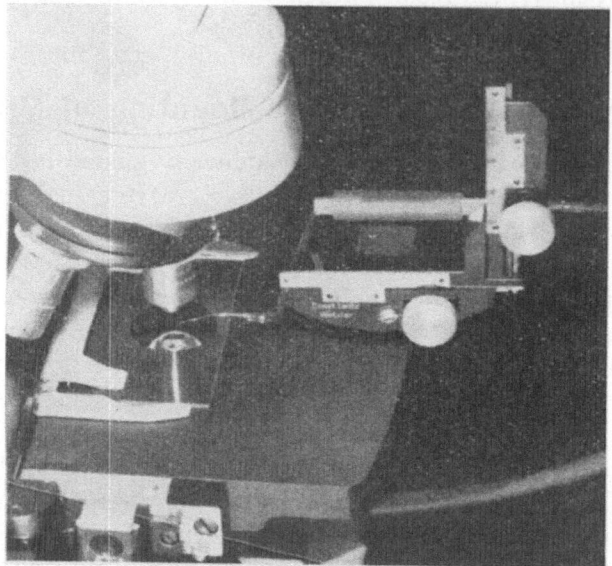

Fig. 162. Assembly for Single-Mounting of Microfossils. Leitz mechanical stage PIRUX mounted vertically on the microscope stage with micropipet in position. Courtesy of R. Y. Anderson.

The pipet is mounted on the mechanical stage to give ample clearance for the revolving microscope objectives. Its wide end is connected with a long piece of rubber tubing of about 3-mm bore. In operation, the free end of the rubber tubing is taken into the mouth of the operator, and the movement of fluid and specimen in the pipet is controlled by blowing or sucking.

For making single-grain preparations, the sample is first dispersed on an object slide in a thin layer of glycerine or oil. Under low power magnification of approximately 100×, a specimen is selected and brought into the center of the field. High power magnification may then be used to verify that the specimen is suitable. All other operations are carried out at low magnifications. By means of the two adjustments on the Leitz mechanical stage, the pipet is manipulated in the 6–12 and the vertical directions to bring the tip into contact with the liquid just in front of the specimen, Fig. 163b, whereupon the fluid, carrying the specimen with it, instantly moves into the pipet. The movement of the fluid, tending to be rapid, may be controlled either by blowing cautiously into the rubber tubing or by taking up into the pipet a long column of the fluid prior picking up of the specimen.

By means of the vertical adjustment of the mechanical stage, the micropipet with the picked up specimen is lifted out of the liquid. The slide carrying the dispersed sample is replaced by a new slide. The tip of the pipet is brought into contact with the new slide and the specimen is discharged from the pipet by blowing into the rubber tubing. The specimen

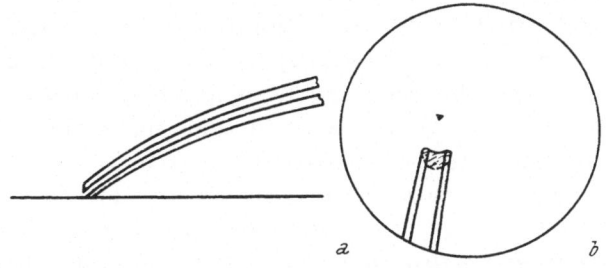

a   *b*

Fig. 163. Micropipet for Picking Up Microfossils. *a*, lateral view showing shape of shaft and grinding of tip; *b*, microscopic view with a specimen in front of tip.   Courtesy of R. Y. ANDERSON.

thus ejected is contained in a droplet of oil or glycerine, expelled with it onto the new slide. A drop of the mounting medium is then placed on top of the droplet and covered with a cover-slip.

In a rapid procedure for making a large number of single-grain preparations, the dispersed sample is scanned under low magnification, and as many as 20 to 30 of good specimens are taken up with the micropipet. The slide bearing the dispersed sample is removed, and several new slides are placed in a row upon the stage of the microscope. Onto each of these new slides, a single specimen is discharged from the pipet until all specimens have been ejected. The specimen on each slide is then covered with the mounting medium and cover slip. The procedure is repeated until all types are fairly represented. Other specimens are then located, examined under high magnification one by one, picked up, and mounted. Paratype slides may be prepared in the same way by scanning and picking up specimens of the same type, mounting them either all together on one slide or on separate slides.

Instead of the LEITZ mechanical stage, the micropipet may be mounted on a simple three-dimensional rack-and-pinion low power micromanipulator as those shown in Figs. 15, 17, and 18. These micromanipulators, permit motion of the micropipet in the three space coordinates instead of the two dimensional movement of the LEITZ mechanical stage. In addition, the microscope stage, would accomodate a greater number of slides for making preparations.

11. Entomology.

a) Brain Research in Insects.

In the field of neurophysiology, general techniques of manipulation under optical microscopy enable the brain physiologist to carry out very interesting experiments on specific areas of the brain of insects in an endeavor to get a better understanding of the origin and the underlying mechanism of instinctive behavior.

Recent research on the behavior of animals showed that many of their actions, the so-called instinctive actions, are species specific, are fixed from birth, are correct from the start and need not be learned. Examples of such actions are the gaits of animals and their modes of conduct during time of propagation and care of the young.

The group of insects, particularly crickets and locusts, are especially suited for the study of the origin and mechanism of instinctive actions, since they are characterized by highly developed modes of behavior. Like birds, many crickets are excellent musicians. Their manifold songs are characteristic of each species and serve to establish understanding between its members. Among the songs of known biological significance are the songs of battles which arise between males of the same species which happen to meet during time of reproduction, and the courting songs summoning the female which hastens to the scene. These sound productions of the males, as well as other behavioral procedures, are generally accompanied by clear and characteristic movements, the execution of which often requires the alternate or simultaneous activity of many elements of movement, each of which performing a specific task well coordinated with the activity of the other elements. For instance, the singing male cricket requires the activity of as many as 28 muscles to raise the wings and rub them against each other.

Since the activity underlying every human movement, as well as those of animals, is regulated through the functioning of the nerve cells in the brain and spinal cord commanding the proper muscles, it is necessary to investigate the neural command posts in order to study the mechanisms underlying various modes of behavior. One of the methods used for the study of the performance of an individual neural center in an animal is based upon the localized electrical stimulation of the required center and the examination of the released behavior.

F. HUBER (124), who wished to carry out such investigations on insects, developed an experimental procedure for crickets and locusts permitting the electrical stimulation of individual portions in the brain of the insect and the simultaneous observation and recording of the response to the stimulus. Though our knowledge of the mode of action of localized brain centers is still far from adequate, this experimental procedure permitted the localization of such brain centers and the study of some of their functions. With crickets and locusts, HUBER was able to demonstrate that all modes of instinctive behavior fail to appear if the brain is put out of action. Stimulation experiments on the brain of crickets proved that specific brain structures are coresponsible for certain performance of the insect. The stimulation of these centers by natural or electrical means releases complex behavioristic actions, and the repeated stimulation of a specific brain structure always produces the same result. It was found, for instance, that an electric stimulation of a specific, sharply defined region of the brain of the male crickets always produced chirping, and the songs thus released cannot, as a rule, be distinguished from the natural ones, even by an expert. Electric stimulus in another definite area of the brain released running and jumping, whereas the stimulation of another spot increased breathing.

The extent of stimulation depends on the intensity of the stimulus, upon duration and sequence frequency of the impulses, and upon the nature or condition, for example the conductivity, of the nervous tissue.

Assembly Used for Brain Stimulation of Crickets. For the successful performance of stimulation experiments for the study of the released behavior, the experimental animal should be well able to endure the necessary operations and have complete freedom of movement during stimulation.

The brain of crickets, housed in the head, has a maximum width of about 1.5 mm, and a volume of only 0.5 to 0.6 mm^3 (μl). Electric stimulation of a localized spot in this tiny brain requires the use of stimulation electrodes less than 40 μm in diameter in order to avoid major injury to the brain on inserting the electrodes, which may render the animal unsuitable for experiment. Electrodes made of lacquer-insulated tungsten and platinum wires, having a diameter of 15 to 20 μm, were found suitable and are sufficiently stiff to be easily introduced through the meninges into the required part of the brain. The only part of the electrode, which is effective in stimulation, is the bright surface of the cross section of the electrode tip.

Unipolar stimulation is carried out by introducing the active electrode into the brain, whereas the neutral reference electrode, soldered to the animal holder, is inserted into the chest cavity. A loop of platinum wire, about 0.1 mm in diameter, may be used as the neutral electrode. Double or bipolar electrodes may be used; the two insulated electrodes are held together by a common lacquer layer and have their tips about 5 to 7 μm apart.

The part of the assembly used for mounting and observing the experimental animal is shown in Fig. 164. A ZEISS stereoscopic binocular microscope is held on a special adjustable structure in such a way that the microscope can be tilted backward and forward or turned sidewise by releasing knurled screw knobs. The microscope is mounted on a base plate which carries also a stand on which both the animal holder b and the electrode holder c are adjustably mounted. Between the microscope and the stand, the common base plate also carries a vertically adjustable reception plate d

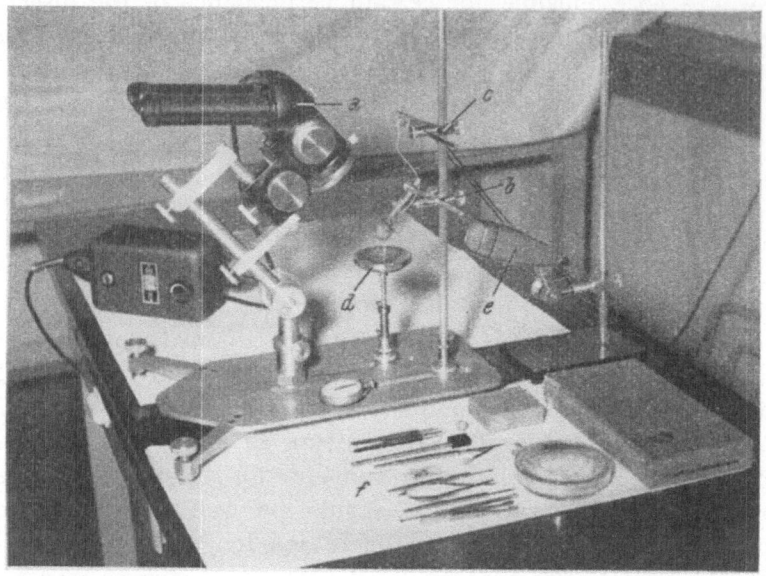

Fig. 164. Set-Up for Mounting and Observing Cricket for Stimulation Experiments. a, stereomicroscope on adjustable mounting; b, animal holder, cricket and cork sphere; c, electrode holder; d, reception plate for the sphere; e, microphone; f, tools used in operation. Courtesy F. HUBER.

for the cork sphere which is held between the legs of the cricket to allow running while the animal remains stationary. Also shown in the figure, held on a separate stand, is a microphone e which is connected to a tape-recorder for registering cricket songs.

The stereomicroscope offers a sufficiently long free-working distance suitable for various manipulations. The magnification can be continuously changed from 6 up to 40 diameters by turning a drum. The assembly should be free, as far as possible, from vibration effects since crickets respond immediately with fright reactions to slight vibrations. A suitable low-voltage lamp is used for illumination, and heat filters or cooling cells may be used to eliminate heat effects which may be disturbing in operations extending over long periods.

The complete assembly including the electrical equipment is shown in

Fig. 165. In addition to the components shown in Fig. 164, the assembly comprises a CO_2 cylinder a for narcosis. The electrical equipment includes a tube generator b, delivering stable voltages with independently variable

Fig. 165. Complete Assembly for Brain Stimulation of Crickets. a, narcosis cylinder; b, tube generator for stimulation; c, potentiometer; d, cathode-ray oscillograph; e, recording camera; f, tape-recorder; also other components shown in Fig. 164. Courtesy of F. Huber.

amplitudes, duration, and frequency and a potentiometer c permitting a finely graduated voltage division with a switch for opening and closing the stimulus. The potentiometer has two outlets, one of which is connected

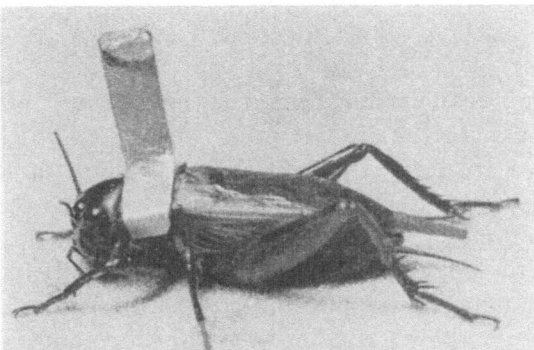

Fig. 166. Male Cricket Fastened to the Animal Holder. A conducting wire soldered to the holder leads to the neutral reference electrode inserted into the chest cavity. Courtesy of F. Huber.

with the stimulation electrode, and the other with the cathode-ray oscillograph d. Also shown in the figure is a recording camera e. This arrangement enables photographic recording of the form and the amplitude changes of the impulses. The tape-recording instrument f is connected with a microphone, shown also in Fig. 164, and serves for registering the cricket songs for comparison with normal songs.

Stimulation Procedure. The cricket is first subjected to a brief CO_2 narcosis, and is then held by a suitable copper holder, shown in Fig. 166, so that the animal is left free to move its head, wings, and legs. The neutral electrode to be inserted into the chest cavity and the conducting wire to the tube generator are soldered to the holder. A small window is cut in the chitin cuirass of the head between the feelers, and while still under narcosis, the animal is fastened to the stand.

Fig 167. Male Cricket and Holder with the Conducting Wire Leading to the Neutral Electrode, a Double Electrode Inserted in the Brain, and the Cork Sphere. Courtesy of F. HUBER.

Observing through the microscope, the muscles and brain tracheae are pushed aside to expose the frontal brain surface. The escaping blood is constantly removed to clean the field of view. The stimulation electrodes hanging down from the electrode holder are then introduced into the required part of the brain. These very fine electrodes are negligible in weight and have sufficient length and flexibility to follow all movements of the head without changing their position inside the brain. Coagulation of blood escaping from the wound forms a thin film and thus protects the brain from drying out. The temporary stimulus which the insertion of the electrodes exerts on the brain subsides a few minutes after the animal comes out of narcosis, and its behavior becomes perfectly normal. The animal is then subjected to stimulation.

Stimulation experiments usually last for several hours. A cork sphere between the legs of the cricket, Fig. 167, permits the animal to run while stationary, turning the sphere as it runs. The selected structures in the brain are stimulated with impulses of various frequencies and intensities, and the released behavior is observed and recorded.

The animal is killed after the experiment, and the position of the electrode inside the brain is finally determined from sections made of the brain. Thus the recorded behavior of the animal is correlated with a specific brain structure.

b) Artificial Insemination of the Queen Bee.

The artificial insemination of the queen honeybee is necessary for carrying out reliable and precisely controlled breeding and inheritance experiments since it permits the strict keeping of pure lines. Controlled matings

Fig. 168.   Abdomen of the Virgin Queen Bee.   *mo*, oviductus medianus or midoviduct; *lo*, oviductus lateralis; *sp*, spermatheca or seminal vesicle; *vf*, vaginal flap; *v*, vagina; *ov*, ovarium.   Courtesy of F. Ruttner.

of the honeybee by establishing isolated mating stations, a procedure still practiced for planned breeding, is not very reliable since, in most cases, such stations are not secure from visits by alien drones.

Fairly recently, the artificial insemination of the queen bees became possible on a large scale and enabled the exact study of the genetics of the bee. Nevertheless, it is already possible to have a clear understanding of the genetics of a number of mutations of body and eye color, whereas several other genetic factors are still under investigation. Examples of the latter factors are those leading to hairless bees, and to the so-called cyclops bees which have a single eye at the center of the brow.

The genital system of the virgin queen is diagrammatically shown in Fig. 168. In natural mating, which takes place in the open air during the nuptial flight, the queen copulates with several drones, and the first eggs are deposited a few days later. On copulation, the semen gets first into the oviducts, *mo* and *lo*, and from there into the seminal vesicle or spermatheca *sp*. Strong backflow of the semen is prevented by means of the vaginal flap or valve fold *vf*. In the seminal vesicle the spermatozoa of the drone are stored and remain alive for about three to four years, during which fertilized or unfertilized eggs are laid according to whether queens and workers, or drones are to be produced depending on the social requirements of the population of the hive.

For artificial insemination, the main problem is the injection of the semen behind the valve fold into the midoviduct which has an opening only about 0.33-mm wide. For this operation, it is necessary to expose the entry to the vagina and to depress the valve fold by means of a suitable probe. This is to permit insertion of the tip of the cannula, used for delivery of semen, into the narrow opening of the oviduct.

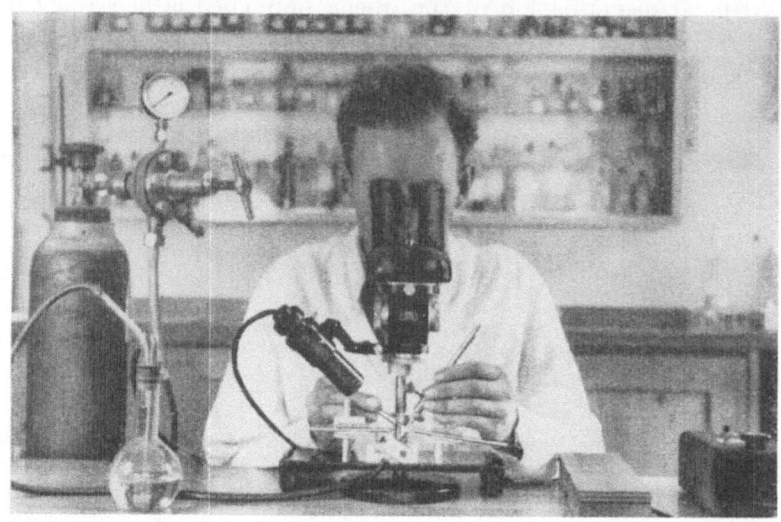

Fig. 169.   Simple Micromanipulator-Microscope Assembly for Artificial Insemination of Queen Bees. Shown is a CO_2 narcosis cylinder, a double micromanipulator device carrying two tractors, and queen mounted under the microscope.   Courtesy of F. RUTTNER.

The assembly used by F. RUTTNER (198) for mounting the queen bee, narcotized by CO_2, in a suitable position for artificial insemination is shown in Fig. 169. It comprises a ZEISS stereoscopic binocular microscope mounted on a column sufficiently long to offer a suitable space for the micromanipulator. The increased depth of field of the stereomicroscope is particularly convenient since the operating region extends for an appreciable depth. The large working distance and the magnification changer with constant working distance greatly facilitate the work. Two very simple micromanipulator units are mounted opposite to each other, one on each side of the microscope. Each unit holds a simple fine tractor for adjusting the rear end of the queen, Figs. 171 and 172.

The queen, narcotized by carbon dioxide, is first fastened in the apparatus with its rear end upwards.

A small syringe, of about 2.5-μl capacity, with a fine cannula of Plexiglass is then filled with semen from drones. For the latter purpose, the drone, under chloroform narcosis, is induced to evert its copulatory organ.

By applying a slight pressure on the abdomen, the semen generally is easily expressed, whereupon it is taken up into the Plexiglass cannula, Fig. 170. The semen from four to five drones is sufficient to fill the syringe.

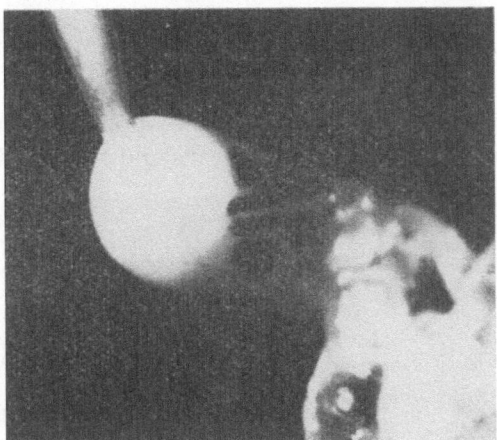

Fig. 170. Drawing into the Cannula the Semen Expressed from the Everted Copulatory Organ of the Drone. Courtesy of F. Ruttner.

This operation is performed under the stereomicroscope at a magnification of about 15× in order to separate the semen from the accompanying viscous mucus.

By means of the two fine tractors mounted on the micromanipulator,

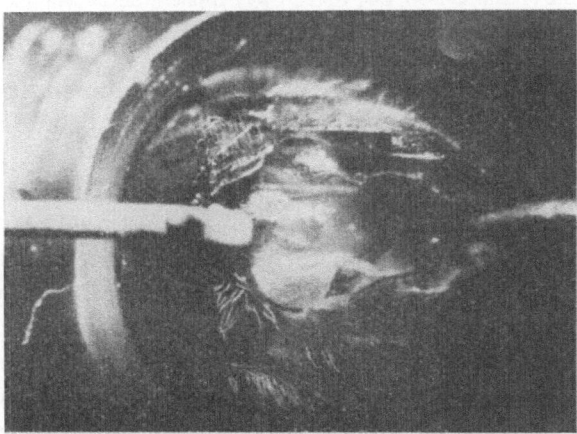

Fig. 171. Queen Bee Mounted in Special Holder under the Microscope. Exposed rear end of the queen adjusted for insemination with the aid of two tractors mounted on the micromanipulator. Courtesy of F. Ruttner.

the rear end of the queen is spread to expose the entry to the vagina, Fig. 171. Using a magnification of about 25×, and good illumination, the valve fold is then depressed, with the aid of a fine probe, to permit the cannula to be introduced into the midoviduct. The tip of the cannula, Fig. 172, is then carefully inserted, and the semen injected behind the vaginal flap. The whole operation is repeated after one or two days to attain maximum filling of the spermatheca.

12. Microbiology.

Micrurgical techniques have been widely applied by microbiologists in the different fields of the investigation of cells and tissues. It is not proposed

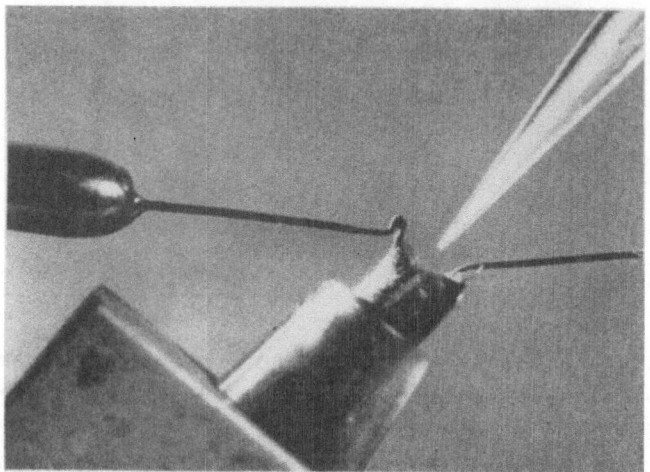

Fig. 172. Rear End of the Queen Bee Spread by Application of the Two Tractors, Tip of Cannula Ready to Enter Vagina. Courtesy of F. RUTTNER.

to outline these applications, many of which have already been mentioned in appropriate places. Many more references are listed in the General Bibliography, p. 315.

Only a few classical experiments carried out by S. L. SCHOUTEN of Utrecht, Holland, and described by WORST (243) will be given briefly.

In the field of bacteriology, micrurgical techniques (8, 58, 80, 104, 131, 206, 208) have long been used for the development of pure cultures. Many kinds of bacteria, which may cause different diseases, appear strikingly similar. In order to study a certain kind of bacteria it is necessary to develop a culture or a colony which is composed only of the species to be investigated. For obtaining such a pure culture, the most reliable method starts with the isolation of a single bacterium. This bacterium can then be grown through many generations to produce a colony known to have originated from one organism.

One of the best and most reliable techniques used for this purpose was developed by SCHOUTEN (208). Use is made of the film of liquid forming across a tiny loop. The photomicrographs, Fig. 173a to d, illustrate the application of the technique for the isolation of an individual bacterium from a mass of bacteria. The bacteria are contained in a hanging drop, and the microloop approaches the specimen from below. The isolated bacterium can then be cultured, and the developing culture is watched closely for the appearance of bacteria showing any hereditary changes. Members which differ from the rest of the population can be isolated by the same technique.

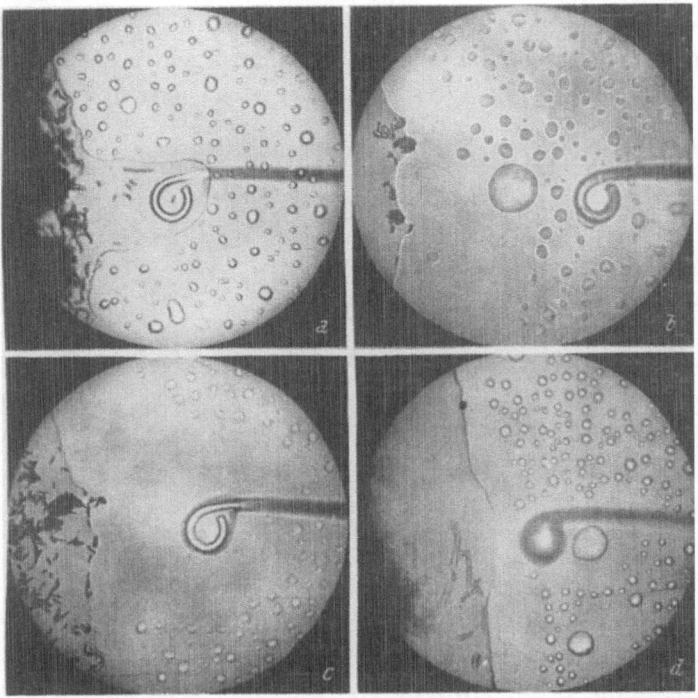

Fig. 173. Isolation of an Individual Bacterium by Means of a Mounted Microloop. *a*, a few bacteria are drawn away from the main preparation; *b*, an individual bacterium is isolated on the microloop; *c*, the isolated bacterium is deposited in a single hanging droplet for inspection; *d*, the isolated bacterium is removed for culture or other investigation. Courtesy of S. L. SCHOUTEN.

In the study of moulds, micrurgical techniques have also proved very valuable since they may be successfully applied in the separation and transfer of individual spores. Moulds, such as the mould penicillium, reproduce by means of spores which form chains on certain branches of the mould. Investigations have shown significant differences between the offspring obtained from apparently similar individual spores which had been separated, labelled, and allowed to reproduce. Similar to the above mentioned method of obtaining a pure culture, application of the procedure to the mould penicillium, for instance, permits selection of the best strains for the production of penicillin. Fig. 174*a*, *b*, *c*, illustrates the separation

of a single spore from *Bacillus Anthracis*. The specimen is contained in
a hanging drop, and the micro operation is performed by means of a mounted
microneedle having a fine tip. The separated spore can then be transferred
for investigation.

In the field of physiology, micrurgy has very important applications.
It may be used, for instance, to study the digestion of certain microscopic
organisms which feed on bacteria, and which may be able to destroy germs
which cause disease. The experiment carried out by SCHOUTEN (243) for
the study of the digestion of *Colpidium*, may serve as example. *Colpidium*
is a single-celled organism feeding on bacteria which may be seen in vacuoles
inside the animal. Fig. 175a to d, illustrates the removal of the bacteria for

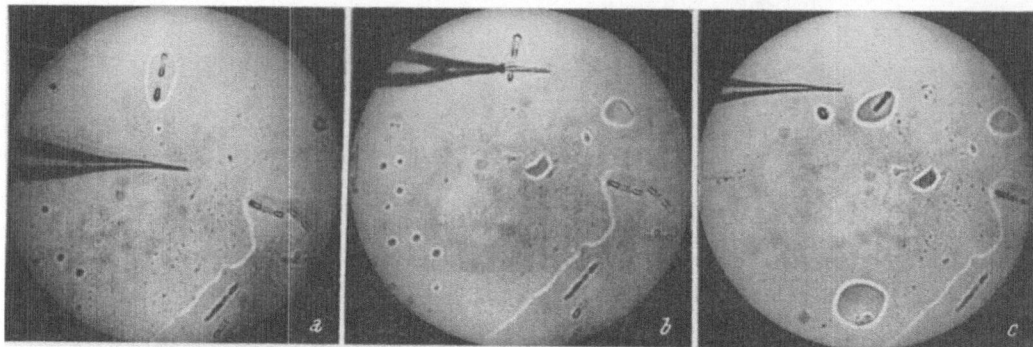

Fig. 174. Separation of an Individual Spore by Means of a Mounted Microneedle. *a*, field
of view before micro operation; *b*, a single spore is cut off by microneedle; *c*, separated spore
seen in a hanging droplet below needle tip. Courtesy of S. L. SCHOUTEN.

investigation. An individual animal, contained in a hanging drop preparation is first caught by means of a mounted loop and transferred to a
single hanging droplet for microdissection. The animal is then dissected
with a fine microneedle to liberate the vacuoles containing the bacteria.
By means of the microneedle, a single vacuole is transferred to a smaller
hanging droplet and is then dissected to liberate the bacteria inside the
vacuole. The bacteria thus liberated can then be investigated for their
ability to grow and multiply. Such investigations may reveal which organisms can destroy harmful bacteria, for instance, in contaminated water.

13. Engineering and Technology.

Mechanical manipulation under optical magnification is finding wide
and diverse applications in the investigation of industrial materials as
well as in many fields of engineering and technology.

In the broad field of testing materials, certain applications of the micromanipulative techniques have already been considered. Among these

applications is the microhardness testing, p. 82, the use of microdrills for taking small samples, p. 191, the study of the microstructures of fibers, p. 291, the investigation of the structure of rubber latex particles, p. 292, as well as the various other micromanipulative methods of handling and investigating small samples which have already been considered in appropriate places.

Modern scientific progress is introducing, among other developments, considerable reduction in size together with increased refinement and com-

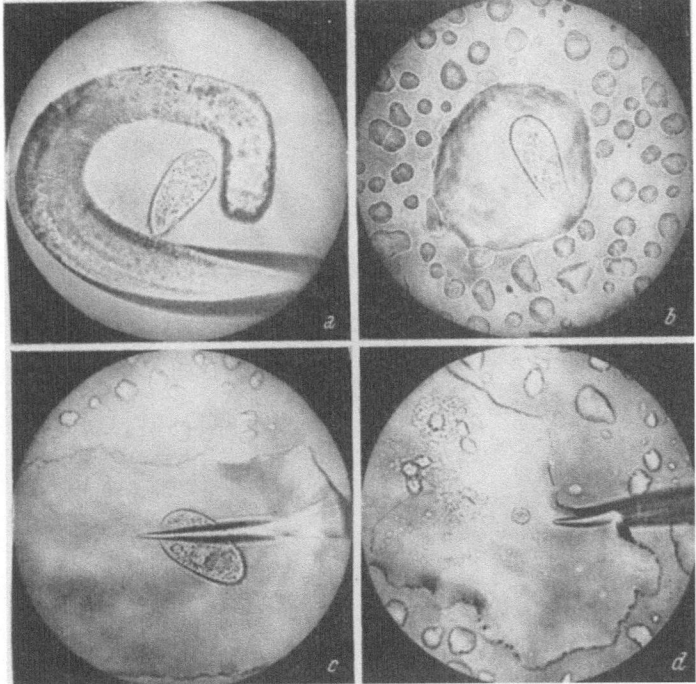

Fig. 175. Removal of Bacteria from Vacuoles Inside a Small Single-Celled Animal. *a*, individual animal caught by a microloop; *b*, animal transferred to a droplet; *c*, microneedle cuts through animal to liberate vacuoles; *d*, needle approaches vacuole filled with bacteria to liberate contents. Courtesy of S. L. Schouten.

plexity of a large and ever increasing diversity of instruments and gadgets of varied functions. Many instruments to be used in aircraft and missiles, for instance, have to be kept as light and as compact as possible. Because of the minute size of the component parts, numerous operations involved in the manufacture and the assembling of complex devices have to be accomplished with the use of mechanical aids and optical magnification. Precise micromachining, microdrilling, and assembling of the precision miniature parts permit the utilization of smaller microcomponents in numerous intricate devices. Among the assembly problems of a transistor, for example, is the soldering and welding of a number of tiny components with considerable accuracy. Beside a number of other optical instruments, the stereomicroscope is now coming into widespread use as a production line tool. Since it offers a sufficiently wide field, a long working distance, and shows tiny objects in their three-dimensional form and orientation, it is used in many assembly shops and in precision mechanical industry. It enables the operator to inspect small objects for the detection and elimination of defects early in the assembly process and helps him to

Fig. 176. Performing Varied Mechanical Workshop Operations in the Field of a Stereoscopic Microscope. Courtesy of C. ZEISS, Oberkochen (244).

considerably reduce final part rejections. It also provides better and more accurate vision for performing a wide diversity of delicate mechanical production operations. Such operations are frequently involved in machining, drilling, aligning, adjusting, mounting, and assembling miniature parts, and these have to be performed in accordance with the exacting precision requirements introduced by modern electronics, radar, aviation, and the manufacture of jet engines, automatic computors, automatic time controls, machine tools, and many other articles of precision mass production.

Commercially available microdrilling machines, p. 72, are becoming of increased importance in industries where tiny holes of microscopic dimensions have to be obtained with considerable precision. Drilling of microscopic precision holes is important in various industries (73). It is needed in the Diesel engine industry for making small multiple holes in the jet nozzles. The synthetic silk and yarn industry uses spinnerettes for the extrusion of synthetic silk and yarn filaments, and the multiple holes, usually of the order of 40-μm diameter, have to be made in hard metals such as stainless steel. Among other industries, in which microscopic precision holes are needed, are the fine watch industry and the med-

Fig. 177. Fastening of a Small Lens in Its Metal Mount under a Stereoscopic Microscope. Courtesy of C. Zeiss, Oberkochen (244).

ical instrument industry. In the medical field, the so-called hypo-spray depends on the use of small precision holes in a device for hypodermic injection of medication.

In mechanical experimental workshops, micromanipulation is frequently used for performing such operations as turning or drilling with an improved precision since small tools and workpieces can be guided in the field of the microscope with added exactness and certainty, Fig. 176.

Among the many other applications of micromanipulation is the mounting of small lenses such as those of microscope objectives which may be as small as 1 to 3 mm in diameter. The process consists essentially of placing the lenses into their metal mounts and then fastening them by turning over a projecting burr. The operations may be performed with the aid of a simple stand magnifier or a monocle, as in older practice, but are best performed under a stereoscopic microscope, Fig. 177.

As a final example may be mentioned the increased use of micromanipulative techniques in the preparation (177) of vitreous silica fibers of various sizes down to a few microns in diameter or less, and for the construction (26, 135, 175, 197) of silica fiber devices which are used in the manufacture of exceedingly sensitive silica fiber balances (26, 122, 135, 178, 179), electrometers, galvanometers, electroscopes, and other instruments.

Appendix.
Literature Cited.

(1) ALBER, H. K., Mikrochemie **14**, 219 (1933/34). — (2) ALOE SCIENTIFIC DIVISION of the A. S. Aloe Company, 5655 Kingsburg, St. Louis 12, Mo., U. S. A. — (3) AMERICAN OPTICAL CO., Instrument Division, Buffalo 15, N. Y., U. S. A. — (4) AMERICAN OPTICAL CO., A.O. Micromanipulator Instruction Manual 1850-101, 1955. — (5) ANDERSON, R. Y., Micropaleontology **4**, 205 (1958). — (6) Private communication (October 1959), Department of Geology, University of New Mexico, U. S. A. — (7) ANDERSON LABORATORIES, Inc., Adrian, Mich., U. S. A.

(8) BARBER, M. A., J. Kansas Med. Soc. **4**, 487 (1904). — (9) J. Inf. Dis. **5**, 379 (1908). — (10) *Ibid.* **8**, 348 (1911). — (11) Kansas Univ. Sci. Bull. **4**, 3 (1907). — (12) Philippine J. Sci. **9**, 307 (1914). — (13) BARER, R., and S. JOSEPH, Quart. J. Micr. Sci. **95**, 399 (1954). — (14) *Ibid.* **96**, 1 (1955). — (15) BARER, R., and A. E. SAUNDERS-SINGER, Quart. J. Micr. Sci. **89**, 439 (1948). — (16) J. Sci. Instr. **28**, 65 (1950). — (17) BARKER, F. J., J. CONVEY, and J. H. OLDFIELD, Engineering **152**, 298 (1941). — (18) *Ibid.* **160**, 481 (1945). — (19) J. Iron & Steel Inst. **18**, 473 (1945). — (20) BAUSCH & LOMB, Inc., Rochester 2, N. Y., U. S. A. — (21) BECKMAN INSTRUMENTS, Inc., Fullerton, Cal., U. S. A. — (22) BÉKÉSY, G. VON, Rev. Sci. Instr. **27**, 690 (1956). — (23) Trans. Amer. Micr. Soc. **71**, 306 (1952). — (24) BELKIN, M., Science **68**, 137 (1928). — (25) BENEDETTI-PICHLER, A. A., Introduction to the Microtechnique of Inorganic Analysis, New York: Wiley, 1942. — (26) *"Waagen und Wägung"* in HECHT und ZACHERL, *Handbuch der mikrochemischen Methoden*, Vol. I, Part 2, Wien: Springer-Verlag, 1959. — (27) Ind. Engng. Chem., Analyt. Ed. **9**, 149 (1937). — (28) *Ibid.* **9**, 483 (1937). — (29) BENEDETTI-PICHLER, A. A., and M. CEFOLA, Ind. Engng. Chem., Analyt. Ed. **14**, 813 (1942). — (30) *Ibid.* **15**, 227 (1943). — (31) BENEDETTI-PICHLER, A. A., and J. R. RACHELE, Ind. Engng. Chem., Analyt. Ed. **12**, 233 (1940). — (32) BENEDETTI-PICHLER, A. A., and S. SIGGIA, Ind. Engng. Chem., Analyt. Ed. **14**, 828 (1942). — (33) BLACET, F. E., and D. H. VOLMAN, Ind. Engng. Chem., Analyt. Ed. **9**, 44 (1937). — (34) BRINDLE, T., and C. L. WILSON, Mikrochem. **39**, 310 (1952). — (35) BRINKMANN INSTRUMENTS, Inc., 115 Cutter Mill Road, Great Neck, N. Y., U. S. A. — (36) BRISCOE, H. V. A., *et al.*, Trans. Min. Inst. & Met., June 1937. — (37) BRISCOE, H. V. A., and J. W. MATTHEWS, Mikrochim. Acta **1**, 266 (1937). — (38) BROSCHKE, H., Microtecnic **6**, No. 1, 1 (1952). — (39) BRYAN, F. R., J. Opt. Soc. Amer. **41**, 1061 (1951). — (40) BRYAN, F. R., and G. NAHSTOLL, J. Opt. Soc. Amer. **37**, 311 (1947). — (41) BRYAN, F. R., and C. H. NEVEU, Metal Progr. **64**, 82 (1953). — (42) BUCHTHAL, F., and C. PERSSON, J. Sci. Instr. **13**, 20 (1936). — (43) BURGER, M. J., X-Ray Crystallography, New York: Wiley, 1942. — (44) BURROUGHS WELLCOME & Co., 1 Scarsdale Road, Tuckahoe, N. Y., U. S. A. — (45) BUSH, V., W. R. DURYEE, and J. A. HASTINGS, Rev. Sci. Instr. **24**, 487 (1953).

(46) CADLE, R. D., Particle Size Determination, New York: Interscience, 1955. — (47) Analyt. Chem. **23**, 196 (1951). — (48) CADLE, R. D., *et al.*, Arch. Ind. Hyg. Occupational Med. **2**, 698 (1950). — (49) CADLE, R. D., A. G. WILDER, and C. F. SCHADT, Science **118**, 490 (1953). — (50) CANON, H. G., J. Roy. Micr. Soc. **61/62**, 58 (1941). — (51) CASSELA, C. F., LTD., London, England. — (52) CEFOLA, M., Microchem. J. **2**, 205 (1958). — (53) CHABRY, L., J. de l'anat. et physiol. **25**, 167 (1887). — (54) CHAMBERS, R., Anat. Rec. **24**, 1 (1922). — (55) Biol. Bull. **34**, 121 (1918). —

(56) *Ibid.* **41**, 318 (1921). — (57) J. Bact. **8**, 1 (1922). — (58) J. Inf. Dis. **31**, 334 (1922). — (59) J. Roy. Microscop. Soc. **60**, 113 (1940). — (60) Science **54**, 411 (1921). — (61) *Ibid.* **54**, 552 (1921). — (62) CHAMBERS, R., and M. J. KOPAC, in C. E. McCLUNG, "Handbook of Microscopical Technique", 3rd ed., pp. 492–543, New York: Hoeber, 1950. — (63) CHAMBERS, R., and H. C. SANDS, J. Gen. Physiol. **5**, 815 (1923). — (64) CHAMOT, E. M., and C. W. MASON, Handbook of Chemical Microscopy, Vol. I. 3rd ed., New York: Wiley, 1958. — (65) *Ibid.*, Vol. II, 2nd ed., New York: Wiley, 1946. — (66) CLIPPINGER, D. R., Ind. Engng. Chem., Analyt. Ed. **11**, 216 (1939). — (67) COCKS, G. G., private communication, 1957. — (68) CROZIER, W. D., and B. K. SEELY, Proceedings First National Air Poll. Symp., p. 45, Stanford Research Institute, Stanford University, Cal., U. S. A. — (69) CUNNINGHAM, B. B., Nucleonics **5**, 62 (1949). — (70) U. S. AECD-1879, Dec. 1947. — (71) CUNNINGHAM, B. B., and L. B. WERNER, "The Transuranium Elements", in SEABORG, KATZ, and MANNING, National Nuclear Energy Series IV-14B, Vol. I, pp. 51–78, New York: McGraw-Hill. 1949. — (72) J. Amer. Chem. Soc. **71**, 1521 (1949). — (73) CUPLER, J. A., Tool Eng. **24**, C-71 (1950).

(74) DANA, E. S., and W. E. FORD, A Text Book of Mineralogy, 4th ed., New York: Wiley, 1932. — (75) DE FONBRUNE, P., Technique de Micromanipulation, Monographies de l'Intistut Pasteur, Paris: Masson et Cie., 1949 — (76) Micromanipulateur pneumatique et microforge, Société Industrielle d'imprimerie, Paris, 1937. — (77) C. r. acad. sci., Paris, **195**, 603 (1932). — (78) Recherches et inventions **16**, 433 (1935). — (79) *Ibid* **16**, 1, 252 (1935). — (80) DICKINSON, S., Phytopathology **23**, 357 (1933). — (81) DORF, H., Analyt. Chemistry **25**, 1000 (1953). — (82) Dow CORNING CORP., Midland, Mich., U. S. A. — (83) DU BOIS, D., Science **73**, 344 (1931). — (84) DUNN, F. L., J. Inf. Dis. **40**, 383 (1927). — (85) DU PONT DE NEMOURS & CO., E. I., Fabrics and Finishes Dept., Wilmington, Del., U. S. A. — (86) DURAM LTD., Thanet House, 231-2 Strand, London W. C. 2, England.

(87) EL-BADRY, H. M., F. R. M. McDONNEL, and C. L. WILSON, Analyt. Chim. Acta **4**, 440 (1950). — (88) EL-BADRY, H. M., and C. L. WILSON, Roy. Inst. Chem. Monog. **4**, 23 (1950). — (89) Analyst **77**, 596 (1952). — (90) Mikrochem. **40**, 141 (1952). — (91) *Ibid* .**40**, 218 (1953). — (92) **40**, 225 (1953). — (93) **40**, 230 (1953). — (94) EMICH, F., and F. SCHNEIDER, Microchemical Laboratory Manual, New York: Wiley, 1932.

(95) FAEGRI, K., and J. IVERSEN, Text-Book of Modern Pollen Analysis, Copenhagen: Ejnar Munksgaard, 1950. — (96) FEIGL, F., Qualitative Analysis by Spot Tests, Inorganic and Organic Applications, 3rd English ed., transl. RALPH E. OESPER, New York: Elsevier, 1946. — (97) FELDMAN, C., Analyt. Chemistry **21**, 1041 (1949). — (98) Private communication, July 1957. Address: Oak Ridge National Laboratory, Oak Ridge, Tenn., U. S. A. — (99) FICKLEN, J. B., Pasadena 7, Cal.. U. S. A. — (100) FITZ, G. W., Science **73**, 72 (1931). — (101) *Ibid.* **79**, 233 (1934). — (102) FRIEDRICH, A., Mikrochem. **15**, 36 (1934).

(103) GAGE, S. H., The Microscope, 17th ed., New York: Comstock, 1943. — (104) GEE, A. H., and G. A. HUNT, J. Bact. **16**, 327 (1928). — (105) GENERAL ELECTRIC Co., Pittsfield, Mass., U. S. A. — (106) GETTENS, R. J., Tech. Studies, Field Fine Arts **1**, 20 (1932). — (107) *Ibid.* **3**, 165 (1935). — (108) GIBB, T. R. P., Optical Methods of Chemical Analysis, New York: McGraw-Hill, 1942. — (109) GILBERT, J.. Econ. Geol. **19**, 668 (1924). — (110) GILBERT, P. T., JR., Science **114**, 637 (1951). — (111) GOETZ, A., J. Amer. Water Works Assoc. **45**, 933 (1953). — (112) GOETZ, A.. R. H. GILMAN, and A. M. RAWN, J. Amer. Water Works Assoc. **44**, 471 (1952). — (113) GOETZ, A., and N. TSUNEISHI, J. Amer. Water Works Assoc. **43**, 943 (1951). — (114) GOOCH, F. A., Proc. Amer. Acad. Arts Sci. **13**, 342 (1878). — (115) *Ibid.* New series **12**, 300 (1884/85).

(116) HAMILTON COMPANY, Inc., Whittier, Cal., U. S. A. — (117) HANSEN, W. W., J. Roy. Micr. Soc. **58**, 250 (1938). — (118) HARDING, J. P., J. Roy. Micr. Soc. **59**, 19 (1939). — (119) HARTSHORNE, N. H., and A. STUART, Crystals and the Polarizing Microscope, 2nd ed., Arnold, 1950. — (120) HAUSER, E. A., Ind. Engng. Chem. **18**, 1146 (1926). — (121) HILSON, G. R. F., J. Gen. Microbiol. **7**, 175 (1952). — (122) HOOD, W. L., and A. J. ROGERS, in H. V. MOYER, "Polonium", U. S. Atomic Energy Commission, TID-5221, 1956. — (123) HOWLAND, R. B., and M. BELKIN, Manual of Micrurgy, New York: New York University Book Store, 1931. — (124) HUBER, F., Zeiss Werkzeitschrift, Oberkochen/Württ. [25], **5**, 76 (1957). — (125) HURWITZ, J. K., J. Opt. Soc. Amer. **42**, 489 (1952). — (126) HYBBINETTE, ANNA-GRETA, and A. A. BENEDETTI-PICHLER, Mikrochem. **30**, 15 (1942).

(127) IMPERIAL CHEMICAL INDUSTRIES, Ltd., Hexagon House, London W. C. 2, England.

(128) JACOBS, M. B., Analytical Chemistry of Industrial Poisons, Hazards, and Solvents, New York: Interscience, 1941. — (129) JOHNSON, D. L., and C. L. SHREWSBURY, Mikrochem. **26**, 143 (1939). — (130) JOHNSON, MATTHEY AND Co., Ltd., 78 Hatton Garden, London, E. C. I., England.

(131) KAHN, M. C., J. Inf. Dis. **31**, 344 (1922). — (132) KEHL, L., H. STEINMETZ, and W. J. McGONNAGLE, Metallurgia **55**, 151 (1957). — (133) KIRK, P. L., Quantitative Ultramicroanalysis, New York: Wiley, 1950. — (134) Ann. Rev. Biochem. **9**, 593 (1940). — (135) KIRK, P. L., and R. CRAIG, Rev. Sci. Instr. **19**, 777 (1948). — (136) KIRK, P. L., R. CRAIG, J. E. GULLBERG, and R. Q. BOYER, Ind. Engng. Chem., Analyt. Ed. **19**, 427 (1947). — (137) KLAUS, W., Microscopie **8**, 1 (1953). — (138) KOCH, W., H. MALISSA, and D. DITGES, Arch. Eisenhüttenwesen **28**, 785 (1957). — (139) KOPAC, M. J., Ann. N. Y. Acad. Sci. **50**, 870 (1950). — (140) Ibid. **63**, 1219 (1956). — (141) Anat. Rec. **111**, 117 (1951). — (142) Int. Rev. Cytol. **4**, 1 (1955). — (143) Science **113**, 232 (1951). — (144) Trans. N. Y. Acad. Sci. **15**, 290 (1953). — (145) Ibid. **17**, 257 (1955). — (146) Ibid. **18**, 22 (1955). — (147) KOPAC, M. J., and J. HARRIS, Anat. Rec. **111**, 116 (1951). — (148) KRIEG, A., Zeiss Werkzeitschrift [19] **4**, 4 (1956).

(149) LAURIE, A. P., The Pigments and Mediums of the Old Masters, London: Macmillan, 1914. — (150) Analyst **55**, 162 (1930). — (151) LEDERBERG, J., J. Bact. **68**, 258 (1954). — (152) LEITZ, E., Wetzlar, Germany. — (153) LODGE, J. P., Analyt. Chemistry **26**, 1829 (1954). — (154) LOSCALZO, ANNE G., and A. A. BENEDETTI-PICHLER, Ind. Engng. Chem., Analyt. Ed. **17**, 187 (1945). — (155) LOVELL CHEMICAL Co., Watertown, Mass., U. S. A. — (156) LUNDE, G., Mikrochem. **5**, 102 (1927).

(157) MACKAY, A. L., J. Sci. Instr. **30**, 140 (1953). — (158) Private communication, 1959. — (159) MADLER, K. A., Micropaleontology **2**, 399 (1956). — (160) MAGNUSSON, L. B., and T. J. LA CHAPELLE, J. Amer. Chem. Soc. **70**, 3534 (1948). — (161) MASON, W. P., and R. F. WICK, J. Acoustical Soc. Amer. **23** (1951). — (162) MATTHEWS, J. W., Analyst **63**, 467 (1938). — (163) MAY, K. R., J. Roy. Micr. Soc. **73**, 140 (1953). — (164) McCLENDON, J. F., Biol. Bull. **12**, 141 (1907). — (165) J. Exp. Zool. **6**, 87 (1909). — (166) McCLUNG, C. E., ed., Handbook of Microscopical Technique, 3rd ed., New York: Hoeber, 1950. — (167) McNEIL, E., and J. F. GULBERG, Science **74**, 460 (1931). — (168) MICROCHEMICAL SPECIALITIES Co., 1834 University Ave., Berkeley 3, Cal., U. S. A. — (169) MICROTECH SERVICES COMPANY, Berkeley, Cal., U. S. A. — (170) MURDOCK, J., Microscopical Determination of Opaque Minerals, New York: Wiley, 1916. — (171) MURRAY, W. M., B. GETTYS, and S. E. Q. ASHLEY, J. Opt. Soc. Amer. **31**, 433 (1941).

(172) NAKHLA, F. M., Econ. Geol. **51**, 811 (1956). — (173) NATIONAL JET Co., Cupler Drive, La Vale Cumberland, Md., U. S. A. — (174) NEEDHAM, J., and D. M.

NEEDHAM, Proc. Roy. Soc. **B 98**, 259 (1925). — (175) NEHER, H. V., in STRONG, Procedures in Experimental Physics, New York: Prentice-Hall, 1939. — (176) NYLEN, C. O., Acta Oto-Laryngologica, Suppl. **116**, 226 (1954).

(177) OLT, G. R., Improvements in Method and Equipment for Drawing Quartz Fibers, MLM-656, Mound Laboratory, Miamisburg, Ohio, U. S. A., Jan. 1952. — (178) OLT, R. G., et al., A Remote Controlled Quartz-Fiber Microbalance: Design, Construction and Characteristics, MLM-1022, Mound Laboratory, Miamisburg, Ohio, U. S. A., Dec. 1954. — (179) A Remote Controlled Quartz-Fiber Microbalance: Fabrication of Quartz Components, MLM-1023, Mound Laboratory, Miamisburg, Ohio, U. S. A., Dec. 1954. — (180) ORDWAY, F., J. Research NBS **48**, 152 (1952). — (181) NBS Tech. News Bull. **36**, 141 (1952).

(182) PALEN, V. W., private communication, 1957. — (183) PAYNE, B. O., Microscope Design and Construction, York, England: Cooke, Troughton, and Sims, 1954. — (184) PEPINSKY, R., Rev. Sci. Instr. **24**, 403 (1953). — (185) PETERFI, T., Pflüger's Arch. Physiol. **208**, 454 (1925). — (186) Naturwissenschaften **6**, 81 (1923). — (187) POWELL, E. O., J. Roy. Micr. Soc. **72**, 214 (1952).

(188) RABINOWITSCH, B., Kolloid-Z. **57**, 203 (1931). — (189) RAWLPLUG Co., Ltd., Rawlplug House, Cromwell Road, London S. W. 7, England. — (190) READ, H. H., Rutley's Elements of Mineralogy, 24th ed., London: Murby, 1949. — (191) RESEARCH SPECIALTIES Co., 2005 Hopkins Str., Berkeley 7, Cal., U. S. A. — (192) REYNIERS, J. A., Anat. Rec. **56**, 307 (1933). — (193) RICHARDS, A. N., J. Biol. Chem. **87**, 463 (1930). — (194) RICHTER, K. M., Science **108**, 119 (1948). — (195) ROHM & HAAS Co., Philadelphia, Pa., U. S. A. — (196) RUBIN, S., J. Atm. Terrest. Phys. **2**, 130 (1952). — (197) RUNGE, E. F., and F. R. BRYAN, Appl. Spectroscopy **13**, 74 (1959). — (198) RUTTNER, F., Zeiss Werkzeitschrift, Oberkochen/Württ. [31] **7**, 5 (1959).

(199) SALVIONI, E., Misura di masse comprese fra g 10^{-1} e g 10^{-6}, Messina: 1901. — (200) SCHAEFFER, H. F., Microscopy for Chemists, New York: Van Nostrand, 1953. — (201) SCHEIBE, G., and J. MARTIN, Spectrochim. Acta **1**, 47 (1939). — (202) SCHMIDT, H. D., Amer. J. Med. Sci., **37**, 2 (1859). — (203) Amer. J. Med. Sci. **73**, 13 (1859). — (204) SCHMITT, F. O., in O. GLASSER, Medical Physics, Chicago: Year Book Press, 1944. — (205) SCHOLANDER, P. F., Rev. Sci. Instr. **13**, 32 (1942). — (206) SCHOUTEN, S. L., Kon. Akad. Wetenschappen Amsterdam, 804 (1911). — (207) Proceedings Fourth International Congress Microbiology **1**, 816 (Rome, Sept. 1953). — (208) Z. wiss. Mikroskop. **22**, 10 (1905). — (209) SCHUSTER, E. H. J., J. Anat. London **81**, 281 (1947). — (210) SEABORG, G. T., Science **104**, 379 (1946). — (211) SEABORG, G. T., and A. C. WAHL, J. Amer. Chem. Soc. **70**, 1128 (1948). — (212) SEELY, B. K., Analyt. Chemistry **24**, 576 (1952). — (213) Ibid. **27**, 93 (1955). — (214) SEIFRIZ, W., and C. W. HOCK, Paper Trade J. **102**, 36 (1936). — (215) SEN, B., Proc. Soc. Exp. Biol. Med. **27**, 310 (1930). — (216) SHORT, M. N., Microscopic Determination of the Ore Minerals, U. S. Geol. Survey Bull. 914, 2nd ed., Washington: Government Printing Office, 1940. — (217) SISCO, R. C., B. B. CUNNINGHAM, and P. L. KIRK, J. Biol. Chem. **139**, 1 (1941). — (218) SONKIN, L. S., J. Ind. Hyg. Toxicol. **28**, 269 (1946). — (219) STRONG, J., Procedures in Experimental Physics, New York: Prentice-Hall, 1939.

(220) TALMAGE, S. B., Econ. Geol. **20**, 531 (1925). — (221) TAYLOR, G. F., Phys. Rev. **23**, 655 (1924). — (222) TAYLOR, C. V., Proc. Soc. Exp. Biol. Med. **22**, 533 (1925). — (223) Ibid. **23**, 147 (1925). — (224) Science **51**, 617 (1920). — (225) University California Publ. Zool. **19**, 403 (1920). — (226) Ibid. **26**, 443 (1925). — (227) THANHEISER, G., and J. HEYES, Arch. Eisenhüttenw. **14**, 543 (1941). — (228) TITUS, R. N., and H. L. GRAY, Ind. Engng. Chem., Analyt. Ed. **2**, 368 (1930). — (229) THOMPSON, J. K., and C. L. WILSON, Mikrochim. Acta (Wien) **1957**, 334. — (230) TREVAN, J. W., Biochem. J. **19**, 1111 (1925).

(231) UNITRON INSTRUMENT DIVISION, United Scientific Co., 204–206 Milk Str., Boston 9, Mass., U. S. A.

(232) VICKERS, A. E. J., ed., Modern Methods of Microscopy, a series of papers printed from *Research*, London: Butterworth, 1956.

(233) WAHLSTROM, E. E., Optical Crystallography, 2nd ed., New York: Wiley, 1953. — (234) WAIT, E., A. E. R. E., Harwell U. K., C/R 1809 (1955). — (235) WAKEFIELD INDUSTRIES, Inc., Skobie, Chicago, Ill., U. S. A. — (236) WENGER, P. E., *et al.*, Second Report ,Reagents for Qualitative Inorganic Analysis, New York-Amsterdam: Elsevier, 1948. — (237) WHITAKER, D. M., Science **70**, 263 (1930). — (238) WHITE, S. S., Bulletin 5205, Industrial Division, 10 E. 40th Str., New York, N. Y., U. S. A. — (239) WIESENBERGER, E., Mikrochim. Acta (Wien) **1957**, 527. — (240) WIGGLESWORTH, V. B., Biochem. J. **31**, 1719 (1937). — (241) WORST, J., Nature **170**, 1129 (1952). — (242) Quart. J. Microscop. Sci. **95**, 469 (1954). — (243) Times Sci. Rev. **9**, 8 (1953).

(244) ZEISS, C., Oberkochen/Württ., Germany. — (245) ZEISS, C., Jena, Germany.

General Bibliography.

This bibliography gives a list of books and articles which deal with micromanipulation but have not been cited in the body of the book.

ALEXANDER, J. T., and W. L. NASTUK, Rev. Sci. Instr. **24**, 528 (1953): Instrument for the Production of Microelectrodes Used in Electrophysiological Studies.

BARBER, M. A., J. Exp. Med. **32**, 295 (1920): Use of the Single Cell Method in Obtaining Pure Cultures of Anaerobes.

BARIGOZZI, C., Arch. exp. Zellforsch. **22**, 190 (1939): Esperienze di microdissezione sui chromosomi delle ghiandole salivari de *Chironomus spec.*

BARON, H., and R. CHAMBERS, Amer. J. Physiol. **114**, 700 (1936): A Micromanipulative Study on the Migration of Blood Cells in Frog Capillaries.

BECKMAN, H., Paleont. Z. **24**, 91 (1951): Hilfsmittel zum Schleifen von Mikrofossilien.

BENEDETTI-PICHLER, A. A., Microchem. J. **2**, 3 (1958): Chemical Experimentation under the Microscope.

— Umschau **52**, 107 (1952): Ultramikromethoden der Chemie.

BENEŠOVÁ, K., Svět Techn. 183 (1953): Operace živicich buněk, Mikrochirurgie ve světě malých rozměrů (Operation on Living Cells, Micrurgy in the World of Small Dimensions).

BOVEY, R., J. Roy. Micr. Soc. **72**, 56 (1952): Electron Microscopy and Microdissection.

BREMER, F., and B. E. GERNANDT, Acta Physiol. Scand. **30**, 120 (1954): A Micro-Electrode Analysis of the Acustic Response and the Strychnine Convulsive Patterns of the Cerebellum.

BRETEY, J., and J. BROWAEYS, C. r. soc. biol. **138**, 465 (1944): Améliorations apportées aux techniques d'isolement de micro-organismes par micromanipulation sur fond noir.

— — Bull. Soc. Path. Expt. **33**, 182 (1945): Présentation d'un appareil de fabrication des microinstruments.

BRIEDIGAM, F. T., and T. M. CHANG, J. Lab. Clin. Med. **9**, 572 (1924): A Simple and Practical Device for Isolating *Bacillus acidophilus* for Single Colonies.

BROWAEYS, J., Bull. Soc. Path. Expt. **36**, 69 (1943): Micromanipulateur à pantographe pour le travail à un grossissement limité.

BUCHTHAL, F., Z. wiss. Mikroskopie **58**, 126 (1942): Ein neuer Mikromanipulator mit Zusatzgeräten (Mikromesser und unpolarisierbare Mikroelektroden).

CAILLOUX, M., Rev. Canad. Biol. **2**, No. 5, 528 (1943): Un nouveau micromanipulateur hydraulique.

CHAMBERS, R., in G. CAMERON: Tissue Culture Technique, New York: Acad. Press, 1950: Micromanipulation of Tissue Cultures.

— Amer. J. Physiol. **43**, 1 (1917): Microdissection Studies, I. The Visible Structure of Cell Protoplasm and Death Change.

— Cellule **35**, 107 (1924): Études de microdissection, IV: Les structures mitochondriales et nucléaires dans les cellules germinales males chez la suterelle.

— J. Exp. Zool. **23**, 483 (1917): Microdissection Studies, II. The Cell Aster: A Reversible Gelatin Phenomenon.

— J. Gen. Physiol. **5**, 189 (1922): A Microinjection Study on the Permeability of the Starfish Egg.

— Lancet Clinic, Cincinnati **113**, 1 and 363 (1915): Microdissection Studies on the Physical Properties of Protoplasm.

— Proc. Soc. Exp. Biol. Med. **17**, 41 (1919): Some Studies on the Surface Layer in the Living Egg Cell.
Ibid., **18**, 66 (1920): Dissection and Injection Studies on the Amoeba.
Ibid., **19**, 85 (1921): Apparatus for micromanipulation and microinjection.

— Protoplasma **12**, 338 (1931): Micrurgical Studies on the Tonoplast of Allium cepa.

— Science **41**, 290 (1915): Microdissection Studies on the Germ Cell.

— and P. MARTENS, Le Cellule **41**, 131 (1932): Études de microdissections. I: Les poils staminaux de Tradescantia.

— and H. POLLACK, J. Gen. Physiol. **5**, 738 (1927): Micrurgical Studies in Cell Physiology. IV: Colorimetric Determinations of the Nucleolar and Cytoplasmatic pH in the Starfish Egg.

— and G. S. RENYI, Amer. J. Anat. **35**, 385 (1925): The Structure of the Cells in Tissues as Revealed by Microdissection.

— and P. REZNIKOFF, J. Gen. Physiol. **8**, 369 (1926): Micrurgical Studies in Cell Physiology. I: The Action of the Chlorides of Na, K, Ca and Mg on the Protoplasm of Amoeba proteus.
Ibid., **10**, 731 (1927): Micrurgical Studies in Cell Physiology. III: The Action of CO_2 and Some Salts of Na, Ca and K on the Protoplasm of Amoeba dubia.

— B. COHEN, and P. REZNIKOFF, J. Gen. Physiol. **11**, 585 (1928): Intracellular Oxidation-Reduction Studies. I: Reduction Potentials of Marine Ova as shown by Indicators.

— M. J. KOPAC, and C. G. GRAND, Ind. Engn. Chem., Analyt. Ed. **9**, 143 (1937): The Living Cell, Physical Properties and Microchemical Reactions.

— H. POLLACK, and B. COHEN, Brit. J. Exp. Biol. **6**, 229 (1929): Intracellular Oxidation-Reduction Studies. I: Reduction Potentials of Amoeba dubia by Microinjection of Indicators.

COMANDON, J., and P. DE FONBRUNE, Ann. physiol. physicochem. **10**, 862 (1934): Nouvelle technique de microinjection.

— — Ann. Inst. Pasteur **60**, 113 (1938): La chambre à huile. Ses avantages pour l'étude des microorganismes vivants, la culture des tissus et la micromanipulation.

— — C. r. 9e reun. ass. physiol., Paris 473 (1935): Observations sur une amibe (Acanthamoeba).

— — C. r. soc. biol. **3**, 999 (1932): Sortie d'une hémogrégarine de son cyste endoglobulaire obtenue par une nouvelle technique de microdissection.
Ibid., **123**, 1072 (1936): Ingestion et digestion de bacilles par une amibe Amoeba phagocytoides (Gauducheau).

Ibid., **129**, 619, 620, 623 (1938): Recherches expérimentales sur les champignons predateurs de Nematodes du sol.

Ibid., **130**, 740 (1939): Greffe nucléaire totale, simple ou multiple, chez une amibe.

Ibid., **136**, 423 (1942): Étude volumétrique comparative *d'Amoeba spheronucleus* et de deux variétés obtenues par l'action de la colchicine.

Ibid., **136**, 746 (1942): Greffes nucléaires croisées entre *Amoeba spheronucleus* et l'une de ses variétés colchiciniques.

— — and J. JOLLY, C. r. soc. biol. **117**, 975 (1934): Étude expérimentale de la division cellulaire nouvel enregistrement cinématographique.

— — and H. VELU, A. Inst. Pasteur **72**, 701 (1946): Perfectionnement aux techniques d'isolement des micro-organismes par micro-manipulation. Application aux spores de Pénicilline.

DE FONBRUNE, P., Ann. physiol. physicochem. **10**, No. 4 (1934): Démonstration d'un micromanipulateur pneumatique et d'un appareil pour la fabrication des microinstruments sous le microscope.

— J. med. de Bordeaux (15. Sept. 1941): La cyto-chirurgie et ses moyens techniques.

DORFMANN, W. A., Protoplasma **25**, 463 (1936): A Simple Type of Micro Electrode for Determination of pH.

DUNN, F. L., Science **64**, 650 (1926): Finely Regulated Movement by Using Hydraulic Devices.

ENSINGER, H., Z. Zellforsch. **28**, 614 (1936): Kurze Mitteilung über Versuche an der lebenden quergestreiften Muskelfaser mit dem Mikromanipulator.

ERIKSON, D., and F. M. MASSON, J. Gen. Microbiol. **11**, 209 (1954): Modifications of Micromanipulative Practice Suitable for Single Cell Isolation and Cultivation of (a) Aerobic and Transiently Chain-Forming, (b) Lipophilic, and (c) Micro-Aerophilic Bacteria.

ETTISCH, G., and T. PETERFI, Pflügers Arch. ges. Physiol. **208**, 454 (1925): Zur Methodik der Elektrometrie der Zelle.

Ibid., **208**, 467 (1925): Elektrometrische Untersuchungen an *Amoeba terricola*.

FARBER, W. P., and C. W. TAYLOR, Univ. California Publ. Zool. **26**, 131 (1926): Fatal Effects of the Removal of the Micronucleus in *Euplotes*.

FISCHER, A., Die mikrurgische Methode in ihrer Anwendung auf Gewebskulturen, in Gewebszüchtung, 3rd ed., Abschn. IV. D. 107—126, Munich: Müller & Steinicke 1930.

FLORIAN, J., Z. wiss. Mikroskopie **45**, 460 (1928): Ein Hebelmikromanipulator.

FRY, H. J., Arch. exp. Zellforsch. **2**, 402 (1926): Zelloperationen ohne Mikrosektionsapparat.

FURTH, R., Kolloidchem. Beih. **28**, 235 (1929): Die physikalischen Grundlagen elektrischer Potentiale im Organismus und die direkten Methoden ihrer Messung (Elektrometrie).

— Physik. Z. **28**, 697 (1927): Über die Messung von elektromotorischen Kräften mikroskopisch kleiner Elemente.

GICKLHORN, J., Kolloidchem. Beih. **28**, 252 (1929): Die Herstellung von Mikroelektroden zu Potentialmessungen.

— and R. KELLER, Methoden der Bioelektrostatik, in Abderhaldens Handb. biol. Arbeitsmethoden, Abt. V. T. 2, Vienna and Berlin: Urban und Schwarzenberg, 1928.

— and K. UMRATH, Protoplasma **4**, 228 (1928): Messungen elektrischer Potentiale pflanzlicher Gewebe und einzelner Zellen.

GOLDACRE, R. J., Nature **173**, 45 (1954): A Simplified Micromanipulator.

HAHN, M., F. SCHUTZ, and L. WAMOSCHER, Z. Hyg. **106**, 746 (1926): Hefe-Einzell-Kulturen mit dem Mikromanipulator.

HARDER, R., Planta **2**, 446 (1926): Mikrochirurgische Untersuchungen über die geschlechtliche Tendenz des homothallischen *Coprinus sterquilinus* Fries.

— Z. Bot. **19**, 337 (1927): Über die Rolle von Kern und Protoplasma bei der Übertragung von Eigenschaften (nach mikrochirurgischen Untersuchungen an Hymenomyzeten).

HAUSER, E. A., Kolloidz. **38**, 76 (1926): Über die Anwendung des mikrurgischen Verfahrens in der Kolloidchemie.

— Z. wiss. Mikroskopie **41**, 465 (1924): Über die Anwendung des Mikromanipulators und anderer neuer optischer Instrumente bei mikroskopischen Studien an Kautschuk-Milchsäften in den Tropen.

HECKER, F., J. Inf. Dis. **19**, 306 (1916): A New Model of Double Pipet Holder and the Technic for the Isolation of Living Organisms.

HEILBRONN, A. L., Jahrb. wiss. Bot. **61**, 284 (1922): Eine neue Methode zur Bestimmung der Viskosität lebender Protoplasten.

HIMMELWEIT, F., klin. Wochschr. **12**, 1909 (1933): Ein neuer Mikromanipulator.

HÖFLER, K., Ber. dtsch. bot. Ges. **49**, 79 (1931): Das Permeabilitätsproblem und seine anatomischen Grundlagen. Mikrochirurgische Versuche zum Hautschichtenproblem.

HOFMEISTER, L., Protoplasma **35**, 65 (1940): Mikrurgische Studien an Borraginoiden-Zellen. I: Mikrodissektion.
Ibid., **35**, 161 (1940): II: Mikroinjektion und mikrochemische Untersuchung.
Ibid., **43**, 278 (1954): Mikrurgische Untersuchungen über die geringe Fusionsneigung plasmolysierter, nackter Pflanzenprotoplasten.

— Zeiss-Nachr. **4**, 1 (1941): Die Vorführung mikrurgischer Experimente in der Projektion.

— Z. wiss. Mikroskopie **57**, 259 (1940): Mikrurgische Studien an Diatomeen.
Ibid., **57**, 274 (1940): Studien über die Mikroinjektion in Pflanzenzellen.

HORT, E. C., J. Hyg. **18**, 361 (1920): The Cultivation of Aerobic Bacteria from Single Cells.

HOWLAND, R. B., J. Exp. Zool. **40**, 251 (1924): Experiments on the Contractile Vacuole of *Amoeba verrucosa* and *Paramaecium caudatum*.
Ibid., **40**, 263 (1924): Dissection of the Pellicle of *Amoeba verrucosa*.

— Proc. Soc. Exp. Biol. Med. **20**, 471 (1923): Notes on the Dissection of *Amoeba verrucosa*.

— and H. POLLACK, J. Exp. Zool. **48**, 44 (1927): Micrurgical Studies on the Contractile Vacuole.

JOHNSON, M. W., J. Bact. **8**, 573 (1923): A Simple Micropipette Holder.

— and L. J. MANHOFF, Science, N. Ser. **113**, 182 (1951): A Simplified Technique for Microelectrocoagulation.

KERR, T., Protoplasma **18**, 420 (1933): The Injection of Certain Salts into the Protoplasm and Vacuoles of the Root Hairs of *Limnobium spongia*.

KIRK, P. L., Density and Refractive Index. Their Application in Criminal Investigation, Springfield, Ill.: C. C. Thomas, 1951.

KITE, G. L., Amer. J. Gen. Physiol. **73**, 282 (1915): The Relative Permeability of the Internal Cytoplasm of Animal and Plant Cells.

— Biol. Bull. **25**, 1 (1912): The Relative Permeability of the Surface and Interior Portion of the Cytoplasm of Animal and Plant Cells.

KOBLMÜLLER, L. O., and R. VIERTHALER, Zbl. Bakt. I. **129**, 438 (1933): Über ein Gerät zum Isolieren von Keimen auf der Oberfläche fester Nährböden (Platten-manipulator).

KOPAC, M. J., Cancer Research **12**, 276 (1952): New Ultramicro-methods for Enzymatic Cytochemistry.

— Trans. Amer. Microscopic Society **48**, 438 (1929): A Micromanipulator for Biological Investigations.

LEDERER, B., Protoplasma **22**, 405 (1934): Färbungs-, Fixierungs- und mikrochirurgische Studien an *Spirogyra*-Tonoplasten.

MALONE, R. H., J. Path. Bact. **22**, 222 (1918): A Simple Method for Isolating Small Organisms.

McCLENDON, J. F., Science **74**, 661 (1931): The Use of Micromanipulators.

MOENCH, G. L., and H. HOLT, Biol. Bull. **56**, 267 (1929): Microdissection Studies on Human Spermatozoa.

NAVILLE, A., Bull. histol. **7**, 273 (sec. 1), 313 (sec. 2) (1930): L'emploi du micromanipulateur de Peterfi: fabrication et montage des microinstruments.

OTTO, L., Feingerätetechn. **4**, 56 (1955): Moderne Mikromanipulatoren.

— VDI-Z. **94**, 754 (1952): Mikromanipulatoren und deren Hilfsgeräte.

PETERFI, T., Die Technik der Zelloperationen (Mikrurgie), in Methodik der wissenschaftlichen Biologie, vol. **1**, 559, Berlin: Springer, 1928.

— Arch. exp. Zellforsch. **4**, 165 (1927): Die Mikrurgie der Gewebekulturen.

— Naturwiss. **11**, 81 (1923): Das mikrurgische Verfahren.

— Z. wiss. Mikroskopie **41**, 263 (1924): Neue mikrurgische Nebenapparate.

Ibid., **43**, 186 (1926): Die Präparierwechselkondensoren und ihre Handhabung bei Dunkelfeldmanipulationen.

Ibid., **44**, 296 (1927): Die heizbare feuchte Kammer.

Ibid., **45**, 56 (1928): Ein Beitrag zur Methode der pH-Bestimmung in Zellen und Geweben.

— and O. KAPPEL, Arch. exp. Zellforsch. **4**, 155 (1927): Die Wirkung des Anstechens auf das Protoplasma der in vitro gezüchteten Zelle. II. Anstichversuche an in vitro gezüchteten Vogelmonozyten.

Ibid., **5**, 341 (1928): III. Anstichversuche an den Nervenzellen.

Ibid., **5**, 349 (1928): IV. Die Pigmentzellen.

— and O. KAPPEL, Z. Krebsforsch. **26**, 89 (1928): Die Wirkung des Anstechens auf das Protoplasma der *in vitro* gezüchteten Zelle. Mikrurgische Untersuchungen an den Geschwulstzellen.

— and H. KOJIMA, Protoplasma **25**, 489 (1936): Die Wirkung mikrurgischer Eingriffe auf den Ruhekern der Pflanzenzelle. I: Anstichversuche. II: Injektionsversuche.

— and S. MOSCHKOWSKI, Arch. Protistenkde. **60**, 492 (1927/28): Mikrurgische Versuche an Leishmanien.

— and A. NAVILLE, Protoplasma **12**, 524 (1931): Die Wirkung des Kernstichs auf das Protoplasma der *Amoeba sphaeronucleus*.

— and O. OLIVO, Arch. exp. Zellforsch. **4**, 149 (1927): Die Wirkung des Anstechens auf das Protoplasma der *in vitro* gezüchteten Zelle. I: Anstichversuche an ... Myoblasten.

— and L. WAMOSCHER, Z. Hyg. Infektionskr. **106**, 191 (1926): Die Isolierung von Bakterien im Dunkelfeld: Einzellkulturen und Tierimpfung mit einzelnen Bakterien.

— and C. S. WILLIAMS, Arch. exp. Zellforsch. **14**, 210 (1933): Elektrische Reizversuche an gezüchteten Gewebezellen. I: Versuche an Nervenzellen.

— and M. W. WOERDEMANN, Biol. Zbl. **44**, 264 (1924): Einfluß der mechanischen Reizung auf die Flimmerzellen.

PFEIFFER, H. H., Chromosoma **1**, 526 (1939/40): Mikrurgische Versuche im polarisierten Lichte zur Analyse des Feinbaues der Riesenchromosomen von *Chironomus*.

PICK, J., Z. wiss. Mikroskopie **51**, 257 (1934): Über ein Mikroskop zur Untersuchung lebenden Gewebes.

PIERCE, J., and H. MONTGOMERY, J. Biol. Chem. **110**, 763 (1935): A Microquinhydrone Electrode: Its Application to the Determination of the pH of Glomerulus Urine of *Necturus*.

PRATT, F. H., and M. A. REID, Science **72**, 431 (1930): A Method for Working on the Terminal Nerve-Muscle Unit.

RANKE, O., Z. wiss. Mikroskopie **45**, 67 (1928): Spannungsmessungen am Mikromanipulator.

REDENZ, E., Z. wiss. Biol., Abt. B. (Z. Zellforsch. u. mikr. Anat.) **4**, 611 (1927): Untersuchungen über die elastische Faser. II: Dehnung der elastischen Fasern und Platten der Aortenwand mit dem Mikromanipulator.

REES, W. M., CHAS., Univ. California Publ. Zool. **20**, 235 (1922): The Microinjection of *Paramaecium*.

REINERT, G. G., Arch. exp. Zellforsch. **22**, 681 (1939): Ein neuer Mikromanipulator für Arbeiten bis 2500 ×.

— Zeiss-Nachr. **3**, 107 (1939): Fortschritte in der Technik des Mikromanipulierens.

— Zeiss-Nachr. **4**, 252 (1943): Die Methode von J. Comandon und P. de Fonbrune zur Ein-Zell-Abimpfung von Bakterien mit Hilfe des Mikromanipulators.

REYNIERS, J. A., J. Bact. **23**, 183 (1932): A New and Simplified Micrurgical Apparatus Especially Adapted to Single Cell Isolation.
Ibid., **26**, 251 (1933): Studies in Micrurgical Technique.

RICHTER, K. M., Science **106**, 598 (1947): A Precision Micro-Pipette Trimmer.
Ibid., **108**, 192 (1948): An Improved Moist Chamber for Use in Micromanipulation.

RIGGLE, G. C., Rev. Sci. Instr. **24**, 402 (1953): An Improved Electromicrocauter Apparatus.

SAUNDERS-SINGER, A. E., Pat. Specif. No. 690494 V. 22. 4. Pat. Off. London (1953): Improvements in or Relating to Microscope Manipulators or Dissection Manipulators.

SCARTH, G. W., Protoplasma **2**, 189 (1927): The Structural Organisation of Plant Protoplasm in the Light of Micrurgy.

SCHOUTEN, S. L., Arch. exp. Zellforsch. **17**, 429 (1935): Untersuchungen mit dem Mikromanipulator.

— Z. wiss. Mikroskopie **22**, 10 (1905): Reinkulturen aus einer unter dem Mikroskop isolierten Zelle.
Ibid., **24**, 258 (1907): Methode zur Anfertigung der gläsernen Isoliernadeln, gehörend zum Isolierapparat für Mikroorganismen.
Ibid., **51**, 421 (1935): Der Mikromanipulator.

SEELIGER, R., Optik **7**, 303 (1950): Sind Oberflächenabdrücke formtreu?

SEIFRIZ, W., Ann. Bot. **35**, 269 (1921): Observations on Some Physical Properties of Protoplasm by Aid of Microdissection.

— Biol. Bull. **34**, 307 (1918): Observations on the Structure of Protoplasm by Aid of Microdissection.

— Brit. J. Exp. Biol. **2**, 1 (1924): An Elastic Valve of Protoplasm with Further Observations on the Viscosity of Protoplasm.

— Bull. Nat. Res. Council **69**, 229 (1929): The Viscosity of Protoplasm.

— The New Physiol. **21**, 107 (1922): A Method for Inducing of Protoplasmatic Streaming.

SEIFRIZ, W., Protoplasma **1**, 1 (1926): Protoplasmic Papillas of *Echinarachnius oocytes*.
 Ibid., **1**, 345 (1926): The Physical Properties of Erythrocytes.
 Ibid., **3**, 191 (1927): New Material for Microdissection.
SIESS, K., Mikrokosmos **28**, 107 (1934/35): Kapillarisolationsgerät zur Einzel-
 präparation kleinster Organismen.
 Ibid., **29**, 93 (1936): Ein Kapillarisolationsgerät für Arbeiten im Bildfeld.
SIMMEL, H., Z. ges. exp. Med. **179** (1931): Mikrurgische Untersuchungen an Erythro-
 zyten und Leukozyten.
TAYLOR, C. V., Univ. Calif. Publ. Zool. **26**, 443 (1935): Improved Micromanipulation
 Apparatus.
— and W. P. FARBER, Univ. Calif. Publ. Zool. **26**, 131 (1923/25): Fatal Effects of
 the Removal of the Micronucleus in *Euplotes*.
TELKES, M., Amer. J. Physiol. **89**, 475 (1931): Bioelectrical Measurements on *Amoebae*.
TOPLEY, W. W., J. E. BERNARD, and G. S. WILSON, J. Hyg. **20**, 221 (1921): A New
 Method of Obtaining Cultures from Single Bacterial Cells.
TSCHACHOTIN, S., Biol. Zbl. **32**, 623 (1912): Die mikroskopische Strahlenstichmethode.
— C. r. acad. sci. Paris **171**, 1237 (1920): La méthode de la radio-piqûre microscopique,
 un moyen d'analyse en cytologie expérimentale.
— C. r. soc. biol. **85**, 137 (1921): Nouveau dispositif de la méthode de la radio-
 puncture microscopique.
— Z. wiss. Mikroskopie **29**, 188 (1912): Eine Mikrooperationsvorrichtung.
VOLKMANN, R. v., and E. v. MARK, Z. Zellforsch. Mikr. Anat. **32**, 543 (1943): Über
 Volumenmessungen an Gewebezellen.
WADA, B., Cytologia **4**, 222 (1933): Mikrodissektion der Chromosomen von *Trades-
cantia reflexa*.
WAMOSCHER, L., Zbl. Bakt. I. (Ref.) **89**, 287 (1928): Tuberkelbazillen-Ein-Zell-
 Kulturen.
— Z. Hyg. **106**, 421 (1926): Infektionsversuche mit einzelnen Pneumokokken.
WATANABE, A., Bot. Mag. **40**, 115 (1926): A New Device of Micromanipulator.
WHITEHEAD, T. N., The Design and Use of Instruments and Accurate Mechanisms,
 New York: MacMillan, 1934.
WOERDEMANN, A. W., Ned. T. Geneesk. **60**, 3 (1925): Micrurgie of microdissection
 des hulpsmiddel voor zytologisch onderzoek.
WORST, J., Amer. J. Ophth. **45**, No. 6 (1958): Method for Removal of Non-magnetic
 Foreign Bodies.
— Nature **168**, 749 (1951): Microelectrodes for Electro-Physiological Work.
 Ibid., **169**, 631 (1952): A Microinjection Needle.
WRIGHT, W. H., and E. F. McCOY, J. Lab. Clin. med. **12**, 795 (1927): An Accessory
 to the Chamber-Apparatus for the Isolation of Single Bacterial Cells.
WURMSER, R., and L. RAPKINE, C. r. acad. sci. Paris, seance 7, Sept. 1931: Micro-
 injéctions quantitatives.

Subject Index

Rules of Thumb

Total Magnification = Product of Objective Magnification (M) and Eyepiece Magnification

Focal Length of Objective = f = 160/M mm; Working Distance of Objective < f

Magnification of Objective = Tube Length/f; Normal Tube length = 160 mm

Diameter of Field of Vision = 20/M mm; Area of Field of Vision = 300/M^2 mm^2

Units of Length, Area, Volume, and Mass

Prefixes: m = milli, μ = micro, n = nano, p = pico, indicating 10^{-3}, 10^{-6}, 10^{-9}, and 10^{-12}

Length

1 mm = 0.1 cm = 0.001 m

1 μm = (1 μ) = 0.001 mm

1 nm = (1 mμ) = 0.001 μm

0.1 nm = (1 Å) = (0.1 mμ)

1 pm = (1 $\mu\mu$) = 0.001 nm

Length continued

Diam. of Red Blood Corpuscle: 7.5 μm

Wave Length of Violet Light: 300 nm

Effective Atomic Radii: 37(H) to 262(Cs) pm

Area

1 mm^2 = (1 sq. mm.) = 0.01 cm^2

1 μm^2 = (1 μ^2) = 10^{-6} mm^2

Volume

1 cm^3 = (1 cu. cm.) = 1 ml = **0.001 l**

1 mm^3 = (1 cu. mm.) = 1 μl = (1 λ) = **0.001 ml**

(0.1 mm)^3 = 0.001 mm^3 = 1 nl = (1 mλ) = **0.001 μl**

(0.01 mm)^3 = 0.001 nl = 1 pl = (1 $\mu\lambda$)

(0.001 mm)^3 = 1 μm^3 = (1 μ^3) = 0.001 pl

Volume of Red Blood Corpuscle: 0.1 pl

Sphere of 1 μm Diameter: 0.0005 pl

Unit Cell of NaCl: 1.5 × 10^{-14} pl

Mass

1 mg = 0.001 g

1 μg = (1 γ) = 0.001 mg

1 ng = (1 mγ) = 0.001 μg

1 pg = (1 $\mu\gamma$) = 0.001 ng

1 E = 0.001 pg

Mass of Red Blood Corpuscle: 60 pg

Sphere, 1 μm Diameter, Filled with Water: 0.5 pg

Sphere, 1 μm Diameter, Filled with Hydrogen: 35 E

Hydrogen Atom: 1.7 × 10^{-12} pg = 1.7 n E